T0175019

Nanostructured Ceramics

Nanostructured Ceramics
Characterization and Analysis

Debasish Sarkar

CRC Press
Taylor & Francis Group
Boca Raton London New York

CRC Press is an imprint of the
Taylor & Francis Group, an **informa** business

CRC Press
Taylor & Francis Group
6000 Broken Sound Parkway NW, Suite 300
Boca Raton, FL 33487-2742

First issued in paperback 2020

© 2019 by Taylor & Francis Group, LLC
CRC Press is an imprint of Taylor & Francis Group, an Informa business

No claim to original U.S. Government works

ISBN 13: 978-0-367-57094-1 (pbk)
ISBN 13: 978-1-138-08680-7 (hbk)

This book contains information obtained from authentic and highly regarded sources. Reasonable efforts have been made to publish reliable data and information, but the author and publisher cannot assume responsibility for the validity of all materials or the consequences of their use. The authors and publishers have attempted to trace the copyright holders of all material reproduced in this publication and apologize to copyright holders if permission to publish in this form has not been obtained. If any copyright material has not been acknowledged please write and let us know so we may rectify in any future reprint.

Except as permitted under U.S. Copyright Law, no part of this book may be reprinted, reproduced, transmitted, or utilized in any form by any electronic, mechanical, or other means, now known or hereafter invented, including photocopying, microfilming, and recording, or in any information storage or retrieval system, without written permission from the publishers.

For permission to photocopy or use material electronically from this work, please access www.copyright.com (http://www.copyright.com/) or contact the Copyright Clearance Center, Inc. (CCC), 222 Rosewood Drive, Danvers, MA 01923, 978-750-8400. CCC is a not-for-profit organization that provides licenses and registration for a variety of users. For organizations that have been granted a photocopy license by the CCC, a separate system of payment has been arranged.

Trademark Notice: Product or corporate names may be trademarks or registered trademarks, and are used only for identification and explanation without intent to infringe.

Visit the Taylor & Francis Web site at
http://www.taylorandfrancis.com

and the CRC Press Web site at
http://www.crcpress.com

Debasish Sarkar dedicates this book with great sense of gratitude and love to his parents,

Late Chandidas Sarkar and Mrs. Madhuri Sarkar

Contents

Preface

My decision to write this book is initiated in order to support UG/PG student and researchers who are in front of difficulties to interpret in systematic approach and conclude the research objective. Now-a-days, the science and engineering has become multidisciplinary in nature, and the books in the market are either totally for basic science or complete edition of different classical synthesis, characterizations, and applications of nanomaterials and nanotechnologies. So this book targets to bring forth the paradigm either oriented on fundamental aspects or totally hover onto data interpretations of nanostructured ceramics involved in functional and structural applications.

The book starts with a general overview of ceramics followed by different fundamental aspects and their judicial use in different horizon of nanostructured ceramic research. In this context, Chapter 1 deals with the confined growth of nanomaterials and their influence on fundamental aspects primarily band gap, surface area, surface atom density, surface energy, dangling bonds, Gibbs free energy, nucleation, cohesive energy, lattice parameters, strains, frequency, and particle velocity are discussed. One can correlate the physical properties of nanomaterials with help of these phenomenologies. A classic illustration and future scope in both traditional and advanced ceramics by nanoscale modulation are highlighted. Nanomaterials handling and ethical issues are enlightening to make aware of nanoresearch. Eventually, the bird-eye view has been illustrated in the perspective of case studies like energy, environment, and health issues. Chapter 2 briefs the data analysis and quantification strategy during synthesis of nanoparticles in order to explain the material properties effectually. Not all, but a cluster of inevitable material characterizations are being encountered and discussed. After reading this chapter, quantification of the phase transformation temperature, specific heat capacity, enthalpy, kinetics and activation energy of crystallization, X-ray and electron diffraction pattern indexing, lattice spacing and strain, crystallite size, theoretical density, imaging, particle size phenomena, surface area, and spectroscopic data can be done. The effect of nanoparticle size and processing conditions are also highlighted and eventually explored the probable research scope in nanostructured ceramics. Chapter 3 describes the surface and suspension phenomena that eventually mandate to improve recent nanotechnology evolution for the different sectors including energy, environmental science, medicine, food, transportation, construction, and many others. While discussing the use of nanoparticles, practical performance depends on the concurrence of an optimum solid containing suspension that eventually controls by surface characteristics of particles. In previous chapters, the discussion focused on the plausible physical properties revision when particle size is reducing from bulk counterpart. In this conjecture, the surface of solids and their resultant suspension characteristic variation with nanoparticles are discussed. The influence of particle size on diverse features are analyzed in consideration of recent research data.

Increasing pure water demand and global warming due to exhaustive greenhouse gas emissions from fossil fuel motivates to concentrate on the utilization of abundant solar energy for pollutant degradation and renewable energy by photocatalyst. Thus, Chapter 4 deals with the basic philosophy on the probable use of ceramic photocatalyst to solving the serious environmental pollution and energy problems. Fundamental concept on band gap, selection of band position, electron-hole charge transportation phenomena, role of scavenger, surface modifications, details about regenerative and photosynthetic solar cells, semiconductor–electrolyte interaction and band bending, output, and governing parameters are discussed and compared with nanoscale particle size and characterization parameters.

Ferrimagnetic and ferroelectric material has analogous functional behavior and dominates in recent nanotechnology. Manipulation and modeling of atoms, molecules, and macromolecular structure design promotes unique and enhanced properties to open up a new and exciting research field with revolutionary applications not only in electronic technology, but also in the field of energy sector, environmental science, and medical diagnosis. Thus, precise properties including magnetic moment, superparamagnetism, magnetization and hysteresis, magneto crystalline anisotropic, specific loss powers, magneto resistance change, exchange-spring magnet, dielectric permittivity and loss, and piezoelectric constants are discussed and analyzed their behavioral change with respect to particle or grain size in Chapter 5.

One-dimensional (1D) confined growth nanofiber experience more strain compared to equivalent volume of spherical or cuboid nanoparticles, but the same spherical particle is being consolidated more effectively than other morphologies. In consideration of density, the compact can be classified into transparent, dense, and porous ceramics that eventually enables to alter the mechanical properties under stress with respect to grain size and porosity content. Nanostructured bulk ceramic is competent in high strain rate at elevated temperature, although degree of deformation depends on grain size and morphology, and their distribution. In the perspective of structural ceramic research and development, some fundamental and classical mechanical properties are discussed in Chapter 6.

Chapter 7 discusses about the probable generation of renewable energy and their harvesting through solar energy, solid oxide fuel cell, and waste vibrational energy. Recent research progress has several directions to meet the excessive energy demand without evolving any toxic pollutant and maintaining the safe world. In this regard, it is easy to penetrate and solve the targeted goal when critically involve material scientists, instrumentation engineers, technocrats, and so on. A couple of ceramic oxides are encountered to solve this issue, although unlimited scope exists to use nanostructured ceramics in this perspective. Synthesis protocol, working principle, characterization, and analysis of different modules are discussed to facilitate the renewable energy research and development.

Ceramic is an ancient material and serves as natural air conditioner, filter for toxic gases, and industrial waste water; major concern of global warming. Growing population exponentially using vehicles, civil construction, and industries, and disposes tons of heat, toxic gases, and chemical compounds either in air or water. In preference, metal catalyst along with ceramic substrate facilitate gas filtration, however, chemical engineers and material researchers are effectively solving water pollution with using of abundant solar energy and waste mechanical energy. Several critical issues including smart glass, water purification, and minimization of water usability by nanostructured ceramics are focused in the perspective of environment research in Chapter 8.

Adult human skeletal system including teeth and bone weight is near to 15% of total body weight, out of this, 60% inorganic constituent known as hydroxyapatite. Simply, 60 kg human being carries near to 6 kg ceramics (10 wt%), and thus it is expected to replace our skeleton by artificial implants made of ceramics. Pollution, diseases, and accident either deteriorates or fractures the bone/teeth and necessitates both dense bioimplant and porous scaffold analogous to hard and spongy tissues, respectively. Despite the shaped nanostructured ceramics, photodynamic and magnetic ceramic nanoparticles are in forefront to solve the cancer treatment that also being considered and discussed in Chapter 9.

Chapter 10 deals with the different numerical and basic questions to initiate early stage of translational research that eventually help to develop real component for society. After solving and answering these basic questions, it is assumed that students can get interest in material research and able to analyze independently during development of new class of nanostructured ceramics. Provided numerical questions can be effectively considered for 15 weeks tutorial class in one semester. The book content can be started as a new course work, namely, either "Nanostructured Ceramics" or "Nanosciences and Nanotechnology" for graduate students.

In summary, this book gives a comprehensive account on the solved numerical and brainstorming unsolved (partial hints) problem to boost up the data interpretation skill, and one can envisage the correlation between synthesis process and properties in the perspective of the new material development. It serves as a concise text to answer the basics and achieve research goal for academia and industrial research.

Acknowledgment

I would like to convey my heartfelt thanks to family member son Achyut, wife Shubhra, nieces Diya and Sana, brothers Subhash and Amit and sister Rina for their constant endorsement and motivation throughout the journey of writing the book. I would like to thank my students Sarath, Sushree, Sangeeta, Venkat, Akbar, Megha, Barsa, Monali, Satish, and Sanjaya, for their uninterrupted cooperative actions in manuscript preparation. Their inquisitiveness for advancement in the field of ceramics has consistently simulated me for further research and, finally, putting the efforts into this book.

I would like to acknowledge to NIT, Rourkela; Department of Science and Technology (DST), India; Department of Biotechnology (DBT), India; Board of Research in Nuclear Sciences (BRNS), India; Korea Research Institute of Science and Standards (KRISS), South Korea; Tata Steel Ltd., Jamshedpur; and Tata Krosaki Refractories Ltd., Belpahar for their support.

I would like to thank to all the researchers for their contributions to the scientific society which are the building blocks of the conceptual knowledge of this book.

Lastly, my sincere apology to those whose names are inadvertently not mentioned.

Author

Dr. Debasish Sarkar is currently a professor at the Department of Ceramic Engineering, National Institute of Technology (NIT), Rourkela, Odisha, India. After his undergraduate in Ceramic Engineering from University of Calcutta in 1996, he completed a M.Tech in Metallurgical and Materials Engineering from Indian Institute of Technology, Kanpur, followed by a Ph.D in Ceramic Engineering at NIT, Rourkela. In due course of visiting researcher in Korea Research Institute of Standards and Science (KRISS), South Korea, he has gained extensive experience in synthesis, characterizations, and fabrication protocols of nanostructured ceramics. As Principal Investigator, he has managed several research projects including nanomaterials for functional and structural applications and refractories. The government of India and industry like Tata Steel Ltd., Tata Krosaki Refractories Ltd., etc. sponsored most of the funding.

Debasish is an internationally recognized expert in the academic, industrial, and translational research in ceramics. Several works are highlighted in "Global Medical Discovery, Canada," as well as published as European patent. He has been awarded Materials Research Society (MRSI) of India Medal Award, 2016, for the work on patient-specific orthopaedic implants, is a symbol of his outstanding career.

He has cumulative 22 years of experience and published 60 peer-reviewed international journal papers, 15 papers in conference proceedings, and 2 book chapters in Pan Stanford Publisher, Singapore and Wiley-VCH Verlag GmbH. He has been a lead inventor in one Korean patent, one European patent, and one Indian patent, and co-inventor for two Korean patents. He has published and reviewed in ACS, Blackwell Publishing Inc., Electrochemical Society, Elsevier, RSC, Springer, etc. He is holding positions as a member of different editorial boards and technical committees around the globe.

1

Nanoscale Ceramics

1.1 Introduction

Day to day, magic is gradually becoming logic by the invention of various scientific facts. As we are moving forward in time, we are coming up with new ideas to explore the opportunities in science and technology. The one that is growing interest among scientists and industries for the last few decades is "nanoscience and nanotechnology." Nanoscience implies cumulative knowledge together with the physical, chemical, and structural study of materials of nano (10^{-9} m) range that is 1–100 nm. However, nanomaterials defend as interlink in between nanoscience and nanotechnology. Nanotechnology is in the forefront in the support of various scientific and engineering horizons that invention has made our life quite easy and portable. For an example, in a tiny place, a vast data storage facilitates the data processing technology. Not only mammoth growth in the electronics industry, it is helping to make our life comfortable and user-friendly to a large extent starting from food to health care, packaging to paints, renewable energy to environmental issues, missiles to security application, and so on. In order to develop reliable and effective products, one should be aware of synthesis, properties, and effective characterizations in the perspective of nanomaterial development. Knowledge on lab scale module fabrication made of nanostructured material is eventually going to provide the advancement and luxury of our society.

In consideration of basic electron transfer phenomenon, the material world can be classified in conductors, semiconductors, and insulators. However, a common platform encompasses their existence as metals, ceramics, polymers, or a combination of them. Thus, in consideration of these three, we can achieve any properties in the recent era, although their size controls the resultant appearance and properties of products or devices. Apart from the monolithic single-phase materials, a combination of material results in different composites including metal matrix composite, polymer matrix composite, ceramic matrix composite, or any other alteration. Despite two classic groups of materials like metals and polymers, herein, ceramics are preferentially targeted and discussed, in specific, in the form of nanostructured ceramics. In the perspective of book content, this section starts with a correlation with size and physical properties of materials to intellect and justify the synthesis of nanoscale materials. In the end, details of classification of ceramics are highlighted, and their probable use in modern burning issues like energy, environment, and health treatment is emphasized.

1.2 Confinement and Functional Change in Nanoscale

Continuous improvement toward device miniaturization demands nanoscale particles and coatings for advanced applications. Herein, the focus has been emphasized on the confined growth (0D, 1D, 3D) of nanoparticles and nanostructured bulk ceramics, not nanocoating (2D) material. In a convention, nanoparticles are studied as a solid form, but it represents as an intermediate between the gaseous or molecular structure and the bulk solid state. Consequently, the physical properties of nanoscale materials which consist of any dimension below 10 nm (cluster of ~25 atoms or molecules!) are different from their bulk structures. When the particle size is made to be nanoscale, properties such as surface energy, melting point, catalytic, piezoelectric, magnetic, optical, mechanical properties, and chemical reactivity varies as a function of particle size, but obviously *not linear relationship* in all features. At the same time, a definite window size of particles or grains exhibits best properties, and these aspects have been discussed in subsequent chapters. As nanoparticles are closer to the size of an atom, consideration of quantum mechanics is come across in order to explain the motion and energy of these particles. In this nanoscale dimension, gravitational force becomes negligible, simultaneously, the electromagnetic force dominates and determines the physical properties. Quantum confinement describes the availability of electrons in particular space, rather than free to move in bulk of material, and this allows discrete energy levels to differentiate from bulk materials. Despite quantum confinement, nanoscale chemistry plays an important role to explain the properties influenced by bonding and surface characteristics. In the upcoming section, the reader can concentrate in more detail on the impact of nanoscale size, and why it is different from bulk counterpart.

1.2.1 Density of States

Material has free electrons, and their chance of availability within valence band and conduction band, and this incidence differentiate insulator, semiconductor, and conductor. A selective number of valence electrons very close to the Fermi energy within $k_B T$ (where $k_B = 1.38 \times 10^{-23}$ J/K, and T = absolute temperature) has a probability of shifting to conduction band at a particular temperature. Thus, free electrons are not entirely free, as well as not available in any energy state, whereas they must fit into a specific energy state within the solid. In this context, the Fermi distribution function, $f(E)$, is a statistical approach that describes the probability of finding a free electron in energy state of E [1]:

$$f(E) = \frac{1}{e^{\left((E - E_F)/k_B T\right)} + 1} \qquad (1.1)$$

All materials have at least some free electrons, and, hence, this function applies to illustrate the properties of insulator, semiconductor, and conductor. It is worthy to mention that the Fermi energy (E_F) is in the middle of the highest occupied band in conductors, or in another way, the Fermi energy is the band gap for semiconductors and insulators. This discussion focused in the consideration of a uniform availability of states of electrons in either the valence or conduction bands. Although, in reality, it is not true. Thus, it is necessary to introduce density of state (DOS) function as $D_s(E)$ for real solids, and this function describes the availability of states for electrons to occupy in different states. This 3D DOS function represents the number of available electron energy states per unit volume, per unit energy. It can be applied to bulk 3D materials without any growth direction confinement and is independent of the dimension.

$$D_s(E) = \frac{8\sqrt{2}\pi m^{3/2}}{h^3}\sqrt{E} \qquad (1.2)$$

where m is the mass of an electron (9.11×10^{-31} kg), h is the Planck's constant (6.626×10^{-34} m^2 kg/s). The electron population depends on the resultant product of the Fermi function and electron density of states. The number of electrons per unit volume with energy between E and E + ΔE is given by the multiplication of Equations 1.1 and 1.2:

$$n(E)\,\Delta E = f(E)\,D_s(E)\,\Delta E \qquad (1.3)$$

Thus, the number of free electrons per unit volume is:

$$n = \int_{E_c}^{\infty} n(E)\,dE \qquad (1.4)$$

Figure 1.1a demonstrates the characteristic behavior of conventional Fermi energy function above T > 0, density of states for 3D bulk, and resultant integral of Equation 1.4 as the shaded area. It is worthwhile to derive the Equation 1.2 applying the infinite 3D box model that refers the plausible growth of the particles without any confinement [2]. Figure 1.1b illustrates the growth pattern of the particles with having confinement (shown by arrows) in one or multiple directions that results in 2D nanolayer, 1D nanowire, and 0D quantum dot, respectively. In actual, the D_s varies with respect to dimensional features, and thus it varies with $E^{1/2}$, E^0, $E^{-1/2}$, and $\delta(E)$ for 3D, 2D, 1D, and 0D, respectively. The characteristic behavior of density of states of semiconductors with 3, 2, 1, and 0 degrees of freedom for electron propagation with respect to energy is represented in Figure 1.1c. Systems with 2, 1, and 0 degrees of freedom are referred to as quantum wells, quantum wires, and quantum boxes, respectively.

Virtually, high magnetic susceptibility, X-ray radiated emission, outstanding optical properties, specific heat, electrical properties, and many others depend on the availability and motion of the electrons. This density of state is most relevant on nanoscale and with the support of the aforesaid understanding, a later section describes how a small portion of a bulk solid behaves electronically like an artificial atom, that is the key feature of device miniaturization and advance nanodevices.

1.2.2 Confinement and Band Gap

In an effort to enhance the functionality, comfortability, and reduce energy consumption, the major effort to concentrate on the development of nanoscale materials is followed by effective and portable devices. As a result, electronic devices have changed from few centimeter- to millimeter-sized down to nanometer sized. It was possible because of synthesis of nanoscale materials having atomistic and their tunable properties. In this context,

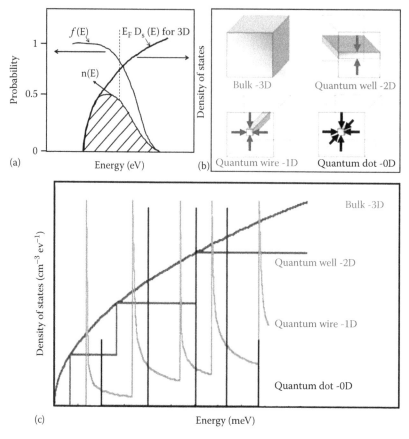

FIGURE 1.1
(a) Probability of finding a free electron and density of states of 3D with respect to band energy; (b) plausible growth confinement of nanoscale materials; and (c) an insight view of DOS and electron energy of 0D, 1D, 2D, and 3D confined nanomaterials.

it is interesting to introduce how a tiny particle either originated from bulk or synthesized nanoscale size can behave as an atom.

In convention, at T = 0 K, the electron distribution goes from zero energy up to the Fermi energy for a conductor and thus taking into account of the Equations 1.1 through 1.4, so, the resultant number of free electrons per unit volume or electron density is:

$$n = \frac{8\sqrt{2}\pi m^{3/2}}{h^3} \int_0^{E_F} \frac{\sqrt{E}}{e^{\left[(E-E_F)/k_B(0)\right]}+1} dE \tag{1.5}$$

Therefore, electron density at 0 K is:

$$n = \frac{8\sqrt{2}\pi m^{3/2}}{h^3} \left[\frac{2}{3}E_F^{3/2}\right] \tag{1.6}$$

A bulk solid has unlimited and thick energy bands, whereas atom has limited thin and discrete energy states, that implies the development of atomistic scale tiny particles compared to bulk, may behave electronically like an atom. Rearranging Equation 1.6, the Fermi energy is dependent on variable "n = number of free electrons per unit volume" (number of free electrons "N" per unit volume, i.e., D^3, D edge length of a cube), and we can obtain:

$$E_F \propto \frac{N}{D^3} \tag{1.7}$$

However, the Fermi energy does not depend on the size of the material, and it is true for conductor, semiconductor, and insulators as well. In order to discuss the total energy state and spacing, it can be justified that the total energy states' span of small volume material is identical with large volume, but the former has wider spacing between energy states to cover the cumulative energy state span. Let total free electrons "N" occupied up to E_F, and thus the average span ΔE is:

$$\Delta E \approx \frac{E_F}{N} \tag{1.8}$$

In consideration of Equations 1.7 and 1.8, we can obtain the average spacing between energy states is inversely proportional to volume:

$$\Delta E \approx \frac{1}{D^3} \tag{1.9}$$

In other words, the spacing between energy states gets larger as the volume gets smaller, and it forms a band in case of the bulk solid. Thus, the average spacing (ΔE) has relation with Fermi energy and consists of total number of valence electrons, N_v (number of valence electron is same with total free electrons "N" for metal, at T = 0 K within E_F), and can equate as:

$$\Delta E = \frac{4E_F}{3N_v} \tag{1.10}$$

When dimension is typically small, it provides fewer number of valence electrons, and, hence, the spacing between energy state ΔE grows inversely, and thus continuous increment of ΔE, or reduction in size, restricts the movement of electrons and needs more energy (temperature) to conduct, so under the condition of $\Delta E > k_B T$, metal exhibits nonmetallic behavior. This feature is very useful to make metal-semiconductor junction to fabricate different electronic devices [3,4].

Confinement concept implies the band-gap energy increment because of size reduction [$E_g(r)$], and thus one metal can behave as semiconductor. In another way, the quantum confinement prediction for the band-gap energy increment for ceramic oxides by [5]:

$$E_g(r) = E_g(bulk) + \frac{2\pi^2 h^2}{D^2 \mu} \tag{1.11}$$

where $1/\mu = \left(1/|m_e| + 1/|m_h|\right)$.

in which m_e, m_h, μ, and D are the effective mass of electron, effective mass of hole, reduced mass, and diameter of nanoparticle, respectively. In preference, the quantum

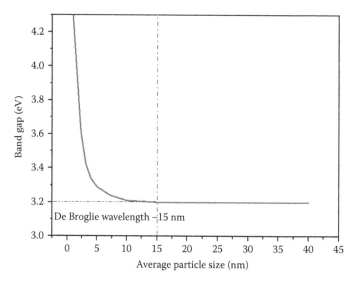

FIGURE 1.2
The change in band gap of BaTiO$_3$ with respect to particle size. Band gap becomes constant beyond the De Broglie wavelength (~15 nm), it varies from material to material. (From Supasai, T. et al., *Appl. Surf. Sci.*, 256, 4462–4467, 2010.)

confinement effect is predominant if particle size is less than the De Broglie wavelength ($D_{DB} = 4\pi\varepsilon_o\varepsilon_r h^2/\mu e^2$). The band gap of BaTiO$_3$ approaches to only bulk value when particle size is beyond $D_{DB} = 15$ nm, whether a steep change in band gap below this critical limit is found, as described in Figure 1.2, 'a similar pattern of Figure 1.3b'. This is attributed to minimize the particle diameter and their effective mass of electrons and holes, as described in Equation 1.11.

Apart from the different behavior in nanoscale attribute to energy state spacing, it is essential to encounter the surface phenomena including the increment of surface area to volume ratio, surface atom density, and surface energy in order to understand the different physical properties and adequate applications of nanostructured materials.

1.2.3 Surface Area to Volume

Surface area is nothing but the cumulative area of the exposed surfaces of an object. Theoretical estimation of perfect cuboid, sphere, rod, and other geometry is easy to calculate, and one can compare the influence of shape on surface area of identical volume of the different geometries of same material. In fact, this surface-to-volume ratio for individual shape is different, and thus there is chance of different degree of activity with other species. Interestingly, this basic understanding can be established in consideration of smaller fragments of isotropic geometry, say cube, has 24% larger surface area compared to the equal volume of sphere.

Let us consider a piece of cubic particle with having edge dimension of 1 cm and weight 1 gm. Imagine, gradual and uniform fragmentation of the same cube up to several small cubes which consist of 1 nm (size of two atoms!) inside length and thus results in high surface area and eventually high surface-to-volume ratio for identical 1 gm of material. This surface area and their surface-to-volume ratio phenomena can be explained through simple analysis (Table 1.1) and representative plot, as illustrated in Figure 1.3a.

The logarithm of surface-to-volume ratio increases near to four times compared to the logarithm of surface area for cubic particles. The resultant surface area of 1 gm, 1 nm cubic

TABLE 1.1

Theoretical Calculation of Surface Area of Cubic Particle, and Effect of Particle Size on the Surface Area and Surface-to-Volume Ratio

Cubic Particle Edge (cm)	Log (Particle Size)	Volume of Particles (cm³)	Number of Particles	Specific Surface Area (cm²/gm)	Log (Surface Area)	Surface/ Volume (S/V)	Log (S/V)
1	0	1	1	6	0.78	6	0.78
1×10^{-1}	−1	1×10^{-3}	1×10^{3}	6×10^{1}	1.78	6×10^{4}	4.78
1×10^{-2}	−2	1×10^{-6}	1×10^{6}	6×10^{2}	2.78	6×10^{8}	8.78
1×10^{-3}	−3	1×10^{-9}	1×10^{9}	6×10^{3}	3.78	6×10^{12}	12.78
1×10^{-4}	−4	1×10^{-12}	1×10^{12}	6×10^{4}	4.78	6×10^{16}	16.78
1×10^{-5}	−5	1×10^{-15}	1×10^{15}	6×10^{5}	5.78	6×10^{20}	20.78
1×10^{-6}	−6	1×10^{-18}	1×10^{18}	6×10^{6}	6.78	6×10^{24}	24.78
1×10^{-7}	−7	1×10^{-21}	1×10^{21}	6×10^{7}	7.78	6×10^{28}	28.78

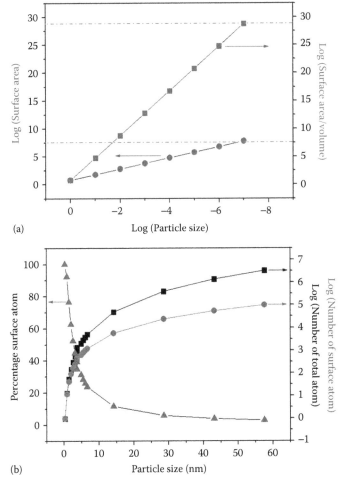

FIGURE 1.3

A representative plot for (a) surface area and surface/volume ratio, and (b) percentage surface atom and total atom as a function of particle size. The details calculated data for the plot is tabulated in Table 1.1.

particle envisages equivalent a soccer ground (118 × 92 m) area. In this context, one can understand why chopped potato softens faster than uncut one! Thus, when it comes to material science, a single parameter that is particle size can have a noticeable impact on properties of the material.

Fundamental material science defines macro-sized material properties depend upon its chemical composition and atomic structure. The bulk properties of material are negligibly associated with the surface atoms, as the number of surface atoms is quite less in comparison to that of core atoms. Hence, the properties of the bulk material are not very much dependent on the surface atoms. Thus, the question is what benefit can be obtained due to high surface-to-volume ratio? The answer is a dramatic increment in surface atom density that results in surface energy due to formation of unsaturated bond and exposed atoms. Section 1.2.4 illustrates how available surface atom density changes in consideration of simple example of FCC metallic silver particle below 100 nm.

1.2.4 Surface Atom Density

How number of surface atoms enhances with nanoscale and thus surface atom density? In order to understand this circumstance, a simple FCC metallic Ag has been considered and calculated the total number and surface atoms with increment of dimension from 1 to 60 nm, thus surface atom density (ρ_s) defines the ratio of number of surface atoms to total number of atoms. Let, total number of atoms N, so, N_{total} is being calculated by the formula [6]:

$$N_{total} = (1/3)\left[10p^3 - 15p^2 + 11p - 3\right] \qquad (1.12)$$

and their number of surface atom

$$N_{surf} = \left[10p^2 - 20p + 12\right] \qquad (1.13)$$

where p = number of layers, assume consecutive layers are grown on reconstructed spherical unit cell, and consider "p" = 1, 2, 3, 4, ... 100, respectively. The equation for total number of atom (N_{total}) is valid for $p \geq 1$, however, N_{surf} is more feasible when $p > 1$. In the same time, the diameter of particle can be calculated by the equation D = (2p − 1)d, "d" is the distance between the centers of nearest-neighbor atoms and, d = a/(2)$^{1/2}$, where a = lattice constant. Let, lattice constant for Ag, a = 4.079 Å. Calculation in each step provides three pieces of information on total number of atoms, surface atoms of each individual particle, and percentage surface atom, followed by surface atom density, and their logarithmic data are plotted with respect to particle diameter in Figure 1.3b.

A clear difference is noticed in between total number of atoms and surface atoms beyond 10 nm particles, whereas below 10 nm size, there is no significant difference that exists. Indeed, such difference clearly predicts the availability of more number of exposed atoms in nanoscale that eventually responsible for the change in different properties compared to bulk. Thus, as we move toward lower dimensions, more entities are present on or near surface than bulk. In fact, questions are coming to mind, how those surface atoms appear and their consequences.

In bulk, valence of atoms is satisfied by its coordination number, whether surface atoms have unsaturated bonds and lower coordination number compared to bulk entity comprises higher energy, resembles to surface energy, and it would be discussed in Section 1.2.5.

1.2.5 Surface Energy and Dangling Bonds

Nanoscale synthesis protocol follows two basic philosophies: top-down and bottom-up approach. Nevertheless, bottom-up approach preferentially develops tailor-made shape and morphology including 0D, 1D, 2D, or 3D nanoscale particles, where synthesis of such confined growth by top-down approach is limited. Reduction in smaller particle size facilitates new surfaces or crystal faces that involves high surface defects. In any synthesis process, existence of unsaturated surface atom is common phenomena that result in unsaturated (green circle) valences, refers to "dangling bond," and experience an inward acting force due to underlying atoms.

These surface atoms possess excessive energy as compared to bulk (orange circle) known as surface energy. This is the (thermodynamically unfavorable) energy of making "dangling bonds" at the surface, as illustrated in Figure 1.4. This surface energy is always associated with surface, regardless of the type of bonding whether covalent, ionic, or non-covalent. Every system wants to get rid of this excess energy or thermodynamically stable by some means. In general, the surface energy is often denoted as "γ," that is defined as the free energy required to create a unit area of new surface of a material; so, $\gamma = (\delta G / \delta A)$, where G = Gibbs free energy and A = surface area. It implies, more surface energy attributed for smaller area, that is 'NANO.'

This surface energy depends on available surface atoms, and thus it is very high preferentially below 10 nm particles (Section 1.2.4). Again, consider the one centimeter cubic particle, and cut into two equal halves, by ignoring the interactions of neighboring atoms, the surface energy density or energy per unit area further can be represented as:

$$\gamma = \frac{1}{2} N_b \, \rho_s \, \sigma \tag{1.14}$$

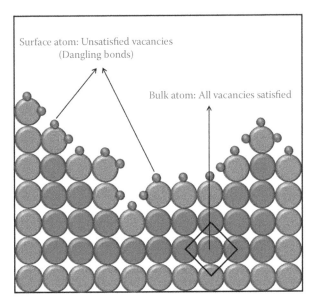

FIGURE 1.4
A schematic representation of the bulk core atom (orange color) satisfied with coordination number, and unsaturated valence atom (green color), refers as dangling bond, responsible for surface energy. High S/V ratio comprises more unsaturated bonds; surface energy.

where, N_b is number of broken bonds by forming new surface, ρ_s is surface atomic density, σ is half of bond strength. This surface energy of each crystal plane is different due to presence of different number of broken bonds. Close-packed plane has the highest coordination number or closest neighboring atoms that comprise fewest dangling bonds, and thus broken-bond model estimates the surface energy order for FCC $\gamma(110) > \gamma(100) > \gamma(111)$, but BCC $\gamma(111) > \gamma(100) > \gamma(110)$ metals, respectively [7]. However, the surface energy per unit area may alter this sequence. The crystal plane has different surface energies and thus thoughtful synthesis parameter can effectively determine the possibility of preferred plane growth and subsequent performance of nanoscale materials. In convention, a material is thermodynamically more stable at its lowest Gibbs free energy, and it controls the nuclei formation during synthesis of nanoparticles that in time related with surface energy as well.

1.2.6 Gibbs Free Energy and Critical Radius

High surface-to-volume (S/V) ratio comprehends high surface energy that contributes maximum Gibbs free energy of nanoparticles. A brief illustration of how Gibbs free energy and surface energy per unit area result in the critical radius during synthesis of nanoscale is discussed. Basic thermodynamics describe a new phase that appears either exceeding the solute solubility limit or reduction in temperature below the phase transformation point. Herein, this new phase says nanoparticle formation mechanism through wet chemical method is considered. Let, C is the concentration of solute, and C_o is the solubility or equilibrium concentration. The change in Gibbs free energy per unit volume of the solid phase, ΔG_v, is dependent on the concentration of solute:

$$\Delta G_v = -k_B T \frac{\ln(1 + \sigma_s)}{v} \tag{1.15}$$

where, "k_B," Boltzmann constant, "T" process temperature, "σ_s" supersaturation of the solution, i.e. $(C-C_o)/C_o$ and "v" molar volume. Under this condition, no nucleation occurs if $\sigma_s = 0$, whether a spontaneous and favorable nucleation occurs when $C > C_o$, hence the, Gibbs free energy per unit volume of the solid phase (ΔG_v) is negative. Supersaturate solution possesses a high Gibbs free energy, thus the overall energy of the system is like to reduce, that is the driving force to form new phase from the solution. Assuming a spherical nucleus of solid particle with a radius "r" develops from liquid phase, and this total change of chemical potential is the summation of volume free energy $(\Delta \mu_v)$ and interfacial energy $(\Delta \mu_s)$. So, overall Gibbs free energy (ΔG) is:

$$\Delta G = \frac{4}{3} \pi r^3 \Delta G_v + 4\pi r^2 \gamma \tag{1.16}$$

where γ is the surface energy per unit area.

In order to favor the formation of nucleus, the process has to overcome critical energy (ΔG^*) barrier, in other way the formation of critical radius (r^*). It is only favorable when $d\Delta G/dr = 0$. Under this condition, $r^* = -2\gamma/\Delta G_v$; and $\Delta G^* = [16\pi\gamma^3]/[3(\Delta G_v)^2]$; and plotted in Figure 1.5. Thus, with help of Equation 1.15, the

$$\text{critical radius } (r*) = \frac{2\gamma v}{[k_B T \ln(1 + \sigma_s)]} \tag{1.17}$$

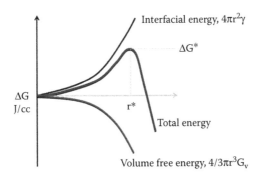

FIGURE 1.5
Theoretical representation of change in Gibbs free energy as a function of particle radius.

From Equation 1.17, one can reduce critical radius during synthesis either decrease the required surface energy per unit area for nanoparticle or increase T and σ_s. Below the nucleation of critical radius, ΔG is positive and dissolved back to the parent phase, nevertheless, beyond critical radius, addition of further atom ($r > r^*$) encompasses growth followed by Ostwald ripening with time. This philosophy is effective for homogenous nucleation, however, most often a system follows heterogeneous nucleation at preferential sites, such as phase boundaries or impurities like dust and requires less energy than homogenous nucleation. At such preferential sites, the operative surface energy is lower and results in reducing the free energy barrier and accelerating the rate of reaction. Surface energy promotes advance nucleation sites because of variation of wetting phenomenon, and thus contact angle (θ) determines the ease of nucleation by reducing the energy needed. A simple relation between ΔG_{hete} and ΔG_{homo} can be expressed as Equation 1.18:

$$\Delta G_{hete} = \Delta G_{homo} \cdot f(\theta) \tag{1.18}$$

where $f(\theta) = 1/2 + 3/4 \cos\theta - 1/4 \cos^3\theta$.

Eventually, the nanoparticle synthesis protocol follows competitive nucleation and growth process and results in different size distribution that facilitates surface energy difference of particles and association together to form agglomeration and change the particle size distribution. With judicious selection of process parameters, one can control the particle size distribution.

1.2.7 Cohesive Energy

Particle size influences the melting and Debye temperature, and both depend on cohesive energy. Cohesive energy defines the amount of energy required to make free atoms from crystal, and thus cohesive energy per mole [kJ/mol] is conventionally lower for nanosolid (E_{ns}) when number of free surface atoms is relatively high in nanosolids compared to bulk ones. In consideration of N_{surf} and N_{total}, the total cohesive energy (E_{total}) is the result of both interior and surface atoms. Hence, number of interior atoms is ($N_{total} - N_{surf}$), and:

$$E_{Total} = E_o \left(N_{total} - N_{surf} \right) + \frac{1}{2} E_o N_{surf} \tag{1.19}$$

where E_o is the cohesive energy per atom, so, Equation 1.19 becomes:

$$E_{ns} = E_{bs} \left[1 - \frac{N_{surf}}{2N_{total}} \right] \tag{1.20}$$

where $E_{ns} = AE_{total}/N_{total}$ and $E_{bs} = AE_o$; A = Avogadro number. As cohesive energy and melting temperature both describe the bond strength and follow the linear relationship, the melting temperature of nanosolid (T_{ns}) can relate with melting temperature of bulk solid (T_{bs}) and are represented as [8]:

$$T_{ns} = T_{bs} \left[1 - \frac{N_{surf}}{2N_{total}} \right] \tag{1.21}$$

Debye temperature (θ) describes the highest temperature achieved due to single mode of vibration or in another way beyond this temperature, the thermal vibrations are more important than quantum effect. In consideration of Lindemann criterion on melting for small particles, a crystal prefers to melt when the root mean square displacement of atoms in the crystal exceeds a certain fraction of the interatomic distance [9]. Under this situation, the simple relation between Debye temperature of nanosolid (θ_{ns}) and Debye temperature of bulk solid (θ_{bs}) can be represented as analogous to Equation 1.21 [10]:

$$\theta_{ns} = \theta_{bs} \left[1 - \frac{N_{surf}}{2N_{total}} \right]^{1/2} \tag{1.22}$$

Apart from size phenomenon, identical material may have different cohesive energy and melting temperature because of different shape and morphology. In this context, simple 3D (spherical), 2D (thin film), and 1D (nanowire) morphology have been considered and highlighted their competitive values, as every size and shape have definite interest of application. Let, "d" is the diameter of the atom, and D is the diameter of spherical solid, then N_{total} is expressed as the volume ratio of nanosolid and atom, and thus $N_{total} = (\pi D^3/6)/(\pi d^3/6)$. Similarly, total surface area of nanosolid is $4\pi(D/2)^2$, and the contribution of each surface atom to surface area of nanosolid is the area of the highest circle of the atom, $\pi(d/2)^2$, so, $N_{surf} = (4\pi D^2/4)/(\pi d^2/4)$, so the ratio of $N_{surf}/(2N_{total}) = 2d/D$ and the Equations 1.20–122 becomes:

$$E_{ns} = E_{bs} \left[1 - \frac{2d}{D} \right] \tag{1.23}$$

$$T_{ns} = T_{bs} \left[1 - \frac{2d}{D} \right] \tag{1.24}$$

$$\theta_{ns} = \theta_{bs} \left[1 - \frac{2d}{D} \right]^{1/2} \tag{1.25}$$

Confined nanofilm (2D) with thickness (h) and nanowire (1D) with diameter (Ø) and length = 1 of same atomic diameter "d" provide the ratio of $N_{surf}/(2N_{total})$ is $2d/3h$ and $4d/3Ø$, respectively. From this information, the cohesive energy, melting temperature, and Debye temperature of 2D and 1D nanostructured materials can also be calculated.

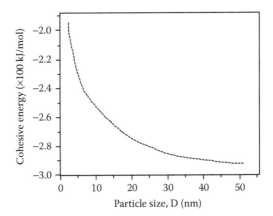

FIGURE 1.6
Variation of cohesive energy with respect to particle size of Ag. (From Zhu, Y. F. et al., *Appl. Phys. Lett.*, 95, 083110, 2009.)

In consideration of E_{bs} = –295.9 kJ/mol, the variation of cohesive energy with respect to particle size of Ag, preferably below 50 nm, is illustrated in Figure 1.6. A dramatic change is noticed beyond 10 nm and followed consistent plateau when particle size is bigger than 40 nm, this is further well matching for both experimental and theoretical concepts [11].

The aforesaid fundamental concept is also true for ionic bonding, where the surface atom has lower binding energy than the bulk and contributes lower cohesive energy per mole.

1.2.8 Lattice Parameters and Strains

Small nanoscale consists of high percentage free surfaces and comprises lattice contraction induced by high surface-to-volume ratio. In mechanical point of view, hydrostatic pressure on the surface induced by intrinsic surface stress results in lattice contraction or lattice strain. Consider a perfect liquid sphere having diameter 2R, that experiences excess pressure ΔP toward inside the sphere because of curvature, and thus Laplace relation depicts:

$$\Delta P \, dV = \gamma dA \tag{1.26}$$

or

$$\Delta P = \frac{2\gamma}{R} \tag{1.27}$$

where dV (V = 4/3πR³) is the volume change corresponding to a change dA (A = 4πR²) in the area of the droplet, and γ is the surface energy per unit area. In this context, the nanosolid experience extensive elastic strain (ε), and thus encounters this phenomenon to understand and evaluate the influence of particle size on the resultant lattice parameters. The pure metallic nanoparticles experience contraction with decreasing particle size, however, above critical size limit, the lattice parameters do not vary with particle size [12].

Usually, the increment of surface energy of nanoparticles inclines to contract their sizes by distorting their crystal lattice elastically, although it is very small compared to the entire particle size. Thus, the cumulative energy (F) variation is the result of both *increased surface energy* and *increased elastic energy*. Taking the equilibrium condition, and minimizing the total energy at dF/dε, the elastic strain becomes:

$$\varepsilon = \frac{1}{1+(\pi G/2\gamma_0)D} \tag{1.28}$$

where, ε = elastic strain, D = particle diameter, γ_o = surface energy per unit area at room temperature, G = shear modulus of material. Interesting to note that the lattice strain increasing with decreasing particle size, whereas above critical size, say 10 nm, the lattice parameter of Ag does not contract anymore [13]. In ideal crystal lattice, the parameter contraction is proportional to the particle diameter, so:

$$\frac{\Delta a}{a_o} = \frac{(1-\varepsilon)D-D}{D} \tag{1.29}$$

Elastic contraction results in $D - \varepsilon D$ and $\Delta a = a_{particle} - a_o$, where $a_{particle}$ and a_o are the lattice parameter of nanoparticle and bulk material, respectively. Considering Equations 1.27 through 1.29:

$$\frac{\Delta a}{a_o} = -\frac{1}{1+MD} \tag{1.30}$$

where M = πG/(2γ_o), and both estimation of G and γ_o encompass a correlation with percentage contraction with respect to particles size, and such behavior for Ag is represented in Figure 1.7.

The degree of contraction may vary with respect to particle morphology. This aforesaid relation is valid within a particle range of 1–15 nm of pure metallic spherical nanoparticles that experience more strain compared to metal oxide particles, however, this concept fails to predict the contraction behavior of 1D or 2D morphologies. However, some particles exhibit dilatation of lattice parameters with smaller particle size because of

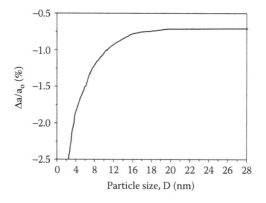

FIGURE 1.7
Contraction behavior of FCC Ag with respect to nanoscale particle size. An optimum ~10 nm particle size is noticed in this case. (From Wasserman, H. J. and Vermaak, J. S., *Surf. Sci.* 22, 164–172, 1970.)

pseudomorphism, substitution of oxygen, carbon, and hydrogen atoms [14]. Recent research shows interesting experimental data and their explanation on ceramic oxide (e.g., Ceria, CeO_2), emphasizing better insight of this phenomenon. Availability of oxygen atom and vacancy alter the oxidation state of Ce in between Ce^{3+}(1.034Å) and Ce^{4+}(0.92Å) that expedite different catalytic properties. In a recent study, both lattice parameter (α) and elastic strain (ε) of nanocrystalline CeO_2 enhances with reduction of particle size (D) during synthesis through the wet chemical method [15]. This phenomenon can be explained in the consideration of crystal structure of CeO_2. The pure CeO_2 has fluorite structure, where Ce^{4+} surrounded by eight O^{2-} ions, but soft chemical reaction facilitates the reduction to Ce^{3+} and creates oxygen vacancy that leads to distortion of local symmetry, enhancing the resultant bond length of Ce – O, and overall lattice parameter, but loss of oxygen ion in nanoscale (<3 nm) results in high strain. However, extensive experimental data and theoretical simulation on the effect of particle size and shape on the lattice parameters and lattice strain for different crystal structures and materials including metal oxides, perovskites, carbides, and nitrides are expected to explain the strain assisted phenomena like sintering to cell interaction with nanoparticles.

1.2.9 Frequency

Every motion and position can be predicted exactly by classical mechanics, and this is an inevitable consequence for larger objects. However, Newton's revelation is occasionally insufficient when object is very small, preferably nanoscale. This group of material exhibit some weird and unusual phenomena that are advantageous for the new generation nanotechnology. The classical mechanics cannot explain their behavior, rather quantum mechanics can do it. Thus, oscillation is most concerned in consideration of motion of small-scale objects. An oscillation or vibration follows repetitive motion in time with a particular frequency that depends on the mass and stiffness of the object. Prior to explaining such phenomena, let us consider a simple ball with having mass "m" is attached to a spring with spring constant of "k." This spring follows simple harmonic motion when it is slightly displaced in the direction of "x" from its equilibrium position, and this force F = –kx. From Newton's 2nd law of motion:

$$F = m\left(\frac{d^2x}{dt^2}\right) = -kx \tag{1.31}$$

So,

$$\frac{d^2x}{dt^2} = -\left(\frac{k}{m}\right)x = -\omega^2 x \tag{1.32}$$

The angular frequency $\omega(rad/sec) = 2\pi f = (k/m)^{1/2}$ represents sinusoidal behavior of the previous equation with expression of $x = A\cos(\omega t + \phi)$, where A is the amplitude of oscillation, and ϕ is the phase of oscillation. In the same time, the frequency of oscillation 'f' is the inverse of the oscillation period "τ." Thus:

$$f = \frac{1}{2\pi}\left(\frac{k}{m}\right)^{\frac{1}{2}} \tag{1.33}$$

The resultant or effective mass is $m_{eff} = 0.24$ m for rectangular beam when beam is subjected to fix in one end. The stiffness further depends on the dimension (length-L and width-w, thickness-t) and elastic modulus (E), for an example, the stiffness of a rectangular beam (k_{beam}) is:

$$k_{beam} = \frac{Ewt}{4L^3} \tag{1.34}$$

Nevertheless, this frequency is very high for nanoscale objects and enables to detect and like to store the information rapidly (may be in nanosecond!) in the world of miniaturization devices. In order to support this phenomenon, let's consider a 10 μm long, 0.5 μm width and 0.2 μm thin Si beam having density ($\rho = 2.33$ g/cc) and elastic modulus (E = 110 GPa) is attached in one end, and it enables to switch over within nanoseconds from downward to upward direction during vertical vibration [16]. Thus, frequency of 1D nanoscale objects inversely proportional to two important parameters mass and length of the wires, whiskers, or tubes, etc.

1.2.10 Velocity and Motion in Fluid

Take a spherical (5 mm diameter) small iron ($\rho_{Fe} = 7.1$ gm.cm^{-3}) bearing ball and allow to fall under gravitational force within a jar of water (viscosity $\eta = 0.89$cP), and, hence, it will slowly fall down under in a particular terminal velocity and direction. Now reduce the particle size from millimeter to micron and, finally, nanoscale, and what will be the expected velocity and motion? Importance of such small-scale object is very important in consideration to design micron size nanorobots for imaging and surgery or make suspended drug-loaded nanoparticles for targeted medication. While discussing this aspect, it is worth to remember the influence of surface by volume ratio of nanoscale particles that expose and adsorb more molecules compared to bulk entities. Furthermore, particle surface roughness and annular space dimension may alter the velocity and motion of the solid objects during flow in a fluid. Either inertial force or viscous force is the predominant factor to control the velocity of fluid. In the same fluid, velocity and characteristics may change due to the presence of foreign particles. The ratio of such forces is known as Reynold's number (R_e), it is nondimensional quantity and defined by:

$$R_e = \frac{inertial\,force}{viscous\,force} = \frac{\rho vD}{\eta} \tag{1.35}$$

where, ρ = density of the fluid (kg/m³), v = characteristics velocity (m/s), D = characteristics diameter or dimension of the object (m), η = coefficient of the dynamic viscosity (N.S/m²). It seems that a different degree of horizontal movement is expected when 2 m long human swims in water ($\rho = 1000$ kg/m³, $\eta = 0.89 \times 10^{-3}$ N s/m²) at the speed of 1 m/s compared to 2 μm long bacteria at 10 μm/s. A distinct difference R_e in the order of 10^{12} is observed in two cases that encompass the larger object dominates by inertial momentum, whereas smaller objects prefer viscous force only. In this context, one can think about to equate the Reynold's number by alteration of density, velocity, and viscosity, however, daily activity of human life at very low Reynold's number is tough, but nanoscale prefers! In order to estimate the settling velocity of nanoscale object, an interesting force balance among viscous drag force (F_v), buoyancy of liquid (F_b), and gravitational force (F_g) can be balanced, and it depicts:

$$F_v + F_b = F_g \tag{1.36}$$

where $F_v = 6\pi R v_s \eta$; $F_b = \rho_f V_p g$; and $F_g = \rho_P V_p g$; in which R = radius of particle (D/2), v_s = settling velocity, ρ_f = fluid density, ρ_P = particle density, V_p = volume of the particles, and g = gravitational acceleration. Substituting the parameters and solving the Equation 1.36, the following relation can obtain, where the settling velocity is square proportional of radius (R) of the particle:

$$v_s = \frac{2R^2 \rho_f}{9\eta}\left(\frac{\rho_P}{\rho_f} - 1\right)g \tag{1.37}$$

In consideration of the previous equation, 0.2 μm, Fe_3O_4 magnetic particle (density = 5.6 g/cc) experience settling velocity of few micron/sec in water and maintain suspension more than a day when crossed 30 cm long columnar passage. The earlier equation is explicit that the velocity is proportional to the square of the diameter of particles and enhance the retention time to travel that further increases the probability of the motion across jig jag path, known as Brownian motion. This particle motion can be represented by Gaussian distribution, where the diffusivity (D′) of particle is inversely proportional to the particle radius (R), and described as [17]:

$$D' = \frac{kT}{6\pi\eta R} \tag{1.38}$$

k = Boltzmann constant, T = temperature, R = particle radius, η = fluid viscosity. However, three-dimensional nanoscale particle motion is a complicated issue and need to solve the differential equations with concern of both diffusive motion and viscosity of the fluid in describing the particle's motion. Discussed data and plots imply both metal and ceramics have several different physical properties when size reduces in bulk, and thus nanomaterials are an interesting choice to fabricate components for different horizons. In the perspective of book content, however, a brief classification and synopsis on the ceramics are emphasized in consecutive sections.

1.3 Classification of Ceramics

Materials selection, and understanding their properties, is often a challenging task, and argument is there to select either metal or ceramic or polymer for structural and functional applications. Nevertheless, it is indeed that each and every material has different degree of advantages and disadvantages to target better and extensive performance of a specific application for prolong duration, some are highlighted in Table 1.2. A brief material classification along with the influence of nanoscale for both traditional and advanced ceramics is illustrated in Figure 1.8.

Metals have characteristics like a decent mechanical strength, hardness, high thermal and electrical conductivity, ductility, etc. These are often prone to corrosion damage, as the metals react with their environment to re-form those compounds. They have a very good property to be shiny due to nonlocalized electron clouds on the surface, which reflect

TABLE 1.2

Comparative Properties of Metal, Ceramic, and Polymers

Properties	Metals	Ceramics	Polymers
High temperature strength (MPa)	Moderate	High	Very poor
Ease of fabrication	Good (ductile)	Poor (brittle)	Very good (ductile)
Conduction (Thermal or electrical) (W/m.K or S/m)	Good conductor	Insulator	Insulator
Resistance to chemical attack (mm/year)	Poor	Inert	Inert
Dimensional stability (stiffness) (N/m)	High	High	Poor
Density (gm/cc)	Very high	Low	Very low
Lustre (GU)	Excellent	Poor	Poor
Elastic modules (MPa)	High	Very high	Low
Melting point (°C)	Moderate	High	Low
Heat capacity (J/K)	High	Moderate	Low
Hardness (GPa)	Moderate	High	Low
Toughness (MPa)	High	Moderate	Low
Coefficient of thermal expansion (K^{-1})	Moderate	Low	High
Compressive strength (MPa)	Moderate	High	Low
Tensile strength (MPa)	High	Moderate	Low
Dielectric constant (unit-less)	Infinity	High	Very low
Magnetism (T)	High	Low–high	Very low
Band gap (eV or J)	Very low	Moderate	High
Wear rate (mm^3/Nm)	Moderate	Low	High
Coefficient of friction (unit-less)	High	Low	Very low

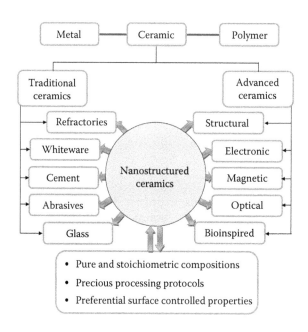

FIGURE 1.8

Classification of materials in specific ceramics. Nanostructured ceramics has multivariant scope to develop both traditional and advanced ceramics, although precious control demands to achieve the desirable properties.

back the light rays. To the contrary, the ceramics are generally nonmetallic, inorganic compounds, preferentially oxides, nitrides, and carbides that either exist in nature or are synthetic, preferentially at high pressure and/or temperature [18].

Most of the ceramics have electrical and thermal insulation, high temperature resistance, and corrosion resistant behavior. In comparison to metals, these materials have lower electrical and thermal conductivity, higher stiffness, good resistance to corrosive environments, and lower fracture toughness, but exceptional thermal conductivity is also noticed, for example, some nonoxide ceramics such as SiC [19]. Polymers are generally organic compounds. This group of material has very large molecular structures, consisting of unit structures known as monomers. Usually, these are light weighed, low density materials, and not stable at high temperatures. From a day-to-day activities point of view, polymers are mostly used materials due to their easy moldability, in turn, can be formed into complex shapes. Their strength, stiffness, and melting temperatures are generally much lower than those of metals and ceramics. Their lightweight, low cost, and ease of forming make them the preferred material for many engineering applications.

Among all materials, prime emphasize is given to ceramics, and this can be further classified into two types: traditional ceramics and advanced ceramics. High purity and surface chemistry, precious stoichiometric composition, and processing protocol are more popular to fabricate the advanced ceramics compared to traditional ceramics, and thus component cost is relatively higher in the former case. However, recent trends are converging to synchronize the traditional ceramics through nanoscale materials. Thus, it is worthy to highlight about the gigantic potentiality of nanostructured ceramic in both traditional and advanced ceramics.

1.3.1 Traditional Ceramics

From ancient age, the use of traditional ceramics in preference clay-based components and their multimode uses are well known to our society. In advance, the traditional ceramics including refractories, whitewares, tiles, cement, abrasives, and glasses have significant interest and regular use for socio-economic benefit. Apart from the use of conventional raw materials and fabrication protocol, insertion of nanoscale materials is now a new horizon of research to enhance the performance and user-friendly of such traditional ceramics.

Literally, refractories are the ceramic materials that can withstand very high temperature, have high thermo-mechanical strength, and have high corrosion resistance. Several shaped and unshaped refractories lining are essentially required for the metal and alloys manufacturing vessel. Despite conventional refractories, high surface area nanoscale carbon particles facilitate synergetic strengthening effect and reduce the carbon content in shaped refractories, and, thus, results enhance slag resistance and low carbon pick-up during manufacturing of steel [20]. Recently, nanobonded refractory castable is also attempted to develop unshaped refractories through using nanocolloidal silica mullite and alumina [21].

Whitewares are often referred to as the triaxial bodies owing to clay, quartz, and feldspar. This broad class of ceramic products is favorably white to off-white in appearance, but not necessarily white always. Ware is the technical name for a ceramic object that is sold in commerce. These includes porcelain, sanitary ware, earthenware, stoneware, bone china, and vitreous tiles. Regardless of the decorative and aesthetic appearance, researchers have tried photocatalytic nanotitanium dioxide (TiO_2), anatase-coated functional tiles to make self-cleaning, self-sterilizing, and air purification under irradiation of ultraviolet sun light [22].

Cements are the binding agent generally used for the building and construction purpose, having a composition of calcium silicate, referred as Portland cement. Another type of cement is also there which can be used at a very high temperature, particularly used in refractory industries, known as refractory cement, and generally has a composition of mixture of various calcium aluminates. However, uncontrolled microvoids in the cement matrix of cementitious material drastically reduce the load-bearing capacity because of weak interface between aggregates and mortar that induces microcrack initiation and promote crack propagation. It, therefore, becomes imperative to reduce the voids in the mortar and promote a better hydration degree of cement. A method that has been proven to be effective is the use of nanocement particles. The smaller cement particles result in an effective hydration process and create smaller ettringites and calcium hydrates [23].

Abrasives are preferentially ceramics that can withstand very high mechanical abrasion without any wear and tear. These are used to cut, grind, and polish other softer materials and shape or finish a work-piece through systematic and precious grinding and polishing. Nanoscale abrasives can reduce the average roughness of surface up to the extent of a few nanometers. In this aspect, Saint-Gobain develops a wide range of nanoabrasive slurries that are capable to withstand outstanding polishing speed and excellent surface finish on a variety of substrates [24].

Glass can be defined as amorphous solid produced by the cooling of molten state of material, where the cooling rate should be faster compared to the crystallization time of their atoms. In earlier days, glass was considered as ceramic material, but now-a-days with the inventions of science, metal and polymer can also be transformed to metallic glass and amorphous-polymers in a suitable heating profile. Interestingly, nanocrystallized transparent glass provides very high hardness and strength compared to ordinary glass or even standard crystallized glass results much harder than ordinary glass. High temperature up to 1800°C and prolonged time of more than a day prefers molecular alignment of the glass on a nanoscopic scale and results in hardening [25]. Furthermore, the advancements of nanotechnology applied in the field of glass technology are resulting in smart glasses, self-cleaning glass, etc.

1.3.2 Advanced Ceramics

From prehistoric time to modern technical era, humans are dependent on ceramics. This demand is more and more in the recently advanced technological world. With advancement of technology, the understanding level of composition, structure, properties, and application of ceramics has been improved in wide horizons. These assists help to grow their applications in space, automobiles, electronics, renewable energy, environment, paints, textile, water filtration, food storage, medication, etc. Thus, advancement of ceramics leads us to a smart life with progressive models in various interdisciplinary broad groups, like structural ceramics, functional ceramics, and bio ceramics. In convention, structural ceramics is predominately designed for load bearing application, although ceramic hip prosthesis is an example of structural ceramics as well as bioceramics. So, one successful material has to cover different range of criteria.

Moreover, functional ferrite particles are being used for cancer treatment and imaging, and thus researchers are encountered as bioceramics as well. However, all structure ceramics like Si_3N_4 piston, zirconia scissors, alumina cutting tool insert, and so much more are no longer related to bioceramics. In preference, functional ceramics including magnetic, dielectric, optical, photocatalytic, mechanoelectrical and more have versatile applications and enormous potential in recent days. Although, one can debate on this classification, however, any modern advance ceramics that are not included in traditional ceramics will get nearer within this class of advanced ceramics. It is commendable to mention that current research progress has enough upcoming thrust to convert the traditional ceramics to advanced mode by the judicious use of nanomaterials and nanotechnology.

Interestingly, ceramics exhibit extensive range of properties including conductors (In_2O_3, SnO_2, YSZ, etc.), semiconductors (WO_3, ZnO, TiO_2, $BaTiO_3$, $CoFe_2O_4$, etc.), and insulators (Al_2O_3, ZrO_2, $Ca_{10}(PO_3)_6(OH)_2$, etc.) in variation of crystal structure and dimension. In addition, the composite materials containing nanoceramics have captured attention of many technologists. By inclusion of nanostructured ceramics in varying volume fractions, the properties of the composite materials can be tailored well, resulting according to the desired application fields. In order to develop dense ceramics for advance application, it is necessary to consider the granulometry and their packing density that eventually reflects in density and porosity. However, the controlling over microstructure is desired for particular applications. These materials also cover in order to understand the structural, functional, and biological applications in the perspective of energy, environment, and health issues (Figure 1.9). Prior to analyzing these burning zones, a brief understanding on the

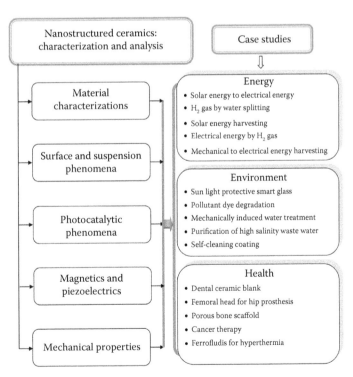

FIGURE 1.9
Bird-eye view of the book content in the perspective of characterization and analysis of nanostructured ceramics.

different characterizations of nanostuctured ceramics, surface and suspension phenomena, photocatalytic behavior, magnetics, piezoelectrics, and mechanical properties have been discussed in consecutive sections, followed by case studies that are enlightened to understand the importance of the subject and future development.

1.4 Rationale of Energy, Environment, and Health Materials

Nanotechnology has become the most demanded arena that is providing us a blessed lifestyle for having the potential to improvise renewable energy sources, make pollution free environment, and advance medication. Energy storage has been greatly enhanced by utilizing nanomaterials as both energy sources and energy storage. Nanomaterials are also being intensely used for the conventional photovoltaic cells. In the case of silicon crystal solar cells, an anti-reflecting coating made of nanomaterials is adhered for higher light yield that ultimately gives a better efficiency. It leads to a path to develop other alternative solar cells, consists of silicon, copper, selenium, indium, or gallium, and many more with very high absorption coefficient and flexibility in use. An innovative recent development in energy application is organic solar cells. It is a two-type, dye-sensitized solar cell consisting of liquid electrolyte and polymer solar cell made of solid electrolyte [26]. In fact, the latter is more flexible to fabricate for different modes of utility including military tents, bag packs, and portable electronic devices and so on, but conversion efficiency is less compared to counterpart dye-sensitized solar cell. Apart from the invention in renewable energy sources, batteries, super-capacitors, and nanocatalyts are tremendously growing in energy sectors. The gift of nanotechnology does not end at this. It has also provided us safe and effective water purification, filtration, and desalination through cheap and portable systems, which can lead the world into a better future where everyone will have possibility of their drinking water resources. In this aspect, several nanomaterials are seeking attention, for example, TiO_2, $BaTiO_3$, ZnO, WO_3, etc. as effective adsorbents and nanoscale photocatalyst to remove contaminants from the waste water of textile industries through several mechanisms. Nanosilver impregnated ceramic filters have already marketed due to their anti-bacterial and anti-viral action, for example, the nanosilver is used in infant soaps and many more cosmetic products to avoid anti-bacterial treatment. Synthetic biomolecules are better known as nanomedicine and have massive opportunity for unprecedented breakthroughs in medicines to damage monitoring, repairing, construction, and immune human biological systems at molecular level. A different class of synthetic nanoparticles (using as direct) and nanostructured bulk ceramics (as indirect application of nanoparticles) may be used as a bridging material within nanoscience and nanotechnology that eventually resolve different successive applications including drug delivery, medical imaging, cancer treatment, porous scaffolds, bioimplants, etc. In theoretical perspective, bottom-up nanomaterial synthesis resembles a natural phenomenon and thus controls synthesis, and device fabrication can target more precise and effective components. Thus, successive synthesis materials and their judicious selection provide renewable energy, clean environment, and effective medical treatment that in due course are beneficial to maintain healthy life for ourselves and next generation as well. In brief, this class of materials research is rationalized in the view of both socio-economic advantages.

1.5 Handling and Ethical Issues

Development of nanoscience and, hence, nanotechnology is like a double-ended sword. Undoubtedly, it has the capability to enhance the status of human life. At the same time, it can destroy its origin, if not used wisely. Miniaturized electronics and molecular machines sound amazing, but it also arises many potential threats to mankind by violation of basic human rights. Molecular microphones and cameras can peep into our freedom and privacy. Unregulated use of nanotechnology leads to issues related to health, safety, social, philosophical, environmental, educational, and other legal issues. Nanoethics is a discipline which deals with the study of societal and moral implications of advances in nanotechnology and handling of such enormous advancements without disturbing the universal balance. It provides a set of guidelines that should be followed while using these nanomaterials. Issues that nanoethics takes into consideration are broadly classified as:

1. Health and safety issues
2. Societal and philosophical issues
3. Environmental issues
4. Educational issues
5. Other issues (moral and legal etc.)

Thus, proper handling of nanomaterials involves:

- Warning signs at the entrance of labs using nanomaterials
- Labeling of storage containers
- Identification of people who are potentially exposed to nanoparticles and hazardous equipment; provide them adequate medical care regularly and necessary nanosafety training
- People working with nanoparticles may inhale them or get them on their skin and that may be carcinogenic or cause dermal issues, and, thus, we need to protect them from skin disease

Usually, the influence of nanoparticles on human body can be done three ways:

1. *Inhalation*: Gases and vapors are the most common substances, although inhaled mists, aerosols, fine dust in preference nanoparticles, nanoobjects, their agglomerates, and aggregates may directly injure the pulmonary epithelium at various levels of the respiratory tract, leading to a wide range of tissue damage and disorders from tracheitis and bronchiolitis to pulmonary edema. For example, silicosis is a common disease for industrial workers who are frequently exposed to nanoscale silica-based particles [27].

2. *Dermal exposure*: Skin comprising of epidermis and dermis, wherein, hair follicles and sweat glands provide pathways across these layers and peripheral blood flowing into the dermis. Intact stratum corneum (protective epidermis layer) provides an effective barrier against viruses, bacteria, and atmospheric toxic components. Although, this barrier is not absolutely impervious and theoretically very small

particles can diffuse across the stratum corneum via cellular and/or intercellular pathways. The damaged barrier can facilitate permeation of foreign particles through skin. For example, the subchronic skin exposures to TiO_2 (in sun protection cosmetics) could induce inflammation of the epidermis, leading to effects such as focal parakeratosis and spongiosis, whereas chronic exposures to TiO_2 may accelerate skin aging [28].

3. *Ingestion*: Use of biomedicine made of polymeric, solid lipid, hydrogels, metal, and ceramic nanosystems are growing exponentially, as these enhance the treatment efficacy and reduce the side effects. However, unwanted ingestion of nanoparticles, overdose, and noncompeting nanomedicine may be the reason of gastrointestinal diseases that affect any part of the gastrointestinal tract including acute, chronic, recurrent, functional disorders, or liver damage [29]. Thus, the parameters such as shape, size, surface chemistry, and geometry of nanoparticles are also important to encounter in the designing of a nanocarrier.

The handling of nanomaterials and careful utilization is a civic concern during synthesis of nanomaterials and manufacturing of nanodevices.

1.6 Concluding Remarks

The last two decades continuous advent in spectroscopic and imaging technique provides us deeper insight on how small object functions in the atomic scale and these up growing approach allowing quantum level understanding of materials. Combined with advanced physics and chemistry, scientists have been enabled to develop continuous nanostructured products consisting of unique properties, often tailored to meet specific applications in diverse areas such as agriculture, information technology, defense, energy (harvesting, conversion, and storage), environment, medicine, etc. Despite existence technologies, we common people consume fresh air, water, food, medicine, energy, wood, and minerals in our daily lives, as well as cropping food, pumping groundwater, harvesting wood, mining minerals, and burning fuel exhaust our resource base and produces pollution. In fact, this trend turns exponentially as population is highly correlated with consumption of energy, domestic water, and medicine, as when diseases increasing day to day in a polluted atmosphere. Resource use and their impact in environment are critically measured by the energy consumption. *This causes* increasing prevalence of chronic diseases and tissue damages that contribute toward the increase in total healthcare expenditure on medicine and medical devices. *Because of our continuous intervention, either natural energy resources are decreasing or water getting polluted that gradually asks for more dependency on synthetic medicine, and thus nanotechnology is in upfront to supply low cost renewable energy, protect environment, and cure critical health disorder by several classes of nanomaterials. Among material worlds, nanostructured ceramic oxides encompass astonishing properties and are opening up* attractive horizons for these highlighted and challenging applications. In this context, a brief discussion and understanding on the different material characterization and their analysis are highlighted in Chapter 2.

References

1. K. F. Brennan, *The Physics of Semiconductors*, Cambridge University Press, New York, 1997.
2. G. W. Neudeck and R. F. Pierret, *Advanced Semiconductor Fundamentals*, Peterson Education, Upper Saddle River, NJ, 2003.
3. R. Dalven, Metal-semiconductor and metal-insulator-semiconductor devices, In: *Introduction to Applied Solid State Physics*, Springer, Boston, MA, 1980.
4. Y. Cui and C. M. Lieber, Functional nanoscale electronic devices assembled using silicon nanowire building blocks, *Science* 291(5505), 851–853, 2001.
5. T. Supasai, S. Dangtip, P. Learngarunsri, N. Boonyopakorn, A. Wisitsoraat, and S. K. Hodak, Influence of temperature annealing on optical properties of $SrTiO_3$/$BaTiO_3$ multilayered films on indium tin oxide, *Appl Surf Sci* 256, 4462–4467, 2010.
6. F. J. Owens and C. P. Poole Jr, *The Physics and Chemistry of Nanosolids*, John Wiley & Sons, Hoboken, NJ, 2008.
7. J. Mackenzie, A. Moore, and J. Nicholas, Bonds broken at atomically flat crystal surfaces—I, *J Phys Chem Solids* 23, 185–196, 1962.
8. W. H. Qi, Size effect on melting temperature of nanosolids, *Physica B* 368, 46, 2005.
9. F. A. Lindemann, The calculation of molecular vibration frequencies, *Physik Z* 11, 609, 1910.
10. L. H. Liang and L. Baowen, Size-dependent thermal conductivity of nanoscale semiconducting systems, *Phys Rev B* 73, 153303, 2006.
11. Y. F. Zhu, W. T. Zheng, and Q. Jaing, Modeling lattice expansion and cohesive energy of nanostructured materials, *Appl Phys Lett* 95, 083110, 2009.
12. W. H. Qi and M. P. Wang, Size and shape dependent lattice parameters of metallic nanoparticles, *J Nanopart Res* 7, 51–57, 2005.
13. H. J. Wasserman and J. S. Vermaak, On the determination of a lattice contraction in very small silver particles, *Surf Sci* 22, 164–172, 1970.
14. C. Goyhenex, C. R. Henry, and J. Urban, In-situ measurements of lattice parameter of supported palladium clusters, *Philos Mag A* 69, 1073–1084, 1994.
15. F. Zhang, S.-W. Chan, J. E. Spanier, E. Apak, Q. Jin, R. D. Robinson, and I. P. Herman, Cerium oxide nanoparticles: Size-selective formation and structure analysis, *Appl Phys Lett* 80(1), 127–129, 2002.
16. B. Rogers, J. Adams, and S. Pennathur, *Nanotechnology: Understanding Small Systems*, 3rd ed., CRC Press, Boca Raton, FL, 2014.
17. C. W. Shong, S. C. Haur, and A. T. S. Wee, *Science at the Nanoscale: An Introductory Textbook*, Pan Stanford Publishing, Singapore, 2010.
18. G. Murray, C. V. White, and W. Weise, *Introduction to Engineering Materials: Behavior: Properties, and Selection*, Marcel Dekker, Inc., New York, 1993.
19. E. Ruh and J. S. McDowell, Thermal conductivity of refractory brick, *J Am Ceram Soc* 45(4), 189–195, 1962.
20. S. Behera and R. Sarkar, Low-carbon magnesia-carbon refractory: Use of N220 nanocarbon black, *Appl Ceram Technol* 11(6), 968–976, 2014.
21. M. Nouri-Khezrabad, M. A. L. Braulio, V. C. Pandolfelli, F. Golestani-Fard, and H. R. Rezaie, Nano-bonded refractorycastables, *Ceram Int* 39, 3479–3497, 2013.
22. M. Hasmaliza, H. S. Foo, and K. Mohd, Anatase as antibacterial material in ceramic tiles, *Procedia Chem* 19, 828–834, 2016.
23. P. Sabdono, F. Sustiawan, and D. A. Fadlillah, The effect of nano-cement content to the compressive strength of Mortar, *Procedia Eng* 95, 386–395, 2014.
24. https://www.surfaceconditioning.saint-gobain.com/products/classic-cmp-slurries-sic-polishing.
25. http://www.countertopresource.com/nano-crystallized-glass-for-countertops/.

26. P. Wang, S. M. Zakeeruddin, J. E. Moser, M. K. Nazeeruddin, T. Sekiguchi, and M. Grätzel, A stable quasi-solid-state dye-sensitized solar cell with an amphiphilic ruthenium sensitizer and polymer gel electrolyte, *Nat Mater* 2(6), 402–407, 2003.
27. D. Napierska, L. C. Thomassen, D. Lison, J. A. Martens, and P. H. Hoet, The nanosilica hazard: Another variable entity, *Part Fibre Toxicol* 7, 39, 2010.
28. J. Wu, W. Liu, C. Xue, S. Zhou, F. Lan, and L. Bi, Toxicity and penetration of TiO_2 nanoparticles in hairless mice and porcine skin after subchronicdermal exposure, *Toxicol Lett* 191, 1–8, 2009.
29. I. L. Bergin and F. A. Witzmann, Nanoparticle toxicity by the gastrointestinal route: Evidence and knowledge gaps, *Int J Biomed Nanosci Nanotechnol* 3(1–2), 10, 2013.

2

Material Characterizations

2.1 Introduction

Development of nanomaterials in specific nanostructured ceramics is a challenging and fascinating expanse for the material research community. While such activity is in the mainstream, one can come across several classic characterization protocols, including phase transformation, crystallography, imaging, size phenomena, and spectroscopy to fulfill the research and development objective. At the same time, obtaining authentic data from different characterization tools is only a secondary concern until those are not properly analyzed and correlate to each other in a systematic approach. Several classic books described the working principle, mode of data acquisitions, and characterizations of concerned techniques [1–5]. Herein, the brief discussion is primly enlightening to interpret the data and their analysis to promote the nanoscale research. In this perspective, different sections are chronologically describing the brief characterization protocols and data analysis for their plausible utilization. Section 2.2 describes the mode of phase change while preparing nanoparticles and advanced industrial materials at elevated temperature from organic or inorganic precursors with an emphasis on the decomposition, amorphous state, crystallization, specific heat, enthalpy change, and reaction kinetics. Section 2.3 briefs how to index the crystallographic phase, miller indices, crystallite size, true density, and strain related to nanoparticles. Apart from morphological behavior, crystallinity, the lattice spacing of particles, and plane indexing of nanostructured ceramics are discussed in Section 2.4. In this connotation, quantification of particle size distribution, surface area, and porosity of nanoparticles are emphasized in Section 2.5. Despite particle synthesis phenomena, other important clusters of material characteristics known as crystal disorder, binding energy, band gap, unknown concentration, existence of functional groups, and bond behavior in order to develop nanoscale ceramics are described in Section 2.6.

2.2 Thermal Analysis

Functional and structural application demands wide choice of nanoparticles either undoped or doped or composites that are usually synthesized starting from different organic and inorganic precursors. The oxide and nonoxide ceramic particle synthesis at elevated temperature is in the forefront, and thus selection of processing temperature of the targeted phase consisting of either amorphous or crystalline character is a critical issue to achieve the resultant properties. A simple thermal analysis can do it.

Differential thermal analysis (DTA) is an early days (1899) technique for ceramic industries to evaluate the temperature transitions in materials. In advance, differential scanning calorimetry (DSC) is a relatively modern (1964) instrument consisting of a separate heater for tested sample and reference material under dynamic heating treatment. Because of the single furnace design in DTA, heat flux is less sensitive to small transitions and contributes relatively less accurate values for specific heat (C_p) and enthalpy (ΔH). Although, the additional heat flow difference (DSC) compared to only temperature difference (DTA) allows precious determination of the process enthalpy or heat of crystallization during transitions. The measured μV (thermal voltage, microvolt) signal associates with temperature difference in between sample and reference, and eventually converts into a heat flux difference in mW (milliwatt, mJs^{-1}) through appropriate calibration. In the same time, another important thermal analysis module refers to thermogravimetric analysis (TGA) to deliver the mass change information when material is subjected to experience a constant rate of heating. Thus, combination of DSC-TGA is an excellent tool to study and understand the several physical properties as a function of temperature when the factors vary, like particle size, packing density, sample weight, heating rate, and atmosphere. With the synthesis of ceramic particles, one can effectively determine thermal decomposition behavior of precursors, formation of amorphous state, activation energy for crystallization, heat capacity, enthalpy, degree of crystallization, etc. However, first derivative of DSC and TGA plots often provide an inflection point for accurate and more in-depth analysis. Thus, DSC-TGA curve interpretation is therefore obvious for the synthesis of ceramic nanoparticles as a starting point.

2.2.1 Differential Scanning Calorimetry and TG Patterns

To begin understanding, we need an equipment (thermal analyzer) to obtain a representative DSC and TG plot, as shown in Figure 2.1a and b. Obtaining a combined plot is common practice during dynamic thermal analysis. A calorimeter measures the heat into or out of a sample, but differential calorimeter estimates the heat of sample relative to a reference. In advance, the differential scanning calorimeter measures both reference and sample heated up with a linear temperature at a particular heating rate.

The process refers to endothermic when heat flows into the sample and exothermic when heat flows out of the sample. Since the DSC is at constant pressure, heat flow is equivalent to enthalpy changes:

$$\left(dq/dt\right)_p = dH/dt \tag{2.1}$$

Here, dH/dt is the heat flow and the difference between the sample and reference (α-alumina):

$$\Delta dH/dt = (dH/dt)_{sample} - (dH/dt)_{reference} \tag{2.2}$$

and it can be either positive (endothermic) or negative (exothermic). Thus, DSC plot represents the heat flow (exo or endo) with respect to temperature variation (Figure 2.1a), but baseline correction endures precise information. This plot exclusively provides endo and exothermic peaks related to different transformation including glass transition, crystallization, melting, decomposition, oxidation, evaporation, and sublimation, but it does

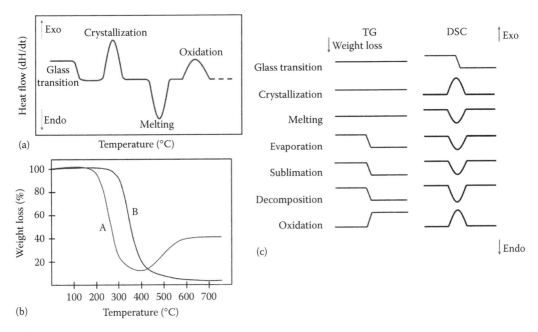

FIGURE 2.1
(a) A typical DSC plot consists of probable exothermic and endothermic peaks related to different transformations, (b) weight loss and gain phenomenon during TGA, and (c) cumulative representation of both DSC-TG plot characteristics for any particular transformation.

not contribute any information about weight loss or gaining during dynamic heating process, eventually that encompasses the completion of the specific transformation. The TGA plot represents two different characteristic features during thermal treatment that finally implies either weight loss or gain or remaining constant. The plot "A" follows early stage of weight loss compared to plot "B," although a substantial weight gain is noticed after 400°C in former case and follows constant plateu beyond 600°C. In consideration of Figure 2.1c, weight loss is accomplished by evaporation/sublimation/oxidation, whereas the weight gain is due to oxidation (e.g., oxidation of nonoxide ceramics). Combined heat flow and weight change characteristics of aforesaid processes are summarized in Figure 2.1c. After performing the thermal analysis, one can pick up the temperature in order to synthesize the targeted phase, followed by confirmation by phase analysis, say X-ray diffraction (XRD) technique (see Section 2.3). In convention, this dynamic analysis describes a particular temperature, but certain degree of less temperature (~25°C less from estimated dynamic DSC temperature) can develop the identical phase during soaking in prolong time or less time at relatively high temperature (~25°C high from estimated dynamic DSC temperature). This hypothesis facilitates to synthesize pure phase of nanoscale particles and avoids unwanted growth of the particles. However, this process demands extensive XRD analysis to ensure the formation of pure phase. In a recent study, a flash calcination (high temperature less time) protocol was used in the synthesis of 120 nm pure and tetragonal $BaTiO_3$ nanoparticles from solid oxide mixture of $BaCO_3$ and TiO_2 [6]. Apart from the understanding of phase transformation and different chemical reactions, several important thermodynamic parameters in the perspective of the development of ceramic particles, such as specific heat capacity (C_p), enthalpy (ΔH), and activation energy for crystallization (E_a) can be estimated, as in the following discussion.

2.2.2 Specific Heat Capacity

Heat capacity is the amount of heat required to raise one kelvin temperature of a material, unit is J/K. It is not normalized. Thus, specific heat capacity defines the amount of heat required to raise one kelvin temperature of one gram material, unit is J/g.K, and designated as C_p. Crystalline material has more ordered atomic structure, and, thus, less molecular motion results in lower specific heat compared to amorphous material. This can be quantified from DSC plot, and their change provide information about phase change. As DSC analysis is performing under constant pressure and uniform heating rate, the specific heat capacity and enthalpy can be related as:

$$C_p \equiv \left(\frac{\partial H}{\partial T} \right)_P \tag{2.3}$$

The displacement "h" within baseline and sample is proportional to heat capacity of material (Figure 2.2a):

$$h = B\beta C_P \tag{2.4}$$

The proportional factors depend on the heating rate (β) and calibration factor (B), but elimination of these factors by a comparative study with standard sample (sapphire) under identical conditions provide a better insight to quantify the specific heat capacity. Figure 2.2b illustrates the schematic representation of DSC curves with an empty sample holder (pan), a tested sample, and standard sample recorded on the identical condition, and represented in same chart. The specific heat capacity of the tested sample can be calculated with the specific heat capacity of the standard (C_{ps}) and displacements measured from Figure 2.2b:

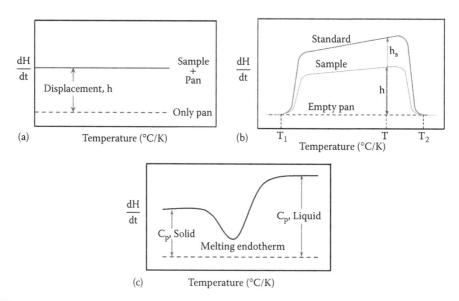

FIGURE 2.2
Enthalpy change with respect to temperature (a) sample, (b) sample and reference, and (c) change in specific heat capacity in different states, say solid and liquid.

$$C_P = C_{Ps}\left(\frac{hM_s}{h_sM}\right) \qquad (2.5)$$

M_s and M are the mass of standard and sample, respectively, h_s is the displacement between the baseline (i.e., empty pan) and standard line. Both the displacements h and h_s can be determined graphically, however, instrument baseline correction is important criteria to establish representative C_p measurement. It is worth to mention that the instrument baseline is like to consider after performing the blank experiment, and it may not be perfectly horizontal. In the same time, the temperature zone T_1 and T_2, as mentioned in figure, is not like to encounter for the estimation of C_p. The C_p difference is noticed for an interesting incidence, say endothermic melting, where a high C_p is recorded after melting compared to solid, this is obvious because of less atomic motion in solid (Figure 2.2c).

2.2.3 Enthalpy Change

From first principle of thermodynamics, enthalpy is the summation of internal energy and product of pressure and volume. So, H = E + PV, dH = dE + PdV + VdP (dE = dq + W, and W = –pdV, only work done by the reaction is work of expansion). Thus, at constant pressure, $dH = dq_p$. This enthalpy of any particular reaction or transformation can be calculated by using the area of any peak (endothermic/exothermic) present in the DSC curve. But these curves need to be done baseline corrections in order to calculate the area, details can be found elsewhere [1]. The total enthalpy change ΔH should be proportional to the peak area (A). If the DSC plot is dH/dt per mass unit versus temperature, then the change in enthalpy (ΔH) is given by the Equation 2.6.

$$\frac{\Delta H}{m} = \kappa \times A \qquad (2.6)$$

where, m is the sample mass and κ is the calibration factor. This factor has to be determined by relating a known enthalpy change to a measured peak area, in convention, the melting of a pure metal, such as indium, is used for calibration. However, one can get the enthalpy change straightforward to use the software of the DSC instrument.

2.2.4 Kinetics and Activation Energy of Crystallization

Crystallization defines the systematic atomic rearrangement process from their random and discrete orientation in amorphous state. It is an exothermic process that evolves heat during this conversion and unit of the process activation energy E_a is kJ/mol or kcal/mol. In fact, this process follows definite transformation kinetics and, is referred to as "kinetics of crystallization," depends on the unit less constant "n," known as, Avrami kinetics exponent [7]. The kinetics and activation energy for a particular isothermal crystallization process can be estimated by Avrami equation and Arrhenius equation, respectively. However, nonisothermal crystallization in preference of glass follows Ozawa and Kissinger equation to estimate the kinetics and activation energy, respectively [8,9]. Avrami equation defines:

$$1 - X_t = \exp(-Kt^n) \qquad (2.7)$$

where, X_t is the volume fraction transformed after time "t," K is a temperature dependent reaction rate constant, "n" is Avrami kinetics exponent (describes the nucleation rate and growth morphology). In assumption, this temperature dependence reaction follows Arrhenius Equation 2.8, where K_o is a pre-exponential constant, E_a is the activation energy, and R is the universal gas constant.

$$K = K_o \exp\left(\frac{-E_a}{RT}\right) \tag{2.8}$$

or

$$\ln K = \ln K_o - \frac{E_a}{R}(1/T) \tag{2.9}$$

Prior to evaluate the "K" and "n" from Equation 2.7, it is necessary to estimate the fraction transformed as a function of time or relative crystallinity from the DSC curve by partial areas under the peak (at temperature of your interest). In order to produce representative data, the determination uses 20 segments of the curve, evenly spaced by temperature [10]. The fraction can be determined on the basis of area [(dH/dT) dT] and can be defined as:

$$X_t = \frac{\int_{T_o}^{T}\left(\frac{dH_c}{dT}\right)dT}{\int_{T_o}^{T_\infty}\left(\frac{dH_c}{dT}\right)dT} \tag{2.10}$$

From Equation 2.7:

$$\ln\left(1-X_t\right) = -Kt^n \tag{2.11}$$

$$\ln\left[-\ln\left(1-X_t\right)\right] = \ln K + n\ln t \tag{2.12}$$

This equation is in the form of $Y = m\,X + C$. Plotting a graph between $\ln\left[-\ln\left(1-X_t\right)\right]$ and ln t gives a straight line whose intercept provides the K and slope gives the value of n. This exponent constant "n" usually varies within 1–4, based on the rate of nucleation, however, some controversy is there to explain the significance of these values. A fraction value (n < 1) defines diffusion-controlled model with contracting geometry, for example, n = 0.5 for 1D growth, n = 1 for 2D growth, and n = 1.5 for 3D growth of the particles [11].

After getting the values of K as a function of temperature, the activation energy (E_a) can be calculated from Equation 2.9. Plotting a graph between ln K and 1/T results in a straight line, as shown in Figure 2.3. The slope of which gives the activation energy: $E_a = -2.303\,R \times Slope$.

In this context, one can estimate the activation energy for nonisothermal crystallization (preferably in polymer and glass) modified Kissinger equation as:

$$\ln\left(\frac{\varphi^n}{T_p^2}\right) = \ln K + \left(\frac{-mE_a}{R}\right)\frac{1}{T_p} \tag{2.13}$$

where T_p is the peak crystallization temperature, φ is the heating rate, K is a constant containing factors depending on thermal history n and m are constants and have values between 1 and 4, R is the universal gas constant. Slope of the plot gives activation energy.

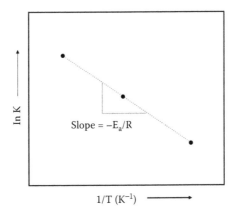

FIGURE 2.3
Arrhenius plot to determine the activation energy of crystallization.

The shape and size of the DSC plot can change with variation of particle size, particle packing density, sample weight, heating rate, and atmosphere, however, extensive data for ceramic systems are limited. Thus, a wide research scope is there to understand the influence of such parameters on estimation of Avrami exponent constant and predict the probable morphology (e.g., clay to mullite), enthalpy, degree of crystallization, and activation energy during transformation of precursors to crystalline ceramic nanoparticles.

2.3 X-Ray Diffraction Analysis

X-rays are a kind of electromagnetic radiation consisting of very short wavelengths and produce high energy when high-speed electrons strike a solid target. The wavelength range is in between 0.01 nm (100 eV) and 10 nm (100 KeV), in other words, it persists within a window of "gamma rays" and "ultraviolet rays." In consideration of energy, it can be classified in two modes, hard X-rays (>10 KeV) and soft X-rays (<10 KeV). The hard X-ray has penetrating ability and widely used to image the inside of the objects, for example, medical radiography, airport security, etc. The wavelength of hard X-ray is too small, similar to atomic size, and thus useful to determine the crystal structure by X-ray crystallography technique. In contrast, soft X-ray is useful tool for critical imaging where material cannot sustain high voltage and low voltage is required for imaging and quantification of chemically distinct nanometer-scale material, for example, soft X-ray spectromicroscopy, with focus on scanning transmission X-ray microscopy [12].

Thus, X-ray diffraction analysis is widely used for not only crystal structure, but also determination of targeted phase (qualitative and quantitative), crystallinity, Miller indices, crystallite size, consist of probable strain, and true density of nanostructured ceramics. Low temperature (selective from dynamic thermal analysis)-assisted isothermal reaction may retain impurity or less crystalline phase, where high temperature promotes wider particle size distribution with heterogeneous strain. At the same time, determination of "hkl" plane for different state of materials may promote to synchronize the functional properties. Usually, particle consists of many fine units called "crystallites," and an appreciable broadening in the X-ray diffraction is obtained when crystallite size is less than 5 nm compared to micron size grain. It should be noted that this crystallite size is equivalent and can be considered as true particle size when particles are really in nanoscale. Apart from details on XRD technique, the mode of data acquisition encompasses the diffraction based on the constructive interference of monochromatic X-rays and a crystalline sample [2]. The interaction between incident rays with the powder or consolidate discs produce constructive diffracted rays with satisfaction the Bragg's law, $n\lambda = 2d \sin\theta$, where n = integer, λ = wavelength of electromagnetic radiation, θ = diffraction angle, and d = lattice spacing of crystalline sample. The result from XRD analysis is a diffractogram that represents the intensity as a function of the diffraction angles, and one can consider their "d" value as a "fingerprint" of any particular material. Deviation of "d" value and thus diffraction peak position consists of homogenous or heterogeneous strain in particles. Homogenous or uniform elastic strain change the d-space and shift the peak position, whereas heterogeneous strain can cause a broadening of the peaks with increment of $\sin\theta$. However, it is worth remembering that the peak broadening is independent with $\sin\theta$ for small crystallite size of a unit cell. Along with crystallite size and strain, determination of true density from crystallographic data is another critical issue to interpret the physical properties.

2.3.1 Indexing and d-Spacing

Powder XRD technique is a common protocol to generate the crystallographic information, and their detail characterization technique is discussed elsewhere [2]. After performing the XRD analysis with a definite step size, scan rate, and a particular angle range of interest, we can obtain a typical XRD pattern that consists of some peaks with indexing at particular angular positions called "2θ" as shown in Figure 2.4. As we know, the simplest crystallographic unit cell is cubic, and thus entire system can be classified as cubic and noncubic system. It is common practice that several analyses including phase identification and their "hkl" indexing can be done by standard software by Joint Committee on Powder Diffraction Standards (JCPDS), but herein objective is to explain how one can perform theoretically indexing of the XRD patterns based on the crystallographic understanding. In this context, first analysis is focused in cubic system, followed by a characteristic sequence of diffraction peaks that when distorted the same cubic into tetragonal and orthorhombic system (for. e.g., BaTiO₃). This cumulative analysis will provide basic insight

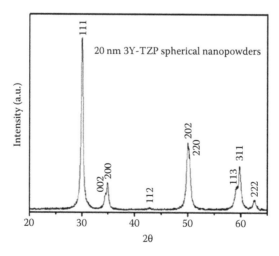

FIGURE 2.4
A typical X-ray diffraction pattern of crystalline 3Y-TZP powder samples (Author Lab data).

on the indexing concept of diffraction patterns, however, a number of complicated issues are discussed in literature.

Before we discuss any further, we assume that reader is fully aware of Miller indices and directions in a unit cell along with knowledge of different crystal structures and Bravais lattices. In order to know the Miller indices of the plane from which the intensity peak has appeared, we need to build a correlation between the interplanar spacing (d spacing), peak position (2θ), and Miller indices (hkl). This process of determining the Miller indices and thereby the unit cell dimensions is called indexing (can avoid this hurdle by direct use of software!). Most materials we come across are cubic crystals, and it is easy to demonstrate by considering a cubic system. Other crystal systems (noncubic) like tetragonal, orthorhombic, monoclinic, hexagonal, etc. include more unknown lattice parameters (a, b, c) and require additional mathematical calculations. By combining Bragg's equation ($n\lambda = 2d_{hkl}\sin\theta$) with plane spacing equation, we can obtain the Equation 2.14:

$$\frac{\lambda^2}{4a^2} = \frac{\sin^2\theta}{h^2+k^2+l^2} = \frac{\sin^2\theta}{S} \tag{2.14}$$

$h^2+k^2+l^2$ is always an integer, and it can be denoted by "S." The cubic crystal system has four lattice types, namely, simple cubic (SC), fcc, bcc, and diamond, and they have S value within $0 \le 2\theta \le 180°$, with having a definite "forbidden plane," as illustrated in Table 2.1.

Certain integers like 7, 15, 23, 28, 31, etc. are not possible because they cannot be formed by adding three integer squares. A typical XRD peak indexing and corresponding lattice

TABLE 2.1

Characteristic Sequence of Diffraction Peaks of Four Common Cubic Lattice Types within a Range of $0 \le 2\theta \le 180°$

hkl	SC	BCC	FCC	Diamond	S
100	√	X	X	X	1
110	√	√	X	X	2
111	√	X	√	√	3
200	√	√	√	X	4
210	√	X	X	X	5
211	√	√	X	X	6
220	√	√	√	√	8
300	√	X	X	X	9
221	√	X	X	X	9
310	√	√	X	X	10
311	√	X	√	√	11
222	√	√	√	X	12
320	√	X	X	X	13
321	√	√	X	X	14
400	√	√	√	√	16
410	√	X	X	X	17
322	√	X	X	X	17
411	√	X	X	X	18
330	√	X	X	X	18
331	√	√	√	√	19
420	√	√	√	X	20

TABLE 2.2

Example of Indexing and Lattice Parameter Estimation for FCC System

Peak No.	2θ	θ	$\sin\theta$	$\sin^2\theta$	Ratios of $\sin^2\theta$	Dividing $\sin^2\theta$ by $(0.11/3^{\#})$	Whole No. of Ratios (S)	Index	$d = \dfrac{\lambda}{2\sin\theta}$ (Å) $(\lambda = 0.154 \text{ nm})$
1	38.49	19.25	0.33	0.11	1 (0.11/0.11)	3	3	111	2.33
2	44.72	22.36	0.38	0.14	1.27 (0.14/0.11)	3.82	4	200	2.03
3	65.16	32.58	0.54	0.29	2.64 (0.29/0.11)	7.91	8	220	1.43
4	78.31	39.16	0.63	0.40	3.63 (0.40/0.11)	10.91	11	311	1.22
5	82.52	41.26	0.66	0.43	3.91 (0.43/0.11)	11.73	12	222	1.17
6	99.12	49.56	0.76	0.58	5.27 (0.58/0.11)	15.82	16	400	1.01
7	112.04	56.02	0.83	0.69	6.27 (0.69/0.11)	18.82	19	331	0.93
8	116.57	58.29	0.85	0.72	6.55 (0.72/0.11)	19.64	20	420	0.91

[#] Trial computation is based on 3 for FCC (111); others can be considered as 2 for BCC (110); 1 for SC (100); and 3 for diamond (111). Here, 0.11 is the lowest $\sin^2\theta$ value.

parameter calculation is demonstrated through following steps, and their brief is discussed in Table 2.2.

What will happen if $\sin^2\theta$ value is divided by 0.11/2 in consideration of BCC or 0.11/1 in consideration of SC system? In first case, the column 7 value becomes; 2, 2.54, 5.27, 7.27, 7.82, 10.55, 12.55, 13.09, and their immediate whole number (S) does not follow the sequence for BCC (Table 2.1). For second case, the value becomes identical with column 5 and also not matching with "S" value for SC system. Similarly, it does not follow diamond system as well. Thus, the obtained XRD pattern and their peaks are associated with FCC system (Table 2.1), only. In order to minimize the trial computation, it is suggested to take a serious look at the measured X-ray diffraction plot and understand how peaks are discrete throughout the pattern. In consideration of Table 2.1, a common inference can be drawn that the simple cubic system has all allowed reflections: bcc has (h + k + l) = even; fcc has unmixed h, k, l value for one plane (i.e., either all even or all odd), and DC has either h, k, and l all are odd or all are even and (h + k + l) is divisible by 4. Apart from the cubic system, manual indexing and lattice parameter for noncubic system can be estimated with help of more complicated equations, and reader can follow the advance literature [2]. Sometime, the synthesis protocol may introduce additional peaks for the same material, and thus it is obvious this is due to the formation of different crystal structure compared to the targeted crystallography. This is attributed to the distortion in the cell and changing of their crystal symmetry. It facilitates the change in lattice parameters and probability of more interaction with diffraction and results in more number of peaks. A simple competitive XRD pattern for BCC, their tetragonal (a = b ≠ c), and orthorhombic (a ≠ b ≠ c) crystal is illustrated in Figure 2.5. From this schematic and appearance of multiple peaks, one can presume why tetragonal $BaTiO_3$ and ZrO_2 is supposed to split, not in their cubic structures [6].

2.3.2 Theoretical Density

X-ray diffraction allows us to determine the lattice parameters of any crystal's unit cell, and, therefore, its volume together with number of atoms in the unit cell. So, the base for the theoretical density determination should be a single unit cell instead of a few cubic

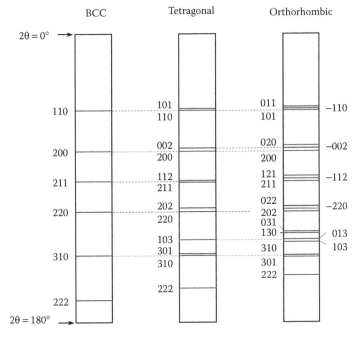

FIGURE 2.5
A characteristic representation of BCC, tetragonal, and orthorhombic crystals of same materials, left to right, the cubic system appears nonsymmetrical that appears more splitting peaks signal, maximum is observed in orthorhombic system.

centimeters, by definition, X-ray density, true density, or theoretical density (ρ) = weight of the total atoms in unit cell/volume of unit cell. So:

$$\rho\left(\frac{gm}{cc}\right) = \frac{\sum A}{N_A V} \tag{2.15}$$

$\sum A$ = sum of the atomic weights of all the atoms in the unit cell; N_A = Avogadro's number, 6.023×10^{23}; V = volume of unit cell (cc). Typical XRD crystallographic data for alumina-zirconia mixed powder are given in Table 2.3. Data reveal that the composite has three

TABLE 2.3

Details Crystallographic Data for Alumina and Zirconia to Determine the True Density

Phase	Crystal Structure	a, α	b, β	c, γ	No. of Atoms in Unit Cell	Atomic Weight of Atoms	ΣA	Unit Cell Volume (Å)³	True Density (ρ)
α-Al$_2$O$_3$	Hexagonal	4.759 Å, 90°	4.759 Å, 90°	12.991 Å, 120°	Al = 12 O = 18	Al – 26.98 Zr – 91.22	611.77	254.8	3.987
t-ZrO$_2$	Tetragonal	3.607 Å, 90°	3.607 Å, 90°	5.176 Å, 90°	Zr = 2 Zr = 4	O – 16	246.65	67.33	6.078
m-ZrO$_2$	Monoclinic	5.150 Å, 90°	5.208 Å, 99.23°	5.317 Å, 90°	Zr = 4 O = 8		492.9	140.76	5.814

different phases namely, α-Al_2O_3, t-ZrO_2, and m-ZrO_2, and we can estimate the true density of individual phase from crystallographic data, as discussed in sequence [13].

The unit cell volume varies with respect to crystal structure, for example, hexagonal $V = 0.866a^2c$, tetragonal $V = a^2c$, and monoclinic $V = abc \sin\beta$.

This is the highest density value, known as X-ray density, theoretical density, or true density for any particular material. The bulk density of material is always lower than theoretical density, and the ratio of bulk density and true density is known as relative density. These values encompass the synthesis process compatibility and explain several physical properties.

2.3.3 Phase Identification and Quantification

XRD pattern consists of the peak positions (determined by 2θ value) based on the d-spacing and definite range of intensity results of crystallinity. At the same time, different ratio of intensities (I/I_1) provides the information about the existence of strongest planes, where I is intensity of particular peak, and I_1 intensity of the highest peak. For phase identification of any unknown sample, Hanawalt method is generally used [14]. In this method, the unknown sample is characterized by comparing "d" values of first three strongest lines (d1, d2, d3) with the already known samples whose data have been tabulated in a card called as JCPDS card. If the unknown sample consists of only one/single phase, it is easy to search and match. If the unknown sample consists of two or more phases in it, then care should be taken while comparing the different "d" spacing values with the standard data. At first, the process is locating a proper d1 group in the numerical search manual, and then the other "d" values are compared to find the closest match with d2 ($\pm0.01A°$). After finding the closest match for all the three strongest peaks, we need to compare their relative intensities with the tabulated values and when the good agreement is made, the identification is said to be completed.

Apart from the phase identification, phase quantification in multiphase material is obvious to understand the synthesis process efficacy that eventually controls the material properties. For this purpose, there are three prime methods of analysis available, namely, external standard method, direct comparison method, and internal standard method. These methods differ from each other on the basis of reference line selection. Of all these methods, "direct comparison method" is the easiest way to use, since, it compares the experimental line intensity from the mixture to a line from another phase in the same mixture [15]. The integrated intensity (peak area) "I" of the diffracted peak from a single-phase powder sample measured by a diffractometer is given in the following form (Equation 2.16), if a sample of the flat plate maintains the infinite thickness (t). So, absorption factor becomes $1/2\mu$:

$$I = \frac{KR}{2\mu} \tag{2.16}$$

again,

$$R = \left(\frac{1}{\upsilon^2}\right)\left[|F|^2 \; P\left(\frac{1+\cos^2 2\theta}{\sin^2\theta\cos\theta}\right)\right]\left(e^{-2M}\right) \tag{2.17}$$

where, K is a constant independent of the kind and amount of diffracting substance, R depends on θ, hkl, and kind of substance μ is the absorption coefficient of material, F is the structure factor e^{-2M} is the temperature factor, θ is the Bragg angle υ is the volume of unit cell, p is the multiplicity factor $\left(\frac{1+\cos^2 2\theta}{\sin^2\theta\cos\theta}\right)$ is the Lorentz polarization factor. For a sample consisting of two phases, this equation can be rewritten as:

$$I_1 = \frac{KR_1c_1}{2\mu_m} \text{ and } I_2 = \frac{KR_2c_2}{2\mu_m} \tag{2.18}$$

$$\frac{I_1}{I_2} = \frac{R_1c_1}{R_2c_2} \tag{2.19}$$

where c_1 and c_2 are volume fractions of phase 1 and 2, respectively. By measuring intensities and calculating the "R" values, we can get the ratios of concentrations from Equation 2.19, and then use $c_1 + c_2 = 1$ to calculate the individual components. This method provides more accurate estimation compared to just taking ratio of summation of all peak areas of particular phase of interest by the total peak areas that appeared during X-ray diffraction. For example, the direct comparison method employed and determined the volume percentage of Si = 48% and Al = 52% for a bimetallic system consisting of integrated intensity (I) of Si(111) – 180.3, Al(111) – 216.2, Al(200) – 93.1, Si(220) – 118.4, however, direct peak area addition provides Si = 100 × (180.3 + 118.4)/(180.3 + 216.2 + 93.1 + 118.4) = 49.13%, and Al = 50.87%, respectively, results, ±3% deviation [2].

2.3.4 Crystallite Size and Microstrain

Prior to discussing the crystallite size phenomena, a basic difference between crystallite size, particle size, and grain size is compulsory to afford a better insight and apposite interpretation of the results. In brief, particle is formed due to combination of several grains separated by grain boundaries. Similarly, the grain is constructed due to the combination of several crystallites with having definite lattice parameter "d." So, XRD measures the crystallite size, neither grain size nor particle size. Crystallite size is the smallest unit among these, and measure as the size of coherently diffracting domains of a material, however, these three sizes have intense relation as crystallite size ≤ grain size ≤ particle size [16]. A brief schematic and their difference is illustrated in Figure 2.6. Crystallite size

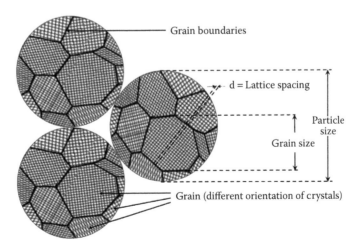

FIGURE 2.6
Schematic representation of crystallite, size, grain size, and particle size.

is equal to grain if the grain is perfectly single crystallite, but grains of sintered sample contain defects and dislocations that interrupt the periodicity, and, thus, an individual grain may contain several numbers of crystallites. Moreover, a particle may be the result of several crystallites or just one crystallite (i.e., particle size = crystallite size). Particles can also be as small as cluster of few atoms (say 2 nm Ag nanoparticles), and it may be polycrystalline, single crystal, or amorphous in state.

In XRD analysis, "the size of a crystal" refers to "size of crystallite," which when becomes less than 0.005 µm, the diffracted peak profile gets deviated from that of the same sample with having standard crystallite size of 0.5–10 µm (diameter). This small crystallite size experience plastic deformation of the sample (lattice planes gets distorted and results in change of spacing from one grain to other or in a grain from one part to other) and stacking faults result in "peak broadening." Usually this kind of nonuniform strain may be the result of either incorporation of additional atom or formation of defects in the crystal lattice during top-down synthesis approach, for example, mechanical grinding, rolling, etc. The effect of lattice strain on the peak profile in specific intensity, shifting, and broadening is shown in Figure 2.7.

In the simplest case where the particles have only homogenous (uniform) strain, the crystallite size (D) can be estimated from a single diffraction peak with employing of Scherrer Equation 2.20:

$$D = \frac{0.9\lambda}{B \cos\theta} \tag{2.20}$$

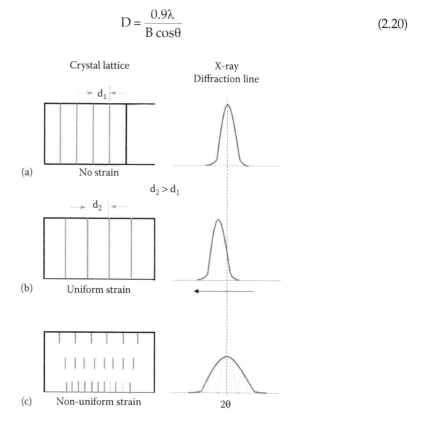

FIGURE 2.7
Effect of uniform and nonuniform strains (left side of the figure) on diffraction peak position and width (right side of the figure). (a) shows the unstrained sample, (b) shows uniform strain, and (c) shows nonuniform strain within the volume sampled by the X-ray beam.

where B is extra broadening over and above the instrumental breadth of line. But in those cases where heterogeneous stress is present, a more robust method involving several diffraction peaks is required, and their crystallite size is estimated by Williamson-Hall equation as given Equation 2.21 [17,18]. This method can also be used to study many other phenomena, such as stacking fault density and nonuniform deformation. Slow scan multipeak XRD data provide more reliable information in order to measure the strain by Equation 2.21:

$$B_r \cos\theta = \frac{\lambda}{D} + 2\eta \sin\theta \qquad (2.21)$$

where B_r is the peak width resulting from the size of crystallites (D) and η is equivalent to inhomogeneous strain. From the XRD peak, a width at half of its peak intensity, known as full width at half its maximum intensity, at every peak is to be determined and denoted by B_{obs} (caused by all cumulative factors). Now apart from crystallite size, instrument and other factors also cause the peaks to broaden, and this can be measured by a standard sample and denoted by B_{std} (generally data taken from Si single crystal). The resultant used peak width is obtained from Equation 2.22:

$$B_r{}^2 = B_{obs}^2 - B_{std}^2 \qquad (2.22)$$

This B_r relates to peak broadening caused by crystallite size, only. Eventually, the angle should be converted to radian to calculate the crystallite size in Å. Now, from the Williamson-Hall equation (Equation 2.21), which is in the form of y = mx + c, a plot between $B_r \cos\theta$ (y-axis) and $\sin\theta$ (x-axis) provides a characteristic graph and their fitting using least-squares method (done easily using software like origin pro) provides a straight line with a slope, so, micro strain and y-intercept give the crystallite size. Alternatively, the microstrain can be calculated from one peak, where one has to first calculate the crystallite size by Equation 2.20, followed by substitution of D in Equation 2.21 provide the strain.

While discussing the particle size, the small angle X-ray scattering is an innovative analytical method performing in the low angle of 0.1°–10° to determine the structure of particle systems in terms of average particle sizes or shapes. The particle or structure sizes can be resolved in a preferential range of 1–100 nm and in a typical set-up, but can be extended on both sides by measuring at smaller (ultra small-angle X-ray scattering) or larger angles (wide-angle X-ray scattering also called X-ray diffraction) [19].

2.4 Transmission Electron Microscopic Analysis

Transmission electron microscopy (TEM) is a high-resolution imaging technique and can be used for wide range of applications, and these are primarily divided in to four sections: conventional, in-situ study, analytical, and very fine details of crystal. Under these different modules, the characterizations are further broadened as [20]:

- Conventional applications: morphology (grain size, orientation) and microstructure, phase distribution and defect analysis.
- In-situ studies: irradiation and deformation experiments; environmental cells (corrosion); and phase transformations at different temperatures.
- Analytical applications: chemical composition analysis by Energy Dispersive Spectroscopy (EDS), Electron Energy Loss Spectroscopy (EELS), etc.
- Very fine details: lattice imaging, atomic structure of defects and/or interfaces.

Highlighted characterizations can be obtained by two different modes: static-beam image mode and diffraction mode. In TEM, illumination source is a beam of high velocity electrons accelerated under vacuum, focused by condenser lenses onto specimen. In static-beam mode, the intermediate lens is required to adjust so that its object plane is the image plane of the objective lens, and thus an image is projected onto the fluorescent viewing screen. In diffraction mode, it is necessary to adjust the imaging system lenses so that the back focal plane of the objective lens acts as the object plane for the intermediate lens. Eventually, it develops diffraction pattern and is projected onto the viewing screen. Electron diffraction pattern can answer about the state of material, for example, amorphous or crystalline and crystallographic characteristics, single or polycrystalline, presence of more than one phase, etc. TEM can develop different electron diffraction patterns including ring patterns (amorphous, polycrystalline), spot patterns (single crystal region), Kikuchi patterns, and convergent-beam electron diffraction. Herein, amorphous, polycrystalline, and single crystal have been considered, and basic privilege is given to plane indexing method from electron diffraction pattern, d-spacing calculation, and eventually correlation with XRD originated crystallographic data. A beginner can get assistance from this information in order to understand and develop nanoscale materials.

In diffraction mode of TEM, the electron beam passing through the selected area of a specimen (powder or bulk) gets diffracted and obtained typical pattern for amorphous, polycrystalline, and single crystal, as shown in Figure 2.8a–c. This is called selected area electron diffraction (SAED) pattern and is useful in crystal structure analysis. In the first step, let us consider a schematic (Figure 2.8d) in order to determine the indexing from

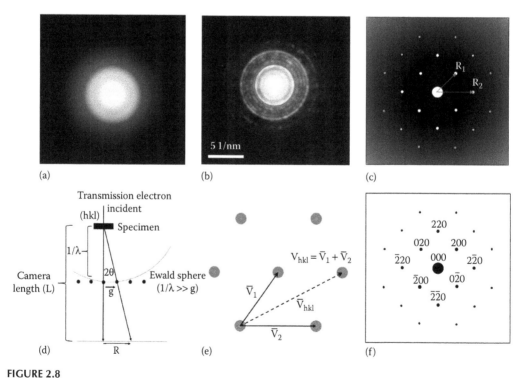

FIGURE 2.8
Photograph of typical electron diffraction pattern of (a) amorphous, (b) polycrystalline, (c) single crystalline nanoparticles, and (d) basic philosophy on the formation of diffraction pattern by TEM. Example for indexing of (e) way to obtain resultant vector by addition of two vectors and (f) single crystal.

diffraction pattern. The Bragg's law is applicable for this electron diffraction, but diffraction angles in TEM are very small, which leads to an approximation of $\sin \theta \approx \theta$, and thus resultant equation becomes:

$$\lambda = 2d\theta \tag{2.23}$$

From Figure 2.8d, $R/L = \tan 2\theta$ as $\theta \to 0$:

$$R/L = 2\theta \tag{2.24}$$

So, the resultant equation becomes (from Equations 2.23 and 2.24):

$$R/L = \lambda/d, \text{ or } \lambda L = Rd \tag{2.25}$$

where, λL is called camera constant of TEM, d = lattice spacing, λ is the wavelength obtained from the operating voltage, say 200 kV has equivalent wavelength $\lambda = 0.00251$ nm. L is the camera length of TEM, R is the distance from the central spot of the transmitted beam to a diffracted spot and can be measured with the method of ratios using the scale or camera constant. On the basis of the accelerating voltage (200 KV), the wavelength (λ) can be calculated from Equation 2.26 as follows:

$$\lambda = \frac{h}{\sqrt{2meV}} \tag{2.26}$$

where h = Planck constant (6.626×10^{-34} J.s), m = electron mass (9.109×10^{-31} kg), e = electronic charge (1.60×10^{-19} C), V = accelerating voltage (say 200 KV). For this conversion, consider the relation Joule (J) = volt (V), coulomb (C) = $kg.m^2.s^{-2}$. Hence, on the basis of constant parameters, the Equation 2.26 reduced to:

$$\lambda = \sqrt{\frac{1.5}{V}} nm \tag{2.27}$$

Thus, the higher the accelerating voltage, the smaller the wavelength of the electrons and the higher the possibility of achievable resolution of nanoscale materials. The diffraction pattern in a TEM represents a reciprocal lattice plane. The diffraction spots are the reciprocal lattice points. This reciprocal lattice plane contains the diffraction of lattice planes belonging to one crystal zone, of which, the axis is parallel to the transmitted beam, for example, if the zone axis is [001] of a cubic crystal, then the reciprocal lattice plane consists of diffraction spots from lattice planes (hk0). The formation of diffraction patterns in a TEM can be explained by Ewald sphere. The spot elongation in diffraction pattern is related to diffraction intensity distribution and can be fully explained by kinematic diffraction theory. However, it can be correlated as analogous to peak broadening in XRD due to small crystallite sizes.

Nonuniform orientation of atoms and consistent hazy electron diffraction pattern results imperfect circular ring in amorphous material, where a perfect ring is noticed for polycrystalline material, as illustrated in Figure 2.8a and b, respectively. The ring pattern in the polycrystal diffraction is obtained because it contains a large number of single crystals that are randomly oriented and at the same time, the rings are often broken, as the specimen may not contain grains in all possible directions. The formation of SAED rings is similar to that of X-ray Debye rings. In the similar fashion, a single crystal is shown in Figure 2.8c. After obtaining the diffraction pattern, it is interesting to know the crystal structure and corresponding plane of the developed material. For this purpose, indexing for both poly-crystalline and single crystal is carried, hence, either each line (polycrystal) or each spot (single crystal) is assigned with a set of h, k, and l. Indexing a pattern without knowing a chemical composition or lattice parameter is very difficult, but a brief explanation is discussed based on a simple crystal structure.

2.4.1 Indexing of Polycrystal

One can consider the following steps to index the SAED pattern of polycrystalline TiO_2, as revealed in Figure 2.8b.

Step 1: The bright central spot in the diffraction pattern is due to transmitted beam. It should be indexed as (0 0 0) of the reciprocal lattice plane.

Step 2: Select any particular ring (polycrystal) of choice and estimate the diameter (2r) of any selective ring (e.g., red color) by image processing software. On the basis of scale bar in Figure 2.8b, the first ring diameter is 2r = 5.625 (1/nm). The radius r (w.r.t. central spot) becomes 2.813 (1/nm), unit 1/nm refers the distance is in reciprocal lattice.

Step 3: Calculate the interplanar distance (d), in real lattice, d = 1/r = 1/2.813 = 0.355 4 nm = 3.554 Å.

The XRD pattern of selective material correlates the d-spacing of particular plane and indexing that corresponding ring. Herein, this prominent (101) plane refers to high intense peak of anatase TiO_2 [JCPDS -21-1272]. In the same way, indexing can be done for all rings and correlates with XRD pattern and their intensity.

2.4.2 Indexing of Single Crystal

Let us consider an example of cubic NaCl single crystal [7], with having lattice constant, a = 0.593 nm, perform the microscopic analysis to estimate the electron diffraction pattern of NaCl at accelerating voltage 200 kV (λ = 0.00251 nm) and camera length L = 1000 mm. Thus, camera constant is λ L = 0.00251 nm × 1000 mm = 2.51 nm.mm. In consideration of the nearest hkl (200) with respect to center hkl (000), the R_1 value can be calculated from following equations:

$$\frac{1}{d^2} = \frac{h^2 + k^2 + l^2}{a^2} \tag{2.28}$$

$$R_{hkl} = \frac{\lambda L(h^2 + k^2 + l^2)^{1/2}}{a} \tag{2.29}$$

Herein, $R_1 = R_{200} = \dfrac{2.51 \text{ nm.mm} (4 + 0 + 0)^{1/2}}{0.593 \text{ nm}} = 8.46 \text{ mm.}$

In order to estimate the R_2 (next intense peak from XRD), for example, hkl (220), the following relation can be employed:

$$\frac{R_1}{R_2} = \frac{(h_1^2 + k_1^2 + l_1^2)^{1/2}}{(h_2^2 + k_2^2 + l_2^2)^{1/2}} \tag{2.30}$$

In a cubic system, all the possible R ratios can be tabulated according to hkl pairs. Hence, R_2 becomes 11.96 mm. Now, with knowing R, λ, L, and "a," the particular dot can be indexed with help of the Equation 2.29, R_1 and R_2 are well matching with (200) and (220), respectively. Actually, there is ambiguity to represent the indexing pattern, as an identical pattern can be obtained after 180° rotation, for convenience herein, both (200) and (220) represent in right hand side. The angle between R_1 and R_2 was measured as 45° during TEM analysis and can be cross checked by Equation 2.31:

$$\cos\phi = \frac{h_1h_2 + k_1k_2 + l_1l_2}{\sqrt{\left(h_1^2 + k_1^2 + l_1^2\right)\left(h_2^2 + k_2^2 + l_2^2\right)}} \qquad (2.31)$$

As we know, any plane (hkl) that is in the zone [uvw] strictly follows $uh + vk + lw = 0$. In this continuation, the crystal zone axis can be identified by vector algebra and is found to be (001). After indexing these two spots, the rest of the spots can be indexed using the rule of vector addition, as shown in Figure 2.8e. The resultant diffraction pattern of NaCl crystal is shown in Figure 2.8f.

2.4.3 Morphology, d-Spacing, and Indexing of WO_3

Synthesis protocol, starting precursors, and processing conditions have influence on the control over the morphology, degree of crystallinity, defect formation, type of crystalline pattern, lattice spacing, and resultant chemical constituent, and all these features can be analyzed by TEM technique. In a recent article, a typical nanocuboid WO_3 was synthesized through hydrothermal method, and its morphology analyzed [21]. A high resolution transmission electron microscopic image was used to measure the lattice spacing and selective area electron diffraction in author's lab. The morphology seems to show the nanocuboid-shaped WO_3 has d-spacing within two crystal planes of 3.77 Å for (002) plane. Can we index the (002) in SAED pattern in Figure 2.9?

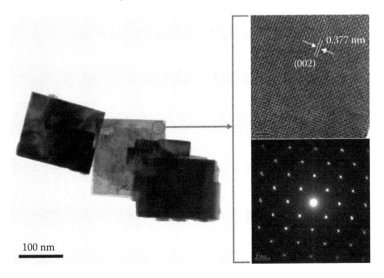

FIGURE 2.9
TEM analysis demonstrates the formation of cuboid morphology of WO_3, d-spacing by high resolution transmission electron microscopic and SAED pattern of selective area, marked by circle. (From Adhikari, S. and Sarkar, D., *RSC Adv.*, 4, 20145–20153, 2014.)

2.5 Particle Size and Surface Area

Ceramic nanoparticles have multidomain applications, thus, a brief subdivision with respect to mode of application can be classified as *direct*: without any morphological alteration of synthesized particles, and *indirect*: utilization after certain degree of consolidation of particles. Hence, the particle size and quantification of their size distribution are one of the important parameters in order to correlate and explain the resultant properties. For example, consistent photocatalytic and hyperthermia activities are expected to narrow size distribution, although a moderate wider distribution commensurate with better packing. Thus, statistical analysis of particle size distribution is performed in order to understand the competency of synthesized nanoparticles for specific interest. As we know, a high degree of surface energy promotes soft or hard agglomeration at elevated temperatures that results in formation of secondary particles as a cluster of primary particles and reduction of surface area. On the other hand, different pore size and content within nanoparticles further increase the surface area, and their adhesion activity that is advantageous for drug delivery and dye synthesized solar cell applications. In this contrary, determination of surface area and pore size phenomena may envisage the justification of adopted synthesis process. Thus, a brief pore size phenomenon is also addressed.

2.5.1 Statistical Analysis of Particle Size

Nanoparticles are subject to experience agglomeration of primary particles and often represent as secondary particles which consist of wider size distribution. Thus, particle size distribution encompasses the particle size and volume fraction with a qualitative prediction only, which is confusing, but a quantitative statistical analysis can envisage the monosize, narrow size, or wider size distribution of particles in order to provide an insight for the selection of nanoparticles for particular applications. In a recent article, a statistical analysis technique has been introduced to analyze nanoparticle size distribution phenomenon, which quantifies the real meaning of the nanoparticle size characteristics [22]. Scanning Mobility Particle Sizer (SMPS) spectrometer is widely used as the standard for measuring airborne particle size distributions, as well as particles suspended in liquids [23]. SMPS spectrometer sizing is a discreet technique in which number populations are measured directly without assuming the shape of the particle size distribution. The method is independent of the refractive index of the particle or fluid and has a high degree of absolute sizing accuracy and measurement repeatability. Table 2.4 illustrates how to approach three parameter Weibull statistical analysis for particle size distribution and their quantification parameters from experimental data (x and N). In order to determine the particle volume fraction in percentage, divide all N_1, N_2, N_3.....N_n values by the highest particle value (N), and multiply by 100, then plot the particle volume fraction (%) versus particle size (x).

A three-parameter Weibull distribution function is considered to quantify the particle size distribution data as obtained from SMPS measurements and is represented by Equation 2.32 [24]:

$$f(x) = 1 - \exp\left[-\left(\frac{x - \gamma}{\beta}\right)^{\alpha}\right]$$

(2.32)

Here, x, α, β, and γ are all positive, and f(x) is the cumulative undersize percent of particle size x present in the distribution. Three parameters α, β, and γ represent shape, scale, and

TABLE 2.4

Quantification of Particle Size Distribution after Obtaining the Data from SMPS

Sample No.	Particle Size (x)	Particle Number (N)	f(x)	$y = \ln\ln[1/(1 - f(x))]$	lnx	$\ln(x - \gamma)$
1	x_1	N_1	$f(x_1) = \dfrac{N_1\,100}{\sum N + 1}$	y_1	$\ln x_1$	$\ln(x_1 - \gamma)$
2	x_2	N_2	$f(x_2) = \dfrac{N_2\,100}{\sum N + 1} + f(x_1)$	y_2	$\ln x_2$	$\ln(x_2 - \gamma)$
3	x_3	N_3	$f(x_2) = \dfrac{N_3\,100}{\sum N + 1} + f(x_2)$	y_3	$\ln x_3$	$\ln(x_3 - \gamma)$
.....	......		....	......	...	
n	x_n	N_n	$f(x_n) = \dfrac{N_n\,100}{\sum N + 1} + f(x_{n-1})$	y_n	$\ln x_n$	$\ln(x_n - \gamma)$

$$\Sigma N = N_1 + N_2 + N_3 ...... N_n$$

location, respectively. A trial and error method is being adapted to obtain best fit within $\ln(x - \gamma)$ and $\ln\ln[1/(1 - f(x))]$, the physical meaning of "γ" is the smallest particle size in the system. Herein, a typical SMPS analysis of spherical $BaTiO_3$ is discussed as an example and illustrated their size distribution and Weibull analysis plot in Figure 2.10. Prior to acquiring data on the particle concentration with respect to particle size, the particles were mixed with ethanol, 1 wt% polyacrylic acid, NH_4OH, and finally mixed through zirconia grinding media in a high-density polyethylene (Nalgene) jar, details found elsewhere [6].

By solving Equation 2.32, a simplified equation represents $y = mx + c$, where $y = \ln\ln[1/(1 - f(x))]$, slope, $m = \alpha \ln(x - \gamma)$, and intercept, $c = -\alpha\ln\beta$. In the present case, a good fit to the SMPS data could be achieved using the above three-parameter distribution function using α, β, and γ equal to 4.6, 124, and 8.6, respectively. The shape ($\alpha = 4.6$) parameter for the synthesized $BaTiO_3$ nanoparticle represents relatively larger value,

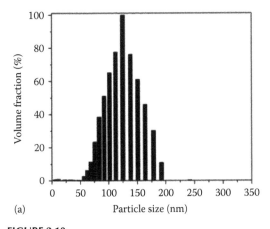

(a)

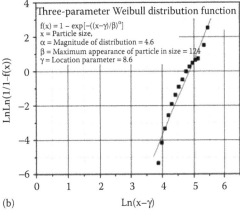

(b)

FIGURE 2.10

(a) Particle size distribution as-obtained from SMPS analysis and (b) three parameter Weibull distribution functions, analyzed from SMPS data. (From Sarkar, D., *J. Am. Ceram. Soc.*, 94, 106–110, 2011.)

indicating narrow size distribution and most of the particles are of 124 nm (β) in size. The, as realized, presence of a smaller ($\gamma = 8.6$) particle is apparently due to the contamination from zirconia grinding media in the time of SMPS sample preparation. In preference, high value of "α" indicates narrow size distribution and vice versa. It is very difficult to achieve narrow size distribution during high temperature synthesis, and, thus, one has to synchronize the process condition selectively for specific interest [22]. Based on maximum particle size (i.e., $\beta = 124$ nm), one can assume the specific surface area (m²/gm) from Equation 2.33, as indicated in the following, however, further surface area measurement and microscopic study are essential to correlate with this data.

$$S = \frac{k}{\rho D} \tag{2.33}$$

where, ρ is particle density (gm/cc), D is particle diameter (micron); an equivalent β (0.124 μm) can be considered for the specific surface area calculation, k is shape factor and varies with particle geometry, for example, 6, 8.94, 8.06, 7.44, 6.59, and 6.39 for sphere, tetrahedron, octahedron, cube, dodecahedron, and icosahedron, respectively.

2.5.2 Surface Area and Particle Pore Size

Usually, the surface area is determined by the nitrogen adsorption isotherms. The adsorption isotherms are classified into five types by IUPAC. One of the most common theories used to determine surface area is Langmuir isotherm that assumes monolayer adsorption on surface and is applicable to type I isotherm. However, the BET (Stephen Brunauer, Paul Emmett, and Edward Teller) method is derived from the Langmuir theory by assuming multilayer adsorption over the adsorbent surface with consideration of following hypothesis [25]:

1. Gas molecules physically adsorb on a solid in layers infinitely
2. There is no interaction between each adsorption layer
3. The Langmuir theory can be applied to each layer

Detail comparisons of these two theories are described elsewhere [26]. Langmuir's equation for type I isotherms is given in the following:

$$\frac{P}{W} = \frac{1}{KW_m} + \frac{P}{W_m} \tag{2.34}$$

A plot of P/W versus P will give a straight line of slope $1/W_m$ and intercept $1/KW_m$, from which both K and W_m can be calculated. The sample surface area can then be calculated from:

$$\text{Specific surface area of the adsorbent } (m^2/g) = \frac{W_m \left(\frac{mg}{g}\right) \times N \times A \left(\frac{m^2}{molecule}\right)}{M} \tag{2.35}$$

where W_m = Maximum amount of adsorbate adsorbed per gram of adsorbent for monolayer formation, M = Molecular weight of the adsorbate (mg/molecule), N = Avogadro's number and A = Contact surface area by each molecules (m²).

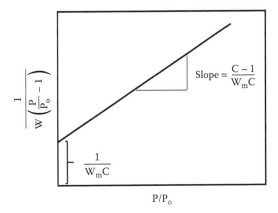

FIGURE 2.11
A typical plot after performing BET analysis.

The BET equation in order to estimate the surface area is given in the following:

$$\frac{1}{W((P/P_o)-1)} = \frac{1}{W_m C} + \frac{C-1}{W_m C}\left(\frac{P}{P_o}\right)$$ (2.36)

where W is the weight of gas adsorbed at a relative pressure, P/P_o, W_m is the weight of adsorbate constituting a monolayer of surface coverage. The term C, is constant related to the energy of adsorption in the first adsorbed layer, and, consequently, its value is an indication of the magnitude of the adsorbent/adsorbate interactions. To determine the surface area from BET theory, a plot between $\frac{1}{W((P/P_o)-1)}$ and $\left(\frac{P}{P_o}\right)$ gives a straight line, as illustrated in Figure 2.11.

This plot will have slope $s = \frac{C-1}{W_m C}$, and intercept, $i = \frac{1}{W_m C}$, and obtain the W_m and C. The specific surface area can be calculated by using the Equation 2.37:

$$S_{BET} = \frac{W_m N\ A_o}{M\ X}$$ (2.37)

where, M = molecular weight of the adsorbate, A = average occupational area of an adsorbate molecule (nitrogen is the most popular adsorbate gas, and it has an average occupational area of 16×10^{-20} m²), N = Avogadro's number, and X = mass of the sample. BET specific surface area can be explained with consideration of equivalent spherical diameter of monosized spheres. The BET equivalent for spherical diameter particle is calculated from surface area ($D_{BET} = 6/(\rho\ S_{BET})$, as already described in Section 2.5.1.

While discussing the pore size phenomena in particles, the total specific pore volume is defined as the liquid volume at a certain predetermined P/P_o (usually at $P/P_o = 0.95$). By using the Gurvich rule, the total specific pore volume can be calculated by the Equation 2.38 [27]. The pore volume, V_p is given by:

$$V_p = \frac{W_a}{\rho_l}$$ (2.38)

where W_a is the adsorbed amount in grams and bulk liquid density (ρ_l) at saturation. Assuming, apart from inner walls within pores, there is no additional surface and pore geometry is cylindrical in shape, then the average pore radius, $\overline{r_p}$ can be estimated from the ratio of the total pore volume, and the BET surface area by the equation 2.39:

$$\frac{2V_p}{S_{BET}} = \overline{r_p} \qquad (2.39)$$

If external surface is also included, then the modified Equation 2.39 represents as:

$$\frac{2V_p}{S_{BET} - S_{ext}} = \overline{r_p} \qquad (2.40)$$

Usually, the classical macroscopic thermodynamic methods that are used for calculating the pore size distribution, namely, Dubinin-Raduskevich and related methods, Horvath-Kawazoe, Saito-Foley, and Kelvin equation-based methods (e.g., BJH), assume that pores are filled with a liquid adsorptive with bulk-like properties.

2.6 Spectroscopy Analysis

While analyzing the crystallographic information through XRD technique, additional features including undetected low phase content (<2 wt%), crystal defects, vacancies, implanted atoms, and change in hybridization are critical to encounter and accomplish the nanomaterial research. Usually, the content of secondary or impurity phase and nature of crystal defects nucleated or endured during synthesis that may change from sample to sample, and can be determined by photoluminescence (PL). Engineering semiconductors are extrinsic, where the level of substantial atoms with the wrong valence acts as electron donors or acceptors, and thus appropriate analysis through Raman spectroscopy may assist and facilitate to develop advance materials. Now-a-days, multidomain properties in a single material is a burning issue, and thus core-shell nanoparticles or nanostructured ceramic coatings have high prospectus for functional applications including photovoltaics, catalysis, adhesion, magnetic media, etc. Thus, presence of surface moiety and valuable quantitative composition up to a few nanometers (up to 10 nm) from top of the surface can be determined by photoelectric effect, in convention X-ray photoelectron spectroscopy (XPS). Synthesis of nanostructured ceramic particles by organometallic or organic/inorganic precursors are common practice and remaining surface species may contain preferential organic functional groups that alter the surface charge and energy. Some classic ceramic nanoparticles have extensive functional characteristics and are sensitive to size, shape, concentration, and agglomeration state that makes ultraviolet (UV) spectroscopy a valuable tool to determine the band gap and concentration in a specific wavelength. An estimation and analysis by Fourier transform infrared spectroscopy (FTIR) on functional groups and their bond behavior definitely assist to analyze and obtain pure ceramic phase. Thus, utility of several spectroscopic techniques in order to develop a different class of nanoscale materials are highlighted.

2.6.1 Photoluminescence Spectra

Photoluminescence (PL) is a light emission process initiated by the photon, that is electro-magnetic radiation. Material absorb the photon and impart excess energy as emission of light or luminescence, and this photoexcitation luminescence is called photoluminescence. The excitation facilitates to move into a permissible excited state, followed by returning back to their equilibrium state; during this process, the excess energy is released and may include the emission of light (a radiative process) or may not (a nonradiative process), as shown in Figure 2.12 [28]. In semiconductor system, the most common radiative transition is in between conduction and valence bands, with the energy difference being known as the band gap, a region of forbidden energy. There are some subbands in the band gap, which are closely related to surface defects and surface states.

In concern to the semiconductor materials, when it is excited with light energy greater than band gap energy, four photophysical processes are feasible, as described in the following [28].

1. *Photo-excited process*: the e^- shifting from valence band (VB) to conduction band (CB) results in formation of h^+ in VB. Probable recombination of e^- with h^+ when it returns to VB and releasing energy in form of light and heat. Light energy dissipates as radiation, and this phenomenon is known as PL in the semiconductor. However, the excited electrons with different energy levels

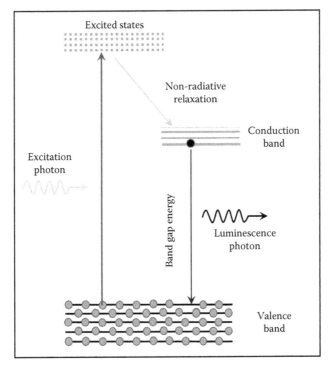

FIGURE 2.12
Principle of photoluminescence spectroscopy; formation of luminescence during photoexcitation.

in the CB easily first transfer to CB-bottom via nonradiative transitions, this follows the occurrence of other three PL processes 2, 3, and 4.

2. *Band-band PL process*: the e⁻ transitions from CB-bottom to VB-top through releasing energy as radiation, this signal is known as band-band PL phenomenon.

3. *Excitonic PL process*: nonradiative transitions of CB-bottom to different subbands and subsequent radiative transitions from the subband to VB-top. This signal is attributed to surface oxygen vacancies and defects present in the semiconductors, thus, it is more prominent in nanoparticles, as the average distance for electron movement becomes less and easily binds the electrons to form excitons in the subband.

4. *Excited e⁻ transitions*: revert back of CB-bottom e⁻ to VB by nonradiative process.

Thus, band-band PL spectrum can reflect the separation of photoinduced charge carriers in the system. A strong band-band PL signal indicates the incidence of higher recombination rate of photoinduced carriers. On the other hand, excitonic PL can inform about surface defects and oxygen vacancies. In presence of photo excitation, the excitation spectrum appears as a graph of emission intensity versus excitation wavelength analogous to absorption spectrum. The greater the absorbance at the excitation wavelength, the more molecules are promoted to the excited state, and the more emission is the outcome. Prior to estimate the excited wavelength, performing UV-visible (UV-vis) spectroscopy is expedient, since UV-vis refers to the absorption spectroscopy or reflectance spectroscopy in the ultraviolet-visible spectral region. In the UV-vis spectrum, an absorbance versus wavelength graph results, and it measures transitions from the ground state to excited state, while PL deals with transitions from the excited state to the ground state.

It is very common about the possibilities of peak behavior alteration in terms of intensity, shape, and position shifting, and these variations are attributed to particle size, excitation wavelength, particle synthesis temperature, dopants, and semiconductor coupling [28,29]. Different classic studies on semiconductors are compiled and highlighted in order to provide better insight and analysis of PL spectra (Figure 2.13). Figure 2.13a and b illustrate the influence of particle size (14, 19, and 26 nm) and excitation wavelength (300 and 350 nm) on the PL spectra of semiconductor ZnO nanoparticles. A wide range of excitonic PL appears within the range of 400–550 nm in presence of 300 and 350 nm excitation light energy, so, both are higher than band gap energy.

High intense excitonic PL peaks at 420 and 480 nm are attributed to surface oxygen vacancies and defects of ZnO nanoparticles. Interestingly, the peak intensity decreases as the particle size increases from 14 to 26 nm, that implies the reduction of surface oxygen vacancy and defect with increasing particle size. Thus, a weak peak, even no signal is expected in the same wavelength window for bulk semiconductor. Herein, the peak characteristics are different with respect to excitation light (300 and 350 nm) because of variation of excited electron energy and their come back from CB-bottom to VB via different sequences. However, peak positions are not changing that encompass the presence of stable energy levels of excitons and surface states.

Dopant has significant influence, and its mode of activity has different results in different PL spectra. Usually, four different types of dopants are most popular to alter the semiconductor properties, these are: dopant type I, stable chemical state possesses a half-filled or full-filled outer electronic structure that is not capable of capturing electrons, such as La^{3+} and Zn^{2+}. Dopant type II, stable chemical state easily captures electrons to become another relatively stable chemical state with a half-filled or full-filled outer electronic structure,

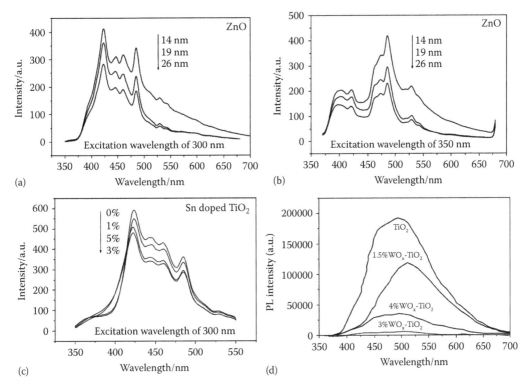

FIGURE 2.13
Effect of (a and b) particle size and excitation wavelength, (c) dopant concentration, and (d) coupling of other semiconductor WO_x. (From Liqiang, J., *Sol. Energy Mater. Sol. Cells*, 90, 1773–1787, 2006; Li, X. Z. et al., *J. Photochem. Photobiol. A*, 141, 209, 2001.)

such as Ce^{4+} and Cu^{2+}. Dopant type III, noble metal dopant, it decreases the excitonic PL intensity due to the capture of noble metal ions. Dopant type IV, semiconductor coupling/ doping.

Influence of semiconductor doping and coupling are interesting aspects to tailor the band gap, followed by photoexcited process. In this circumstance, a Schottky barrier is expected within interface between two semiconductors because of their difference in Fermi energy level and electronic band structure that highly influence the dynamic transport phenomena of photoinduced charge carriers and eventually PL performance. Figure 2.13c shows the PL spectra of different amounts of Sn-doped TiO_2 nanoparticles with excitation wavelength of 300 nm. A continuous decrement of peak intensity with dopant concentration is noticed, but this is not due to surface defect phenomenon, rather the CB-bottom energy level difference between dopant and matrix makes easy transfer of electrons. In consideration of band position, the addition of WO_x (x = 3) to TiO_2 facilitates the photoinduced electron transfer from TiO_2 CB to WO_3 CB that effectively separated electrons and holes and, subsequently, reduces the excitonic PL signal as shown in Figure 2.13d. Apart from the variation of peak intensity, PL blue shifting is attributed to the decrease in resultant particles size and redistribution of surface electronic density [29]. Different classic study of the synthesis of semiconductors, PL spectrum, and correlation on their photodegradation efficiency provides an insight about the utility of such characterizations and analysis [30,31].

2.6.2 Raman Spectroscopy

Raman spectroscopy is an innovative nondestructive technique to identify the chemical composition, phase, structural analysis, and imaging [32]. It resolves most of the limitations of other spectroscopic techniques, and thus it is suitable for versatile applications. Both qualitative and quantitative analysis can be done by this spectra, where the former is analyzed by measuring the frequency of scattered radiations, while quantitative information is measured by the intensity of scattered radiation. Several classic books described the detail insights about Raman spectrum based on theoretical physics, mode of data acquisition for different applications, and their interpretations [33,34]. Herein, a basic understanding, characterization, and analysis of nanostructured materials, say graphene, mostly researched material is discussed.

In Raman spectroscopy, sample is illuminated with a monochromatic laser beam that interacts with the molecules of sample and originates a scattered light. The scattered light having a frequency different from that of incident light (in elastic scattering) is used to construct a Raman spectrum. The Raman spectra arise due to inelastic collision between incident monochromatic radiation and molecules of sample [34]. Thus, atomic and molecular level of study does not affect the nanoscale specimen, high resolution information on microstructure and inner movement of nanostructures is easily obtained directly, thus, this is an excellent tool for the analysis of new nanostructure. While considering the graphene analysis, the micro Raman is supposed to characterize at 514.5 (2.41 eV) laser in backscattering configuration with a consistent laser spot of 400 nm and resolution of ~3 cm^{-1} and scanned area of 200 × 200 μm^2. In order to avoid any defect formation because of heating, low laser power <1 mW is recommended [35].

Theoretical physics explains that the longitudinal optical and transverse optical phonons are responsible for main Raman bands observed in the graphene [36]. A difference between pristine graphene, for example, monolayer flake and damaged graphene flake is shown in Figure 2.14a and b. The competitive peak shifting and splitting comprises the increment of graphene layer from single layer to bi-layer to graphite (Figure 2.14c). Several bands are designated as G, 2D (G′), 2D′, D, D′, and D+D′, and these only appear at specific wavelengths for a particular grade of graphene. The appearance and their significance of individual band at particular wavelength can be summarized as following:

- The G-band at 1500 cm^{-1} is prime Raman band, and it is a doubly degenerate in-plane sp^2 C–C stretching mode. This is present in every carbon system containing sp^2 bond including amorphous carbon, graphite, and carbon nanotubes. But this G mode gets split into G+ and G− when graphene sheet is supposed to roll into a carbon nanotube, and thus the beam width helps in measuring strain and deformation in sample.

- The 2D-band or G′ band at 2700 cm^{-1} is strongest peak and is used to determine the graphene layers in a flake. Without the splitting of 2D band confirms the existence of single graphene sheet, however, 2D peak dramatically splitting single layer to graphite.

- The D-band at 1350 cm^{-1} is disorder induced for laser excitation energies of 2.41 eV. The defect may be sp^3 defects, a vacancy sites, grain boundary, or an edge.

- Similarly, the D′ band at ~1620 cm^{-1} requires a defect for its formation like the D-band. 2D′ band at ~3240 cm^{-1} is a two-phonon process and the overtone of the D′ band. 2D′ and D′ are significantly weaker than the 2D and D-bands.

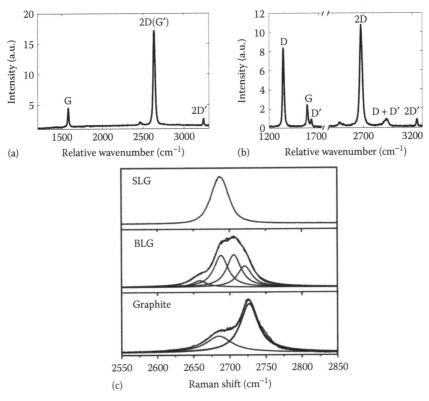

FIGURE 2.14

(a) Raman spectroscopy of defect free pristine graphene sheet, (b) Raman spectroscopy of damaged graphene sheet, (c) spectra of the Raman 2D band single layer graphene (SLG), bilayer graphene (BLG), and graphite. (From Beams, R. et al., *J. Phys. Condens. Matter*, 27, 083002, 26 pp, 2015.)

Despite layer inconsistency, defects can be introduced into the graphene layer by bombarding Ar^+. The ratios of intensities of D to G are conventionally considered to quantify the defect concentration in these samples. Pioneer works have described the mode of defect quantification with respect to the source of bombardment energy [37,38].

Apart from the detail identification of graphene, several materials are characterized for functional applications. For example, atomic vibrational modes identifying the surface species and spatial variations in composition of layered chalcogenides copper sulphide (CuS). Herein, an author reported the presence of an intense peak at 473 cm^{-1} is attributed to the perfect lattice alignment in a periodic array and an additional peak at 322 cm^{-1} assigned to the formation of defective $Cu_{2-x}S$ (x varies from 0, $6 \leq x \leq 1$) under phonon mode of Raman scattering [39]. This nondestructive study assists to get more details about functional behavior of nanoscale materials.

2.6.3 X-Ray Photoelectron Spectroscopy

As we know, the surface constituents and unsaturated dangling bond modify the several physicochemical properties of nanoscale materials, thus, XPS is the right choice to estimate the surface moiety and quantify of those elements that are present within the top of

1–12 nm depth of the sample surface. The basic principle of XPS defines the ejection of an electron from an inner electron shell like K shell of any atom when it absorbs X-ray photon. This K shell electron ejects from the surface as a photoelectron with some kinetic energy (E_K). Knowing this kinetic energy, the binding energy of the atom's photoelectron can be calculated by using the Equation 2.41 [7]:

$$E_B = h\nu - E_K - \Phi \qquad (2.41)$$

"Φ" is the parameter representing the energy required for an electron to escape from a material's surface and depends on material characteristics and spectra, h is the Planck's constant and ν is the frequency. The binding energies (E_B) of each and every atomic electron have characteristic values, and these values are used to identify elements. A typical XPS spectrum of a clean silver surface is expressed as photoelectron intensity versus binding energy (E_B), as illustrated in Figure 2.15. Before XPS peak identification, we often need to calibrate the binding energy. Calibration is particularly important for samples having poor electrical conductivity. The calibration is preferably done with a standard elemental sample like Si that shows little or no chemical shift. Usually, the XPS pattern exhibits three types of peaks formed from photoemission from core electron levels, photoemission from valence levels, and Auger emission spectra excited by X-rays. In Ag-XPS spectrum, the core-level peaks are marked as 3s, 3p, and 3d, the valence-level peak is marked as 4d, and the Auger peaks are marked as MNN [40].

The observed core-level photoelectron peaks are the primary peaks for elemental analysis and observed in mid-range (300–800 eV) of binding energy, whereas low binding energy (0–20 eV) valence-level peaks are usually primarily useful in studies of the electronic structure of materials. The Auger peaks are visible at relatively high binding energy range, arising from X-rays excited Auger process and also useful for chemical analysis. The most intense peak is now seen to occur at a binding energy of 368 eV due

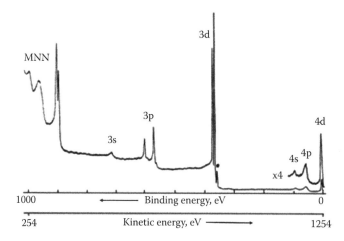

FIGURE 2.15
An XPS spectrum of silver using Mg Kα radiation (hν = 1253.6 eV), (•) at 366 eV indicates the presence of satellite peak. (From Briggs, D. and Seah, M.P., *Practical Surface Analysis*, Vol. 1, John Wiley and Sons Ltd., Chichester, UK, 1990.)

to the emission from the 3d levels of Ag atom. The XPS spectrum can also include extra, or satellite, peaks associated with a main core-level peak, as marked by black circular spot at binding energy (366 eV) in the XPS spectrum. This additional peak results when a photoelectron can excite (shake-up) a valence electron to a higher energy level and thereby lose a few electron volts of kinetic energy. These shake-up satellite peaks are useful for chemical analysis. Sometimes multiplet splitting of a core-level peak may occur in a compound, and it is because of the presence of unpaired electrons in its valence level and is useful in chemical analysis. Plasmon loss defines the energy loss of a photoelectron as it excites collective vibrations in conduction electrons in a metal and is responsible to generate another type of satellite peak. This unwanted information may change the peak shape and thus often needs to concentrate the peak overlapping phenomenon and resolve systematically.

In consideration of peak intensities, the element concentration can be quantified as analogous to X-ray spectroscopy, and most often equipment directly provides this information. Despite surface moiety determination, composition depth profiling for nanoscale material preferentially coating is also done by this technique.

The most commonly used method to obtain a depth profile of composition is sputter depth profiling using an ion gun. The ion beam containing Argon ions bombards a sample, causing it to eject atoms from the surface and thus produces a crater at the surface. For example, an XPS depth profile of titanium surface coated with calcium phosphate is shown in Figure 2.16 [41]. A critical coating thickness (300 nm) and elemental depth profile of coated material are estimated. Herein, the concentrations of Ca, P, and O in the coating gradually decreased, and the concentration of Ti increased in the depth direction, attributed to the interfacial diffusion from the coating into the substrate during fabrication process. In a recent article, the deposition of Ca, P, and O atoms on dental ceramics, say zirconia, in presence of simulated body fluid solution has been confirmed by XPS analysis, and further formation of hydroxyapatite phase was confirmed by XRD analysis [42].

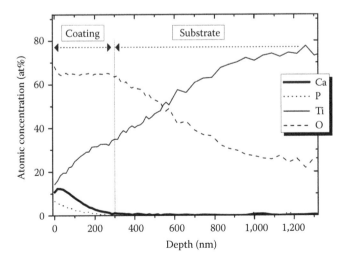

FIGURE 2.16
XPS depth profiling of titanium surface coated with calcium phosphate obtained with Ar etching system. (From Ohtsu, N., *Surf. Coat. Technol.*, 206, 2616–2621, 2012.)

2.6.4 Ultraviolet Visible and Diffuse Reflection Spectroscopy

Ultraviolet-visible spectroscopy is a common technique to determine the unknown concentration of the solute and band gap by UV-vis and UV-diffuse reflection spectroscopy (DRS), respectively. The reflectance spectroscopy is very closely related to UV-vis spectroscopy, both techniques use the wavelength window of 200–800 nm to excite valence electrons to empty orbitals. The difference in these techniques is that UV-vis measures the relative change of transmittance of light as it passes through a solution (Beer-Lambert law), while diffuse reflectance measures the relative change in the amount reflected light off from the surface (Kubelka-Munk theory). In convention, the former is used for solutions, suspensions, or thin films, where the latter is used for solids, usually optically rough films, or powders. Both of the measurements have enormous importance in the perspective of development of new materials, for example, the catalytic efficiency of semiconductor-based nanoscale photocatalyst demands band gap (E_g) calculation to explain the photocatalytic degradation process. Nanoparticles are sensitive to size, shape, concentration, agglomeration state, and refractive index that makes it a promising tool for identifying, characterizing, and studying nanomaterials in different aspects.

Lambert established a relation within transmittance and concentration that states transmission is the ratio of incident intensity (I_o) and transmitted intensity (I) and can be equated as e^{-kb}, where k is a constant and b is the path in length (usually in centimeters). Thus, resultant relation is [43]:

$$T = \frac{I}{I_o} = e^{-kb} \tag{2.42}$$

Beer modified the Lambert concept and introduced concentration parameter in order to estimate the solute content in the solution. The resultant relation states that the amount of light absorbed is proportional to the number of absorbing molecules through which the light passes, and the equation becomes:

$$T = \frac{I}{I_o} = e^{-kbc} \tag{2.43}$$

where c is the concentration of the absorbing species (usually expressed in grams per liter or milligrams per liter). This equation can be transformed into a linear expression in the logarithm form, where a = molar absorption or extinction coefficient, and A is absorbance.

$$\log \frac{I}{I_o} = abc = A \tag{2.44}$$

A simple schematic representation of basic principle of Beer-Lambert law is given in Figure 2.17a and equation follows a linear relation [44]. The absorption coefficient (a) is the characteristic of a given substance under a precisely defined set of conditions, such as wavelength, solvent, and temperature. In practice, the measured extinction coefficient also depends partially on the characteristics of the instrument used. Prior to estimate the unknown concentration of targeted analyte, absorption coefficient is compulsory to determine in identical environment.

One can follow the following process to estimate the coefficient followed by unknown concentration determination of any analyte.

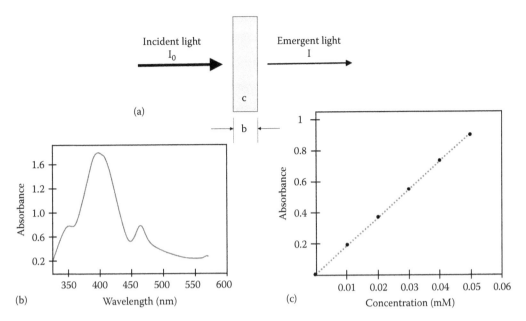

FIGURE 2.17
(a) Schematic representation of Beer-Lambert's law, (b) a typical plot of absorbance and wavelength, highest peak 400 nm, and (c) plot of absorbance against concentration.

Step 1: First measure the UV-vis of targeted sample within a wavelength of 200–800 nm and find out the maximum absorption wavelength, a typical plot can be obtained as indicted in Figure 2.17b. For example, modest water soluble (16 g/l) nitrophenol absorbs well light of about 400 nm, so we can measure the absorbance using light of that wavelength in a cuvette of path length and call the absorbance A_{400}.

Step 2: Prepare some cuvettes containing a range of known concentration of nitrophenol from 0–0.05 Mm with equal interval.

Step 3: Measure the absorbance for each cuvette and plot x-axis as c (mM) and y-axis as absorbance.

Step 4: It exhibits linear plot as shown in Figure 2.17c, calculate the slope from A = abc, and estimate the "a" with known value of b, say 1 cm. Herein, the coefficient is 18.2 mM^{-1}.cm^{-1}.

Step 5: Measure the absorbance (A) of unknown concentration of nitrophenol solution and measure the concentration (c) with help of A, b, a.

However, it is worth remembering that the light scattering is typically very sensitive to the aggregation state of the sample and scattering contribution increasing with particle agglomeration. For example, the optical properties of Ag nanoparticles change when particles combined, and the conduction electrons near each particle surface become delocalized and shared amongst neighboring particles. When this occurs, the surface plasmon resonance shifts to lower energies, causing the absorption and scattering peaks to red-shift to longer wavelength [45].

In another way, this range of wavelength can be used to determine the band gap by diffuse reflection mode as mentioned earlier. This physical property has extensive demand

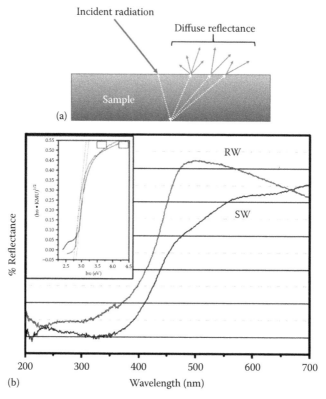

FIGURE 2.18
(a) Schematic representation of diffuse reflectance of a sample upon incident radiation, (b) UV-DRS for spherical (SW) and rod-shaped (RW) WO_3 nanoparticles, inset shows the Tauc plot for the two specimens to measure the band gap. (From Adhikari, S. et al., *Mater. Res. Bull.*, 49, 325–330, 2014.)

in order to develop new class of semiconductors that are effective in UV-vis region. Also, these band gap calculations are very important for those materials which find application in different devices like diodes, transistors, etc. In solids (especially containing transition elements), where electrons jump from localized orbital of one atom to delocalized energy bands, this method can be employed to calculate the band gap. In this experiment, the light is allowed to scatter in all directions from the sample followed by collection of scattered light by an optical detector. Surface reflectance is measured by scanning the sample over a range of wavelengths. The relative change in the amount of light reflected from surface is being measured, as shown in Figure 2.18a. After performing UV-vis spectra analysis of two different morphologies of WO_3 (SW = spherical WO_3, RW = rod shaped WO_3), a typical absorbance with respect to wavelength (UV-vis) is shown in the Figure 2.18b, and band gap calculated by plotting a graph called Tauc's plot (inset).

The reflection probe measures the diffuse reflectance (light scattered to all angles, Figure 2.18a) as a function of wavelength in a close approximation to the signal obtained from an integrating sphere system. The Kubelka-Munk function, KMU or F(R), allows the optical absorbance of a sample to be approximated from its reflectance: $KMU = F(R) = (1 - R)^2/2R$ and allows the construction of a Tauc plot - $(KMU.h\nu)^n$ versus $h\nu$. For a direct band gap semiconductor (WO_3, GaAs, InP, CdS, etc.) the plot $n = 1/2$ exhibits a linear Tauc region just above the optical absorption edge. Extrapolation of this line to the photon energy axis yields the semiconductor band gap, it is a key indicator of its light harvesting efficiency under solar illumination. Indirect band gap materials (Si, Ge, etc.) show a Tauc region on the n = 2 plot. Interestingly, both of the morphologies say SW ~50 nm and RW average length of 140 nm and width of 40 nm exhibits equivalent band gap of 2.82 and 2.75 eV, respectively [46]. However, the band gap may enhance below De Broglie wavelength, for example, below 15 nm $BaTiO_3$ has very high band gap compared to bulk (see Section 1.2.2).

2.6.5 Fourier Transformation Infrared

Different bonds as well as functional groups have different vibrational frequencies, and, hence, the presence of these bonds in a molecule can be detected by identifying the characteristic frequencies as an absorption band in the infrared spectrum. FTIR is a very sensitive and well-established tool for studying the orientation, transformation, and nature of bonds. The investigated sample absorbed certain frequencies in presence of infrared light, while other frequencies are transmitted without being absorbed. The transmittance from the sample is plotted against frequency and is called infrared spectrum. In advance, the FTIR does not separate energy into individual frequencies for measurement of the infrared spectrum. Thus, many scans can be completed and combined on FTIR in a shorter time than one scan on a dispersive instrument and results high signal-to-noise ratio, less error in the range of ±0.01 cm^{-1}, extremely high resolution, wide scan range (14,000–10 cm^{-1}), and minimization of stray light interference. However, most of the oxide particles are analyzed in mid-infrared, approximately 4000–400 cm^{-1} [47]. The transformations involved in the infrared absorption are associated with the vibrational changes in the molecule. Usually, the number of observed peaks is less than the total number of vibrational modes because probable dipole moment change during vibration cycle.

An interesting illustration on the preparation of 15 mol% zirconia (ZrO_2) containing alumina (Al_2O_3) nanopowder through sol-gel protocol has been picked up to discuss as a circumstantial appearance of wide range FTIR spectrum [48]. In this process, the synthesized gel gradually heated to high temperature in target of high crystalline alumina and zirconia phase, and the FTIR studied was carried in several intermediate steps to understand and support the transformation phenomenon, as shown in Figure 2.19.

The FTIR experiment of dried gel and calcined powders was carried out in the wavenumber range 400–4000 cm^{-1} at resolution of 4 cm^{-1} for studying the chemical groups on the surface of the as-dried gel, as well as calcined powder. A small amount of sample (0.2 g) was thoroughly mixed with ground KBr in an agate mortar, and a disc was prepared in vacuum maintaining a pressure of 33 kg/cm^2, prior to performing the FTIR study. After getting the plot from FTIR instrument, the background is required to remove from noise, and then from the standard literature the peaks are assigned to different bonds. Herein, the major peaks appearing in the FTIR spectra of alumina-zirconia hydroxide system could be related to the following:

1. –OH stretching vibration of the surface bonded or adsorbed water
2. –OH bending vibration of the surface bonded or adsorbed water
3. –OH stretching vibration of structural water, corresponding to M–OH bonding
4. –OH bending vibration of structural water, corresponding to M–OH bonding
5. Al–O stretching vibration
6. Zr–O stretching vibration

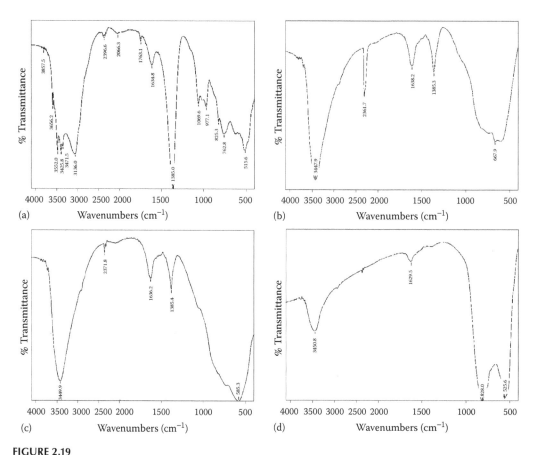

FIGURE 2.19
FTIR patterns of Al₂O₃-15ZrO₂ (mol%); (a) dried gel, (b) 200°C, (c) 400°C, and (d) 1000°C. (From Sarkar, D. et al., *Ceram. Int.*, 33, 1275–1282, 2007.)

While considering these bonding behaviors, several reasonings are ascribed why spectrum patterns are changing with respect to processing conditions, although a detail discussion is elsewhere [48]:

1. Reduced reflectance of the gel powder ascribed to the noncrystalline nature
2. Bending and stretching modes are altered due to change in the charge distribution among molecules and altered frequency of vibration, thus electrical dipole moment
3. Appearance of additional peaks assigned to the coupling effect of stretching and bending vibrations of –OH groups, although same disappears after high temperature calcination

The position and intensity of the IR peaks are also strongly influenced by the zirconia content, crystallization behavior, degree of crystallinity, and particle size. The particle morphology predominately the size and shape of the particles also result in the shift in the position of band. Therefore, in order to identify a mineral correctly, both effects have to be well understood [49].

2.7 Concluding Remarks

Nanoparticle synthesis has to contemplate several steps to control the phase, crystallinity, size, shape, density, and others to synchronize the resultant performance. While considering this fact, the discussed characterizations and analysis can envisage and estimate the phase transformation phenomena, crystallography, microscopic features, size, and spectroscopic behavior of nanoparticles. In brief, thermal analysis ensures to pick up the effective dynamic temperature in order to synthesize nanoparticles and contentment of their physical properties. Although, an isothermal flash calcination temperature can ensure the development of nanoscale particles with having adequate crystallinity and phase. X-ray diffraction analysis can fulfill this demand through analysis of pure phase formation, preferred growth plane, and a probable strain that eventually helps to correlate the properties. Additional electron diffraction of a selective area of synthesized particles confirms the development of amorphous, polycrystalline, and single crystal, and their indexing can assure the reliability of synthesis protocol. Statistical quantitative particle size distribution phenomenology confirmed the real meaning of narrow size distribution that can explain a number of functional incidences. Spectroscopy analysis commensurate together the bonding behavior, band gap, unknown solute concentration in solution, elemental composition profile from surface to certain nanometer depth, as well as the presence of unwanted surface moiety, defects in crystal and composition.

References

1. P. J. Haines, *Thermal Methods of Analysis: Principles, Applications and Problems*, Chapman & Hall, Glasgow, UK, 1995.
2. Y. Waseda, E. Matsubara, and K. Shinoda, *X-Ray Diffraction Crystallography*, Springer, Berlin, Germany, 2011.
3. P. J. Goodhew, J. Humphreys, and R. Beanland, *Electron Microscopy and Analysis*, 3rd ed., CRC Press, Boca Raton, FL, 2000.
4. S. Lowell, J. E. Shields, M. A. Thomas, and M. Thommes, *Characterization of Porous Solids and Powders: Surface Area, Pore Size and Density*, Springer, Dordrecht, The Netherlands, 2004.
5. Y. Leng, *Materials Characterization: Introduction to Microscopic and Spectroscopic Methods*, 2nd ed., Wiley-VCH, Weinheim, Germany, 2013.
6. D. Sarkar, Synthesis and properties of $BaTiO_3$ nanopowders, *J Am Ceram Soc* 94(1), 106–110, 2011.
7. M. Avrami, Kinetics of phase change. I general theory, *J Chem Phys* 7, 1103–1112, 1939.
8. T. Ozawa, Kinetic analysis of derivative curves in thermal analysis, *J Therm Anal* 2, 301–324, 1970.
9. H. E. Kissinger, Reaction kinetics in differential thermal analysis, *Anal Chem* 29, 1702–1706, 1957.
10. D. Feldman and D. Banu, Kinetic data on the curing of an epoxy polymer in the presence of lignin, *J Polym Sci A* 26, 973–983, 1988.
11. A. K. Jena and M. C. Chaturvedi, *Phase Transformations in Materials*, Prentice Hall, Englewood Cliffs, NJ, 1992.
12. J. Maser, A. Osanna, Y. Wang, C. Jacobsen, J. Kirz, S. Spector, B. Winn, and D. Tennant, Soft X-ray microscopy with a cryo STXM: I. Instrumentation, imaging and spectroscopy, *J Microsc* 197, 68–79, 2000.
13. D. Sarkar, Synthesis and thermo-mechanical properties of sol-gel derived zirconia toughened alumina nanocomposite, PhD Thesis, NIT Rourkela, India, 2007.

14. J. D. Hanawalt, H. W. Rinn, and L. K. Frevel, Chemical analysis by X-ray diffraction, *Ind Eng Chem* 10(9), 457–512, 1938.
15. B. D. Cullity, *Elements of X-Ray Diffraction*, 2nd ed., Addision-Wesley, Reading, MA, 1978.
16. A. Guinier, *X-Ray Diffraction: In Crystals, Imperfect Crystals and Amorphous Bodies*, Dover Publications, New York, 1994.
17. G. K. Williamson and W. H. Hall, X-ray line broadening from filed aluminium and wolfram, *Acta Metall* 1(1), 22–31, 1953.
18. D. Sarkar, M. C. Chu, S. J. Cho, Y.-I. Kim, and B. Basu, Synthesis and morphological analysis of TiC nanopowders, *J Am Ceram Soc* 92(12), 2877–2882, 2009.
19. E. G. Vrieling, T. P. M. Beelen, Q. Sun, S. Hazelaar, R. A. van Santen, and W. W. C. Gieskes, Ultrasmall, small, and wide angle X-ray scattering analysis of diatom biosilica: Interspecific differences in fractal properties, *J Mater Chem* 14, 1970–1975, 2004.
20. S. Amelinckx, D. van Dyck, J. van Landuyt, and G. van Tendeloo, *Electron Microscopy: Principles and Fundamentals*, Wiley, Weinheim, Germany, 2007.
21. S. Adhikari and D. Sarkar, Hydrothermal synthesis and electrochromism of WO_3 nanocuboids, *RSC Adv* 4, 20145–20153, 2014.
22. D. Sarkar, M. C. Chu, S. J. Cho, Y.-I. Kim, and B. Basu, Synthesis and morphological analysis of TiC nanopowders, *J Am Ceram Soc* 92(12), 2877–2882, 2009.
23. B. Steer, B. Gorbunov, R. Muir, A. Ghimire, and J. Rowles, Portable planar DMA: Development and tests, *Aerosol Sci Technol* 48(3), 251–260, 2014.
24. W. Weibull, A statistical distribution function of wide applicability, *ASME J Appl Mech* 18, 293–297, 1951.
25. S. Brunauer, P. H. Emmett, and E. Teller, Adsorption of gases in multimolecular layers, *J Am Chem Soc* 60 (2): 309–319, 1938.
26. Y. S. Bae, R. Q. Snurr, and O. Yazaydin, Evaluation of the BET method for determining surface areas of MOFs and zeolites that contain ultra-micropores, *Langmuir* 26, 5475–5483, 2010.
27. S. Lowell, J. Shields, M. A. Thomas, and M. Thommes, *Characterization of Porous Solids and Powders: Surface Area,Porosity and Density*, Springer, Dordrecht, the Netherlands, 2004.
28. J. Liqiang, Q. Yichun, W. Baiqi, L. Shudan, J. Baojiang, Y. Libin, F. Wei, F. Honggang, and S. Jiazhong, Review of photoluminescence performance of nano-sized semiconductor materials and its relationships with photocatalytic activity, *Sol Energy Mater Sol Cells* 90, 1773–1787, 2006.
29. X. Z. Li, F. B. Li, C. L. Yang, and W. K. Ge, Photocatalytic activity of WO_x-TiO_2 under visible light irradiation, *J Photochem Photobiol A* 141, 209–217, 2001.
30. S. Adhikari, D. Sarkar, and G. Madras, Highly efficient WO_3-ZnO mixed oxides for photocatalysis, *RSC Adv* 5, 11895–11904, 2015.
31. N. D. Abazović, M. I. Čomor, M. D. Dramićanin, D. J. Jovanović, S. P. Ahrenkiel, and J. M. Nedeljković, Photoluminescence of anatase and rutile TiO_2 particles, *J Phys Chem B* 110(50), 25366–25370, 2006.
32. http://www.horiba.com/us/en/scientific/products/raman-spectroscopy/raman-spectrometers/.
33. S.-L. Zhang, *Raman Spectroscopy and its Application in Nanostructures*, Wiley-Blackwell, 2012.
34. H. A. Szymanski, *Raman Spectroscopy: Theory and Practice*, Springer, New York, 1967.
35. A. Eckmann, A. Felten, A. Mishchenko, L. Britnell, R. Krupke, K. S. Novoselov, and C. Casiraghi, Probing the nature of defects in graphene by raman spectroscopy, *Nano Lett* 12(8), 3925–3930, 2012.
36. R. Beams, L. Gustavo Cançado, and L. Novotny, Raman characterization of defects and dopants in grapheme, *J Phys Condens Matter* 27, 083002 (26 pp), 2015.
37. A. J. Pollard, B. Brennan, H. Stec, B. J. Tyler, M. P. Seah, I. S. Gilmore, and D. Roy, Quantitative characterization of defect size in graphene using Raman spectroscopy, *Appl Phys Lett* 105, 253107, 2014.
38. L. G. Cancado, A. Jorio, E. H. Martins Ferreira, F. Stavale, C. A. Achete, R. B. Capaz, M. V. O. Moutinho, A. Lombardo, T. S. Kulmala, and A. C. Ferrari, Quantifying defects in graphene via Raman spectroscopy at different excitation energies, *Nano Lett* 11, 3190–3196, 2011.
39. S. Adhikari, D. Sarkar, and G. Madras, Hierarchical design of CuS architectures for visible light photocatalysis of 4-chlorophenol, *ACS Omega* 2, 4009–4021, 2017.

40. D. Briggs and M. P. Seah, *Practical Surface Analysis*, Vol. 1, John Wiley and Sons, Chichester, UK, 1990. [Book - Yang Leng, Materials Characterization, p. 231, 2013 Wiley-VCH Verlag GmbH & Co].

41. N. Ohtsu, Y. Nakamura, and S. Semboshi, Thin hydroxyapatite coating on titanium fabricated by chemical coating process using calcium phosphate slurry, *Surf Coat Technol* 206, 2616–2621, 2012.

42. D. Sarkar, S. K. Swain, S. Adhikari, B. S. Reddy, and H. S. Maiti, Synthesis, mechanical properties and bioactivity of nanostructured zirconia, *Mater Sci Eng C* 33(6), 3413–3417, 2013.

43. J. H. Lambert, *Photometria sive de mensura et gradibus luminis, colorum et umbrae* [Photometry, or, On the measure and gradations of light, colors, and shade] (Augsburg ("Augusta Vindelicorum"), Germany: Eberhardt Klett), 1760.

44. A. Beer and P. Beer, Determination of the absorption of red light in colored liquids, *Ann Phys Chem* 86, 78–88, 1852.

45. M. Ali, R. Rahimi, and S. Maleki. Synthesis, characterization and morphology of new magnetic fluorochromate hybrid nanomaterials with triethylamine surface modified iron oxide nanoparticles, *Synth Met* 194, 11–18, 2014.

46. S. Adhikari, D. Sarkar, and H. SekharMaiti, Synthesis and characterization of WO$_3$ spherical nanoparticles and nanorods, *Mater Res Bull* 49, 325–330, 2014.

47. B. Schrader, *Infrared and Raman Spectroscopy: Methods and Applications*, Wiley, 2007.

48. D. Sarkar, D. Mohapatra, S. Ray, S. Bhattacharyya, S. Adak, and N. Mitra, Synthesis and characterization of Sol-Gel derived ZrO$_2$ doped Al$_2$O$_3$ nanopowder, *Ceram Int* 33, 1275–1282, 2007.

49. J. L. London and C. J. Serna, IR spectra of powder hematite: Effects of particle size and shape, *Clay Miner* 16, 375–381, 1981.

3

Surface and Suspension Phenomena

3.1 Introduction

The dispersion of particles into aqueous or nonaqueous liquid is an industrial practice to make commercial products for daily utility. Selected particles can be hydrophobic, for example, no affinity between particle surface and water, such as organic pigments, agrochemicals, graphene, SiC, Si_3N_4, etc., or hydrophilic solid, having strong affinity between their surface and water, for example, clays ($Al_2O_3.2SiO_2.2H_2O$), alumina (Al_2O_3), zirconia (ZrO_2), silica (SiO_2), etc. Surface energy variation can make both hydrophobic and hydrophilic of an identical material, for example, pyrolyzed TiO_2 nanoparticles are conventionally known as hydrophobic, but in contact with water (or ultraviolet [UV] radiation), they gradually change to hydrophilic, that further revert to hydrophobic after removal of adsorbed water in elevated temperature. Thus, preparation of stable suspension of nanoparticles is a challenging task, as wettability and repellency of solid surfaces are taken into account in practical outlook [1]. Basic dispersion strategy comprises the maximum separation of aggregates and isolation of agglomerated particles into "individual" units through adequate processing and stabilization without further settling in time. Powder wetting, maintaining dispersive state without agglomeration, prevent settling, and workability under optimum shear stress are essential components to use such suspensions made from nanostructured materials. In this context, this chapter starts with an emphasis on the contact angle, surface tension, and critical surface energy that eventually help to analyze the plausible solid–liquid wetting characteristics. Despite different nanomaterial characterizations, a brief illustration on surface analysis brings an impression on recent developments from nanostructured ceramic particle based products, including cosmetics, coating on conductive electrode for alternative energy, consistent pollutant degradation in presence of nanoparticles, drug adsorption and delivery media, nanofluid for energy management, superhydrophobic self-cleaning coating, etc. Discussion on the viscosity with respect to particle size, volume fraction, and their rheological behavior are also enlightened in consideration of casting of high dense nanostructured bioceramics for implants, porous polymer-ceramic composite scaffold for tissue engineering, etc.

3.2 Surface Wetting

Studies on "wetting phenomenon" have focused incredible interest in the field of fundamental and applied sciences since it plays a critical role in understanding of industrial processes like spray quenching, lubrication, liquid coating, printing, oil recovery, and applications involved with superhydrophobic surfaces. The degree of wetting is usually characterized by measuring the contact angle, (θ), at the solid–liquid interface. Usually, larger contact angle, ($\theta > \pi/2$), is indicative of low wettability, but smaller contact angle, ($\theta < \pi/2$), predicts the high wettability or spreading capability of liquid over a solid substrate.

3.2.1 Contact Angle and Hysteresis

The contact angle, (θ), is defined as the developed angle between the tangent to the liquid–air interface and tangent to the solid interface at three phase (vapor–liquid–solid) contact line. The theoretical observations related to the influence of contact angle on the wetting behavior of liquid is compiled in Table 3.1. Figure 3.1 shows a sessile drop of liquid resting on a flat and horizontal solid surface with varying contact angles.

Assuming, γ_S, γ_L, and γ_{SL} are the solid-vapor, liquid-vapor, and solid-liquid interfacial free energies per unit area, respectively, the relationship between surface tension of three phases and contact angle on homogenous surface can be related mathematically using Young's equation [2]:

$$\gamma_S = \gamma_{SL} + \gamma_L \cos\theta \tag{3.1}$$

Despite molecular adsorption and stains, the surface energy and surface roughness play an important role during wettability of solid surface that eventually changes in the contact angle of solid–liquid interface. A metastable droplet can persist with any angle between advancing contact angle, (θ_a), and receding contact angle, (θ_r). However, a combination phenomenon can be noticed when a droplet moves on an inclined surface which consists of both advance (on the downward side) and recede (on the upward side), and these differences between advancing and receding contact angle are known as hysteresis. This hysteresis implies the superhydrophobicity and self-cleaning surfaces. Thus, contact angle hysteresis is a physical phenomenon that usually exists on heterogeneous solid surfaces, it is more common compared to ideal homogenous surface.

TABLE 3.1

Effect of Contact Angle on the Transition of Wetting Behavior of a Liquid

Case	Contact Angle (θ)	Wetting Behavior
A	$\theta < 90°$	Wetting of the surface is favorable, and liquid spreads over the solid surface.
B	$\theta = 90°$	Wetting of the surface is favorable, and hemispherical shape of liquid droplet forms over the solid surface.
C	$\theta > 90°$	Wetting of the surface is unfavorable, and liquid maintains a minimal contact area with solid surface.
D	$\theta = 0°$	This is a special case where the complete wetting of solid surface occurs; superhydrophilicity.
E	$\theta > 150°$	This is another special case representing hydrophobic surfaces, where virtually there is no contact between liquid drop and solid surface; superhydrophobicity.

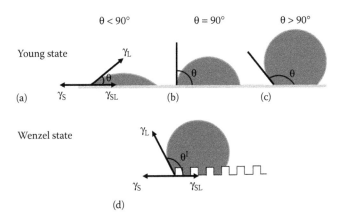

$\theta < 90°$ $\theta = 90°$ $\theta > 90°$

Young state

(a)   (b)   (c)

Wenzel state

(d)

FIGURE 3.1
Schematic illustration showing the wettability of a liquid at three different contact angles: (a) $\theta < 90°$, (b) $\theta = 90°$, and (c) $\theta > 90°$, (d) θ' in heterogeneous surface. a, b, c, drop on homogenous solid surface (Young concept), and "d" on heterogeneous surface, consist of a definite roughness (Wenzel concept).

Let's consider a water droplet on a solid surface, and it formed a sectioned sphere with a measureable contact angle, θ (Figure 3.1). If the droplet volume and angle increase, the droplet advances with a contact angle, θ_a, whereas the reverse incident follows constant receding containing angle, θ_r, and both depend on the surface chemistry and topography. In general, the contact angle measured via Young's equation provides the static contact angle that only validates under ideal conditions (i.e., three-phase equilibrium). In real instances, however, solid surfaces are heterogeneous in nature and three-phase contact line is in actual motion, a dynamic contact angle is defined to characterize the wetting behavior. This dynamic contact angle may be of two types. They are:

1. Advancing contact angle (θ_a): This is formed under expansion of a liquid and approaches a maximum value (Figure 3.2a).
2. Receding contact angle (θ_r): This is formed under contraction of a liquid and approaches a minimum value (Figure 3.2b).

Hence, contact angle hysteresis (H) is mathematically defined as the difference between advancing and receding contact angles:

$$H = \theta_a - \theta_r \tag{3.2}$$

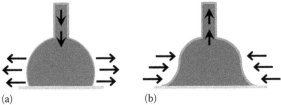

(a)   (b)

FIGURE 3.2
Schematic illustration displays contact angle hysteresis phenomenon using (a) advancing and (b) receding contact angles.

TABLE 3.2

Hysteresis Phenomena of Heterogeneous Surface

Reason of Hysteresis	Incidence during Hysteresis
Heterogonous surface	Consists of preferable hydrophobic domains that create barrier to the liquid motion by pinning the motion of water front and cause contact angle hysteresis.
Surface roughness	Slope change of the surface acts as a barrier to the liquid motion by pinning the water front at three-phase contact line and leads to contact angle hysteresis.

Hysteresis incidence is attributed to heterogeneous surface and surface roughness that alter the liquid movement, and the plausible observations relevant to the generation of contact angle hysteresis are being summarized in Table 3.2.

Wilhelmy plate method is a widely used technique to determine both static and dynamic contact angles (i.e., advancing and receding) that involves the measurement of the force acting in the tensile direction (F) when a plate-shaped solid is held vertically and immersed into the liquid of known surface tension (γ), schematically shown in Figure 3.3 [3].

The static contact angle measurement, (θ_s, Figure 3.3a), follows the immersion of the solid substrate into the liquid where the substrate should be held stationary. When the specimen contacts the surface of the liquid, a force (F) is developed and recorded by the balance, which is equivalent to the weight of the liquid clinging to the specimen. However, measurement of the advancing contact angle, (θ_a), involves recording the force that acts under further advancement of specimen into the liquid for any desired depth (Figure 3.3b). Moreover, the receding contact angle, (θ_r), measurement requires recording the force that acts under retreatment of the specimen from the liquid (Figure 3.3c). Therefore, the contact angles in terms of static, advancing, and receding, ($\theta = \theta_s; \theta = \theta_a; \theta = \theta_r$), can be calculated using the following equation [4,5].

$$\theta = \text{Cos}^{-1}\left[\frac{F}{2\lambda}(L+t)\right] \tag{3.3}$$

where the parameters "L" and "t" are length and thickness of the solid substrate.

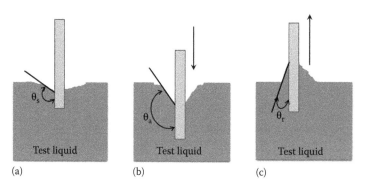

FIGURE 3.3
Schematic illustration of contact angle measurement by Wilhelmy plate method, where (a) static contact angle, (θ_s); (b) advancing contact angle, (θ_a); and (c) receding contact angle, (θ_r).

3.2.2 Critical Parameters for Superhydrophobic Surface

Ideal flat surface without any roughness follows Young's Equation 3.1, and the increment of representative contact angle, θ, by lowering the surface energy encompasses a high degree of hydrophobicity.

$$Cos\theta = \left(\frac{\gamma_S - \gamma_{SL}}{\gamma_L} \right) \tag{3.4}$$

However, any surface is not ideal and consists of a certain degree of roughness that may be responsible for the resultant wetting characteristics compared to ideal counterpart. In this aspect, Wenzel proposed a model in consideration of rough surface having contact angle, "θ," and the modified equation Young's equation is [6]:

$$Cos\theta' = r\frac{(\gamma_S - \gamma_{SL})}{\gamma_L} = r\, Cos\theta \tag{3.5}$$

In Equation 3.5, "r" is roughness factor, described as the ratio of the real contact area of rough surface (heterogeneous) to the geometric project area (homogenous). Thus, "r" is always more than unity, and surface roughness enhances hydrophobicity of hydrophobic surfaces and the hydrophilicity of hydrophilic surfaces (Figure 3.1d). In consideration of a contact angle of a water droplet on idealized sinusoidal surface, a theoretical calculation established the contact angle of hydrophobic rough surface enhances with the increment of "r," and this is significant beyond the value of 1.7 [6]. Hence, both the lowering of surface energy and adequate increment of surface roughness are essential to prepare the super-hydrophobic surfaces. Apart from these, the thermodynamic concept encompasses that every system desires to be in the state of lowest possible Gibbs free energy to make a stable system. So, a few surface modification methods are listed to reducing the surface energy:

1. Surface relaxation: strong surface atoms or ions interaction with underlying species
2. Surface restructuring: formation of new chemical linkages with surface dangling bonds
3. Surface adsorption: chemical or physical adsorption of terminal chemical species onto the surface by forming chemical bonds or weak attraction forces, such as electrostatic or van der Waals forces
4. Composition segregation or impurity enrichment on the surface through mechanisms like solid-state diffusion

However, long periods of outdoor exposure degrade the excellent hydrophobicity of artificially constructed surface because of the continuous accumulation of stains and their adherence to the surfaces with several modifications. In order to improve the outdoor performance of hydrophobic surface, researchers proposed an addition of a certain percentage of TiO_2 photocatalyst in hydrophobic constituent like polytetrafluoroethylene $[(C_2F_4)_n]$ to fabricate an effective self-cleaning of dirt and maintain high contact angle during long periods of outdoor exposure, details discussed in Section 8.6 [7]. This is attribute to the:

1. Immediate and effective stain adherence by TiO_2 surface and subsequent photocatalytic decomposition or easy washing out by water due to high surface energy or photo-induced hydrophilic property of the surface of TiO_2

2. Reduction of static electricity by the photo-induced hydrophilicity of TiO_2

3. Addition of TiO_2 provides a long diffusion distance of the radical species on the hydrophobic surface

3.2.3 Critical Parameters for Superhydrophilic Surface

The term superhydrophilicity appeared for the first time in the technical literature from Japan [8]. Superhydrophilic surface comprises an apparent contact angle less than 5° and may have many practical applications like anti-fogging, anti-fouling or self-cleaning, and others. It is also important in biological systems, like cell activity, proliferation, signaling activity, etc. In comparison to the superhydrophobic surface, conversely, the superhydrophilic surface possesses high surface energy and nanoscale rough and/or porous surface having a roughness factor r > 1 in order for rapid spreading and formation of very low contact angle and thin water film. Apparently, the formation of zero contact angle, (θ = 0), is common, but needs some strategy to make it. From Equation 3.1, when $\theta \to 0, \gamma_S \geq \gamma_{SL} + \gamma_L$, it suggests that the solid-water (γ_{SL}) interfacial free energy approaches to near zero, and attributes to the water molecular interaction with solid surface, such as hydrogen bonding, and thus thermodynamically more stability of solid-water interface. This implies all solids with γ_S interfacial energy $\gamma_S \geq 72.8$ mJ/m² at 22°C, satisfying the ideal conditions of perfect water spreading [9]. In general, metals, alloys, ionic salts, and ceramics have surface free energy higher than 72.8 mJ/m², whereas a polymer has relatively less free surface energy than that of water. In this context, one can enhance the hydrophilicity of surfaces either molecular or microscopic deposition of new material, more hydrophilic than the substrate, or by modification of surface chemistry of the particles or substrate. Molecular level surface modification or deposition of coating is common practice for inorganic materials, whereas controlling over surface chemistry is extensively used for polymers.

3.2.4 Critical Boundary in Hydrophilic and Hydrophobic Surface

In consideration of equilibrium contact angle, (θ), in Young's Equation 3.1, the window may vary 0°–180°, it represents the extreme boundary of superhydrophilicity to superhydrophobicity, respectively. In case of "zero contact angle," $\gamma_S - \gamma_{SL} = \gamma_L$, it is better to characterize by the work of liquid spreading W_s, that is commonly referred to as spreading coefficient. It defines the work performed to spread a liquid over a unit surface area of a clean and nonreactive solid (or another liquid) at constant temperature and pressure in equilibrium with liquid vapor and represented by:

$$W_s = \gamma_S - (\gamma_L + \gamma_{SL}) \tag{3.6}$$

When liquid forms a definite contact angle, θ > 0, we can combine and rearrange both Equations 3.1 and 3.6 and represents as:

$$W_s = \gamma_L (\text{Cos } \theta - 1) \tag{3.7}$$

Thus, $W_s = 0$, when θ = 0, so, complete spreading and wetting of the surface, otherwise W_s becomes negative for any existing contact angle, this concept as similarly as proposed by van Oss [10]. The hydrophilicity and hydrophobicity can be further explained by free energy of hydration, (ΔG_{SL}), and the absolute value of free energy of hydration equals to work of adhesion (W_a), thus the Dupré equation illustrates:

$$\Delta G_{SL} = \gamma_{SL} - \gamma_S - \gamma_L = - W_a \tag{3.8}$$

Work of adhesion describes reversible thermodynamic work done during separation of the interface from the equilibrium state of two phases to a separation distance of infinity, thus, it determines the strength of the contact between two phases. Conversely, this is the amount of energy released during wetting process. The free energy of hydration, (ΔG_{SL}), has a definite window to explain the hydrophobicity and hydrophilicity, hence, estimated $\Delta G_{SL} > -113$ mJ/m² (hydrophobic molecules) and $\Delta G_{SL} < -113$ mJ/m² (hydrophilic molecules) can differentiate the boundary between hydrophobic and hydrophilic surfaces [11]. In combination of Young-Dupré concept, (Equations 3.1 and 3.8), defines:

$$\Delta G_{SL} = - \gamma_L (\cos\theta + 1) \tag{3.9}$$

In consideration of the Equation 3.9 and water surface energy, (72.8 mJ/m²), free energy of hydration, (–113 mJ/m²), the equilibrium contact angle, $\theta \approx 56°$, and can be classified as boundary contact angle in between hydrophilic and hydrophobic surface. In consideration of molecular level of attraction within liquid–liquid (cohesion) and liquid–solid (adhesion), water adhesion tension, $(\tau, \text{mN/m})$, is defined as:

$$\tau = \gamma_L \cos\theta \tag{3.10}$$

Taking into account the contact angle, work of spreading, free energy of hydration, and water adhesion tension, a classic differentiation within superhydrophilic, hydrophilic, weakly hydrophilic, weakly hydrophobic, hydrophobic, and superhydrophobic has been represented in the Table 3.3.

TABLE 3.3

Specific Parameters for Hydrophilic and Hydrophobic Surfaces

Type of Surface	Measure of Hydrophilicity/Hydrophobicity (20°C)			
	Contact Angle [deg]	Work of Spreading [mJ/m²]	Energy of Hydration [mJ/m²]	Water Adhesion Tension [mJ/m²]
Superhydrophilic (rough r > 1)	~0[a]	≥0[b]	≤ –146[b]	≥73[b]
Hydrophilic	~0	≥0	≤ –146	≥73
Weakly hydrophilic	(56°–65°) > θ > 0	0 > W$_s$ > – (32–42)	–113 > ΔG$_{SL}$ > –146	73 > τ > (30–40)
Weakly hydrophobic	90 > θ > (56°–65°)	– (32–42) > W$_s$ > –73	–73 > ΔG$_{SL}$ > –113	(30–40) > τ > 0
Hydrophobic	120 > θ ≥ 90	–73 > W$_s$ > –109	–36 > ΔG$_{SL}$ > –73	0 ≥ τ > –36
Superhydrophobic (rough r > 1)	θ > 150°[a]	W$_s$ ≤ –136[b]	ΔG$_{SL}$ ≥ –10[b]	τ ≤ –63[b]

Source: Drelich, J. et al., *Soft Matter*, 7, 9804, 2011 [12].
[a] Apparent contact angle.
[b] Estimated based on apparent contact angles using Equations 3.7, 3.9, and 3.10.

3.3 Surface Tension

Consider a sessile drop of liquid resting on a flat, horizontal, and homogeneous solid substrate, where the presence of intermolecular forces among the liquid molecules can be observed, as shown in Figure 3.4.

The net force among the liquid molecules present in the bulk portion of a pure liquid is zero, since its neighboring counterparts pull each molecule in every direction. However, an unbalanced net force can be observed for the exposed molecules at the liquid surface due to the absence of neighbors in all directions that creates an internal pressure and pulls them inward to attain the lowest surface free energy by contracting the liquid structure. The intermolecular force responsible for this phenomenon is known as surface tension, which controls the shape of liquid droplets.

3.3.1 Surface Tension Estimation Protocols

Force tensiometer is used to measure surface tension and interfacial tension of liquid–gas and liquid–liquid interfaces based on force measurement at this interaction zone, however, this estimation depends on the physical characteristics of the probe. Du Noüy ring method is the most commonly used technique for the surface tension measurements [13]. This method consists of a platinum ring held over the selected liquid surface under investigation (Figure 3.5).

This ring is there gradually immersed in the liquid and then withdrawn slowly. At this point, the liquid meniscus is pulled up by the ring and eventually breaks away from it. The series of events in this process is reported in Table 3.4. The surface tension of liquid, (γ), is measured using the following relation:

$$\gamma = \frac{F}{2L \, Cos\theta} \tag{3.11}$$

where "F" is the maximum force, "L" is the wetted length, and $Cos\theta$ is the wetting coefficient.

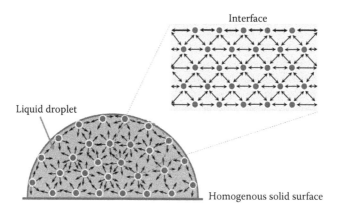

FIGURE 3.4
Schematic illustration of surface tension phenomenon. Inset showing the intermolecular force variation among the liquid molecules present at the bulk and surface.

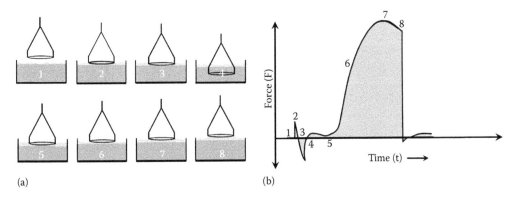

(a)   (b)

FIGURE 3.5
Schematic representation of surface tension measurement by force tensiometer-Du Noüy ring method, (a) movement of Platinum (Pt)- on the tested liquid, and (b) the variation of force with respect to time during conducting the experiment.

TABLE 3.4

Exploring the Series of Events in the Du Noüy Ring Method

Event	Condition	Force at the Interface
1	Ring is just placed above the liquid surface	Zero
2	Ring just touches the liquid surface	Slightly positive (Due to adhesive force between ring and liquid surface)
3	Ring must be pushed through the liquid surface	Negative (Due to surface tension of liquid)
4	Ring breaks the liquid surface	Slightly positive (Due to supporting wires of the ring)
5	Ring is being lifted through the liquid surface to a position 1	Starts increasing
6	Ring is being lifted through the liquid surface to a position 2	Keeps increasing
7	Ring is being lifted through the liquid surface to a position 3	Reaches maximum
8	Ring starts to break away from the liquid meniscus	Slight decrease

The interfacial surface tension can also be determined by pendant drop shape analysis method, where a liquid drop is pushed through the tip of needle and is held under gravity, as described in Figure 3.6. The shape of the liquid droplet helps in determining the surface tension, (γ), that can be calculated using the following relation:

$$\gamma = \frac{\Delta \rho g d_e^2}{H} \tag{3.12}$$

where "$\Delta \rho$" is the density difference between fluids, "g" is the gravitational constant, "d_e" is the maximum diameter of the drop, "d_s" is the diameter of the droplet located at a height equal to the maximum drop diameter "d_e," and "H" is the shape factor ($H = d_s/d_e$).

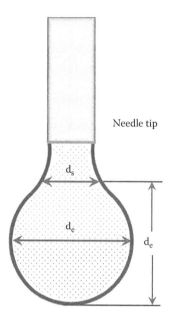

FIGURE 3.6
Schematic representation of surface tension measurement by optical tensiometer-pendant drop shape analysis technique.

3.3.2 Critical Surface Tension

The critical surface tension, (γ_c), of a solid predicts the possibility of the surface wetting. This γ_c is not necessarily the same as surface tension of solid surface, it is true only when $\theta = 0$ and $\gamma_{SL} = 0$ (consider Equation 3.1). The wettability of a particle depends upon the contact angle, (θ), and the observations relevant to the wetting behavior as a function of contact angle is reported in Table 3.3. Surface energy of solid substrates are determined using Zisman plot, in which plot of wetting coefficient, $(\cos\theta)$, versus surface tension, (γ_L), of homologous liquids are used [14]. For example, the surface tension of water, in equilibrium with its vapor at room temperature, is 72 mNm^{-1}. Surface tensions of other liquids are widely available in literature [15]. The critical surface tension equals the surface tension at which the plotted, $(\cos\theta$ versus $\gamma_L)$, data intersect at 1.0. Thus, any liquid having surface tension lower than critical surface tension of solid will wet the solid surface. In a recent article, the critical surface tension for polyethylene particles were estimated and plotted in Figure 3.7 [16].

The Zisman plot of polyethylene particles exhibits the wetting coefficient, $(\cos\theta)$, become 1.0 in the surface tension range of $0 < \gamma < 30$ mN/m that demonstrates the possibility of polyethylene particles wetting by the liquid in this range [16].

3.3.3 Influence of Foreign Elements in Surface Tension

The surface tension of pure water is obvious to change in presence of foreign particles or other substituents like surfactants (although essential for dispersion), and thus several studies concentrated on these issues and discussed their basic reasoning to develop successful nanofluids and suspensions for several applications. In recent studies, the effect of ceramic nanoparticle size and concentration has been studied and illustrated in Figure 3.8a [17].

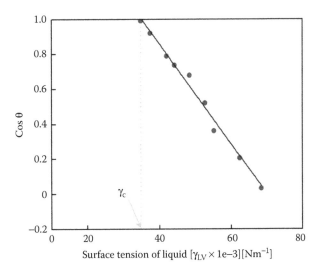

FIGURE 3.7
Zisman plot to determine the critical surface energy of polyethylene. (From Kinloch, A. J., *Adhesion and Adhesives*, Chapman & Hall, New York, 1987, p. 25.)

The change in surface tension behavior of SiO_2, Al_2O_3, ZnO, and TiO_2 nanofluids is primarily attributed to the following factors [17]:

1. Concentration of the nanoparticles in the nanofluid, (0.5–6 Vol.%)

   The surface tension of nanofluids dispersed in propylene glycol/water, (60:40 by mass), is decreasing with the increase of nanoparticle concentration in the range of 0.5%–6% by volume. The density, size, and mass of these nanoparticles (solid) are many times more compared to the liquid molecules. Hence, the addition of nanoparticles to the liquid molecules reduces the surface tension by creating a less dense surface region via attracting large number of liquid molecules on the surface. Consequently, the inward pressure required to pull the liquid molecules on surface reduces.

2. Size of nanoparticles in the nanofluid (15–50 nm)

   The particle size has significant influence on the surface tension behavior that can be emphasized in consideration of the surface tension trend as an example of Al_2O_3 based nanofluids with an average particle size of ~20 nm (plot C) and ~45 nm (plot D). The decrease in particle size cause a reduction in surface tension by increasing the number of particles per unit volume and there by attracting the more number of liquid molecules present on the surface toward itself. This reduces the density of surface region causing a net decrease in the surface tension. Similar kinds of trends have been observed for the case of SiO_2 and ZnO based nanofluids, as well.

Apart from the particle concentration and size effect, surface tension is also being affected by surfactant like sodium dodecyl benzylsulfonate concentration during stabilization of a single wall nanotube nanofluids and illustrated in Figure 3.8b. It is observed that the surface tension trends of water/sodium dodecyl benzylsulfonate (plot-A) fluid mix is constant

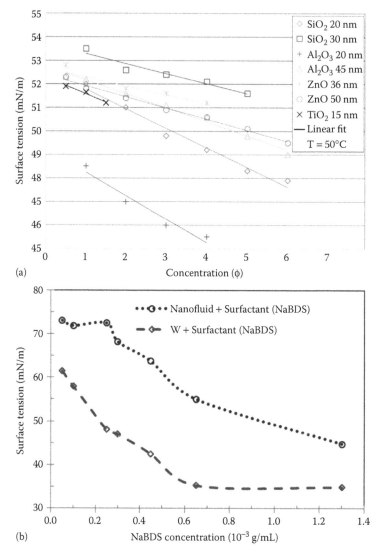

FIGURE 3.8
(a) Surface tension of nanofluids decreases with the increase of nanoparticle concentration. (From Chinnam, J. et al., *Int. J. Commun. Heat Mass Transf.*, 62, 1–12, 2015 [17].) and (b) surface tension of nanofluids decreases with the increase of surfactant concentration. (From Khaleduzzaman, S.S. et al., *Int. J. Commun. Heat Mass Transf.*, 49, 110–114, 2013 [18].)

above a surfactant concentration of 0.62 Vol.% [18]. However, the fluid mixture comprising nanofluids/surfactant (plot-B) exhibits a gradual decrease in surface tension after a surfactant concentration of 0.25–1.3 Vol.%. The decrease trend in surface tension with the addition of surfactant to the nanofluids is primarily attributed to the electrostatic repulsion created by the adsorbed polymer chains over the particle surfaces. Thus, the surface tension is one of the prime factors to control the hydrophilic or hydrophobic behavior of synthesized nanofluids, and, thus, it is essential to optimize the particle size, concentration, and desire dose of surfactant to make a colloidal suspension. However, a stable suspension depends on several factors and discussed in following sections.

3.4 Colloidal Suspension

The terminology "colloidal suspension" describes a substance having permanent solid suspension in a liquid medium, and it is different from solid–liquid solution. In fact, colloid endures in between solution and suspension, and their characteristic features are described in Table 3.5. Colloids have many similarities with solutions; in both cases, particles are invisible through naked eye, do not settle on standing, and pass out through most filters. The particles are colloid (Brownian motion) each other and scattered light in colloids (Tyndal effect), but this incidence is not observed in solution [19]. However, suspension is a heterogeneous mixture of solid particles up to 1 μm in liquid, and two phases may separate (flocculate) if allowed to stand, that is why we need to shake suspended medicine or stir paint thoroughly before use, but colloid do not separate into two phases on standing. In consideration of several applications of stable suspension of solid particles, one can prepare colloidal suspension through smaller nanoparticles in the range of 2–500 nm, however need to consider their surface characteristics and different strategies to make stable without settling in time.

Consider a ceramic particle based colloidal suspension in which the particles undergo Brownian motion and collide against one another. The flocculation behavior of a suspension is characterized by the thermal energy of particles, (kT), where k is Boltzmann's constant and T is the absolute temperature.

Near the room temperature, the thermal energy of particles is insufficient to overcome the van der Waals attractive potential, (V_a), among particles resulting in flocculation of the suspension and makes it unstabilized [20]. Stabilization of these suspensions can be achieved by introducing repulsive forces through several mechanisms, as shown in Table 3.6, and their details is discussed in following sections.

TABLE 3.5

Properties of Liquid Solutions, Colloids, and Suspensions

Type of Mixture	Approximate Size of Particles (nm)	Characteristic Properties	Examples
Solution	<2	Not filterable; does not separate on standing; does not scatter visible light	Air, white wine, gasoline, saltwater
Colloid	2–500	Scatters visible light; translucent or opaque; not filterable; does not separate on standing	Smoke, fog, ink, milk, butter, cheese
Suspension	>500	Cloudy or opaque; filterable; separates on standing	Muddy water, hot cocoa, blood, paint

TABLE 3.6

Stabilization Mechanisms for Colloidal Suspensions

Method	Stabilization Mechanism
(i) Electrostatic stabilization	Repulsive forces are introduced using electrostatic charges on the particle surfaces.
(ii) Steric stabilization	Repulsive forces are created using adsorbed uncharged polymers onto the particle surfaces.
(iii) Electrosteric stabilization	Repulsive forces are introduced via adsorption of charged polymers onto particle surfaces. (Combination of electrostatic + steric stabilizations.)

3.4.1 Electrical Double Layer

Prior to discussing different stabilization mechanisms, a brief on the electrical double layer is highlighted. Let's consider a colloidal suspension consisting of ceramic oxide particles dispersed in water that acquire negative charge over their surfaces via preferential adsorption of ions. Consequently, an electrical double layer is expected and formed around the particle surface that consists of a large number of counter ions (+ve charged) and a small number of co-ions (−ve charged). Figure 3.9 schematically demonstrates the electrical double layer formed around a negatively charged surface that is a combination of stern layer and diffuse layer. Several steps and phenomenology are associated with such model and these are summarized as:

1. Surface charge and surface potential: It refers to the charge of ions adsorbed on the particle surface. The electrical potential acquired by the particle surface due to this charge is known as surface potential.

2. Stern layer, stern plane, and stern potential: This is the first layer comprising counter ions attached to the negatively charged particle surface by the electrostatic forces. The boundary pertaining to the stern layer is known as stern plane. The electrical potential at the stern plane is known as stern potential.

3. Diffuse layer: This is the second layer that constitutes free ions with a higher concentration of positively charged counter ions. The electrostatic potential at the boundary of this layer is zero, indicating that the double layer is electrically neutral.

4. Slipping plane and zeta potential: Slipping plane or shear plane is the boundary of a layer of the liquid surrounding the moving colloidal particles in the dispersion medium. The electrical potential at the slipping plane is known as zeta potential [21].

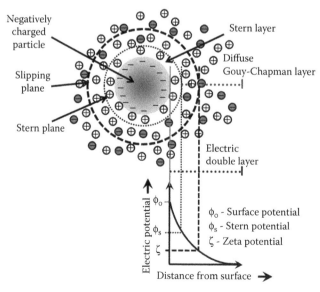

FIGURE 3.9
Schematic illustration of electric double layer and zeta potential.

3.4.2 Electrostatic Stabilization

Assume oxide particles having radius "r" are dispersed in an electrolyte solution, such that their surfaces get negatively charged. At any instant, the total potential energy, (V_T), between two colloidal particles are separated by a distance 'S_o' can be given as:

$$V_T = V_A + V_R + V_G \qquad (3.13)$$

V_A is the attractive potential energy arising from van der Waals forces

V_R is the repulsive potential energy arising from electrostatic and osmotic forces

V_G is the gravitational potential energy

The minima in the potential energy curve represents the flocculation of suspension due to the dominance of van der Waals attractive potential over thermal energy barrier (Figure 3.10).

For particle size ranges, r <<< 1 µm.

The effect of gravity becomes negligible and the total potential energy equation reduces to the following form.

$$V_T = V_A + V_R \qquad (3.14)$$

The DLVO (Derjaguin, Landau, Verwey, and Overbeek) theory describes the electrostatic stabilization mechanism of a ceramic suspension as a function of interparticle distance of separation, (S_o), and two essential cases of this theory is discussed in the following.

Case-1: Interparticle separation distance is greater than the thickness of two double layers, $(S_o > 2d)$

At this instant, both van der Waals attractive and repulsive forces reduce to zero, since the two particle surfaces are far apart, there would be no interaction of diffuse double layers (Figure 3.11a).

Case-2: Interparticle separation distance equals the thickness of electrical double layer, $(S_o = d)$

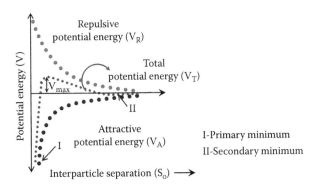

FIGURE 3.10

A schematic illustrating DLVO theory using potential energy curve as a function of interparticle separation distance. The minima in the potential energy curve represents the flocculation of suspension due to the dominance of van der Waals attractive potential over thermal energy barrier.

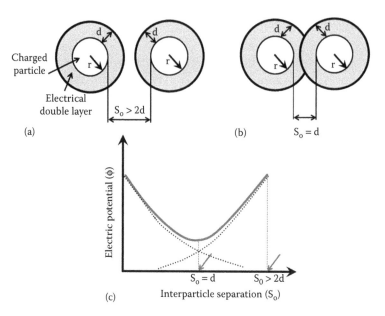

FIGURE 3.11
(a) Schematic illustration of condition for no interaction between two double layers, (b) overlap of two double layers, and (c) interparticle separation and electric potential relation.

At this instant, electrical double layers of two particles overlap each other and generate repulsive force (Figure 3.11b). The cumulative electric potential and interparticle separation distance is represented in Figure 3.11c. This repulsive force can be two types.

1. Repulsive force derives from the overlap of two double layers instead of surface charge on solid particles.

2. Repulsive force derives from the osmotic flow of the solvent into the overlapped region of two double layers to restore the original equilibrium concentration profiles of the individual double layers. The maxima shown in the potential energy curve (Figure 3.10) represents the potential energy barrier, where the repulsive potential dominates the attractive potential.

3.4.3 Steric Stabilization

Steric or entropic stabilization of a colloidal system is a thermodynamic stabilization technique in which repulsive forces are predominantly originated from the change in configurational entropy of the interacting uncharged lyophilic polymer chains adsorbed onto the particle surfaces (Figure 3.12). Consider a colloidal ceramic suspension comprising two oxide particles with an interparticle separation distance of "S_o" and covered with terminally anchored polymers having a thickness of "δ" (Figure 3.12a). The adsorbed polymer chains like to begin interacting when the interparticle distance, (S_o), is less than twice the thickness of the polymer sheath (2δ, Figure 3.12b). However, the interparticle forces at this instance can be attractive or repulsive, which typically depends upon temperature of the solvent, and they significantly influence the Gibbs free energy of the system. The force is said to be repulsive in nature, ($\Delta G > 0$), and the solvent becomes ideal when solvent temperature is higher than the critical flocculation temperature, (θ). Above this temperature, a dispersion is sterically stabilized by separating the

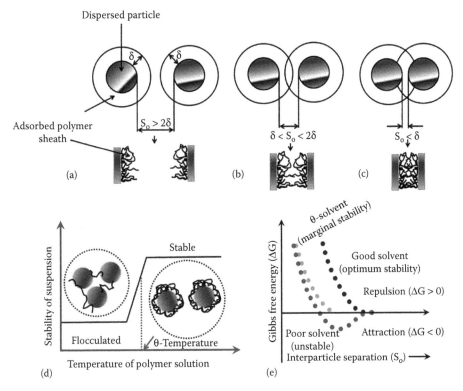

FIGURE 3.12
Steric repulsion of a colloidal ceramic suspension as a function of interparticle separation (a) $S_o > 2\delta$, (b) $\delta < S_o < 2\delta$, (c) $S_o < \delta$, and (d) dramatic change in the stabilization behavior of a suspension at CFT, and (e) change in free energy as a function of interparticle separation.

dispersed particles apart via an osmotic pressure effect that causes flow of solvent ions from bulk to the intervening region in order to offset the established chemical potential gradient (Figure 3.12d and e). Moreover, there would be an attractive force, ($\Delta G < 0$), resulting from the coil up, or collapse of the adsorbed polymer chains at a solvent temperature below the critical flocculation temperature may limit the colloidal stabilization and lead to bridging flocculation.

But the interparticle forces that are purely repulsive with further decrease in interparticle distance, (S_o), below to the thickness of a single polymer sheath, (δ, Figure 3.12c), can be attributed to increase in the Gibbs free energy, ($\Delta G > 0$), of the colloidal system. It losses in configurational entropy, (ΔS), are owing to the compressive deformation of adsorbed polymer chains on particle surface actuated by interacting with the rigid surface of the other particle. This deformation is purely elastic in nature, and this contribution to the steric repulsion can be described as volume restriction effect or entropic effect. Therefore, the total energy of interaction, (V_T), for a sterically stabilized dispersion can be expressed as:

$$V_T = V_A + \Delta G_{Steric} \tag{3.15}$$

where "V_A" is the van der Waals energy of attraction, and "ΔG_{Steric}" is change in free energy of the steric interaction. It has been known that free energy change in steric repulsion is sum of the change in free energies due to osmotic effect, ($\Delta G_{Osmotic}$), and entropic effect, ($\Delta G_{Entropic}$). So that:

$$\Delta G_{Steric} = \Delta G_{Osmotic} + \Delta G_{Entropic} \tag{3.16}$$

But, the free energy change due to the entropic effect is given as:

$$\Delta G_{Entropic} = -T\Delta S \tag{3.17}$$

From Equations 3.15 through 3.17:

$$V_T = V_A + \Delta G_{Osmotic} - T\Delta S \tag{3.18}$$

Thus, cumulative interaction can minimize through high entropy of the system or increment of temperature above the critical flocculation temperature, (θ).

3.4.4 Electrosteric Stabilization

Electrosteric stabilization of a colloidal dispersion involves the use of electrosteric stabilizers in order to stabilize the colloid against flocculation through a combination of electrostatic and steric repulsions. The most common electrosteric stabilizers are polyelectrolytes, which can be cationic or anionic in nature [22,23]. These are lyophilic polymers, having ionizable groups distributed along their back bone, develop a net charge on adsorbed polymer chain upon their dissociation in solvent. Figure 3.13 schematically represents a speculative electrosterically stabilized colloidal ceramic suspension comprising polymer chains anchored to the negatively charged particles surrounded by a positively charged species formed via dissociation of an anionic polyelectrolyte.

The effectiveness of an electrosteric stabilization is typically influenced by the interparticle forces, which are primarily a function of distance of separation between particles, concentration of the polyelectrolyte, and concentration of the solvent (electrolyte). The electrostatic component offers repulsive forces through electrical double layer formation, and dominant at a larger separation, however, steric repulsion becomes more prominent as the distance of separation decreases. Therefore, the total energy of interaction, (V_T), for an electrosterically stabilized colloidal dispersion as a function of the interparticle separation distance can be written as:

$$V_T = V_A + V_R + \Delta G_{Steric} \tag{3.19}$$

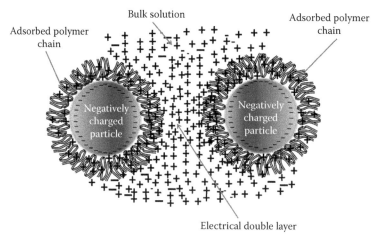

FIGURE 3.13
Schematic illustration of electrosteric stabilization.

where "V_A" is the van der Waals energy of attraction, "V_R" is the potential energy of repulsion, and "ΔG_{Steric}" is the change in free energy of the steric repulsion.

Substituting Equations 3.16 and 3.17 in Equation 3.19 yields:

$$V_T = V_A + V_R + \Delta G_{Osmotic} - T\Delta S \tag{3.20}$$

Moreover, a relatively higher concentration of the adsorbed polyelectrolyte is beneficial for an effective stabilization, since a lower concentration of the polymeric species can lead to flocculation due to weaker interparticle forces engendered by the charge neutralization phenomenon. However, the electrosteric stabilization through electrostatic component is much pronounced in presence of low electrolyte concentrations, whereas a breakdown of the electrical double layer at higher electrolyte concentration significantly weakens the resultant electrostatic repulsion.

3.4.5 Zeta Potential and Isoelectric Point of SiO$_2$ and MgO: Influence of pH

As discussed early, the zeta potential is the charge on a particle at the shear plane that is important to understand and predict the particle interaction behavior in suspension. In convention, a large magnitude (either positive or negative) above 25 mV zeta potential, (ζ), generally indicates the high degree of electrostatic stability of particle suspension.

Thus, measurement of zeta potential and its analysis can envisage to make stable colloids. To brief this incidence, the zeta potential was estimated with respect to different pH variation of suspended acidic (SiO$_2$) and basic (MgO) oxide particles and represented in Figure 3.14 [24]. The Isoelectric point (IEP) is an important parameter to prepare the suspension, as it describes the point at which molecule does not experience net electrical charge in the statistical mean, in other ways, it is defined as the pH where zeta potential is zero. Experimental data reveal different characteristic features of ζ – pH plots, where acidic oxide surfaces have a relatively low IEP value compared to basic oxide surfaces since the acidic surfaces have a low density of hydroxyl ions. Ceramic suspensions are unstable and flocculated at a pH value near or at the IEP since the repulsive forces are absent due to

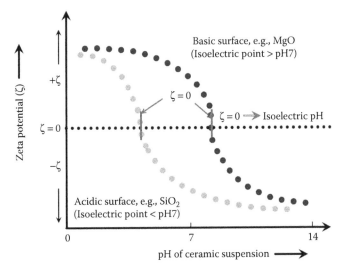

FIGURE 3.14
Zeta potential and isoelectric point of acidic and basic surfaces as a function of pH of the solution.

zero net electrical charge. In this perception, one can maintain an optimum pH of particle suspended solution to achieve repulsive force and thus stable colloids.

3.4.6 Zeta Potential and Isoelectric Point of ZrO$_2$: Influence of pH and Particle Size

Despite the influence of pH, additional effect of particle size is also important aspect as nanoscale particles experience high surface energy and easily get to agglomerate and shift toward flocculation behavior. Recently, the variation of ζ and agglomerated size analogous to average particle size of ZrO$_2$ with respect to pH was studied and represented in Figure 3.15 [25].

The commercial grade free flowing starting ZrO$_2$ particle size <100 nm was used in this experiment and observed the agglomeration (average particle size, nm) behavior at different pH level without addition of surfactants or dispersants. IEP is noticed at approximately pH 6. However, the change in zeta potential and average particle size can be described in two ways:

1. pH of the solution equals to the isoelectric point, (pH = IEP).

   The average particle size distribution is symmetrical, and a significant increase in the average particle size can be observed at a pH near IEP. From DLVO theory, the increase in average particle size is indicative of a flocculated suspension since the repulsive forces at or near IEP (zero charge) are zero. The maximum zirconia particle size at the IEP is 1200 nm.

2. pH of the solution shifted away from the isoelectric point, (pH > IEP or pH < IEP). When pH of the suspension is shifted away (either increasing or decreasing) from IEP, the average particle size markedly decreased from 1200 to 100 nm, which reflects the stability of the suspension.

3.4.7 Packing Fraction and Sedimentation

Higher agglomerated size comprises flocculation and continuous sedimentation, as discussed in Section 1.2.10. Sedimentation velocity, (ϑ, m/s), defines the velocity at which

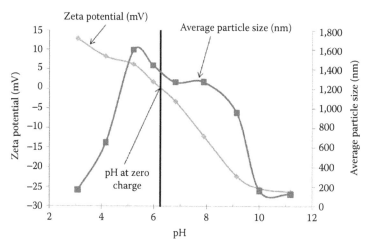

FIGURE 3.15
Zeta potential and flocculation behavior of ZrO$_2$ nanoparticle dispersed nanofluid as a function of pH [25].

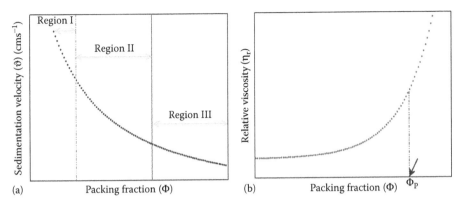

FIGURE 3.16
(a) Variation of sedimentation velocity, (ϑ), and (b) relative viscosity, (η_r), as a function of particle packing fraction.

settling of sedimented particles occur. Generally, unstable suspensions have higher sedimentation velocities compared to the stable suspensions due to their higher flocculating tendencies. Figure 3.16a and b shows the variation of sedimentation velocity, (ϑ), and relative viscosity, (η_r), with respect to packing fraction, (volume fraction of solids, Φ), of the suspension. Apparently, sedimentation velocity decreases exponentially, and the corresponding relative viscosity increases with the increase of particle packing fraction [26,27]. However, sedimentation velocity becomes zero and the corresponding relative viscosity of the suspension approaches infinity when packing fraction, (Φ), approaches the critical value, refers as, maximum packing fraction, (Φ_p). This observation reflects complete sedimentation of the suspension occurs if the packing fraction (Φ) equals to the maximum packing fraction (Φ_p), and the system is unstabilized. The transition in sedimentation behavior of the suspension can be explored more critically by dividing the sedimentation velocity and relative viscosity plots into three regions as a function of packing fraction.

Region I: This region refers to the very dilute suspensions, ($\Phi < 0.01$), where the sedimentation is controlled by the density difference between particles. These suspensions are observed to be more stabilized since the decrease in sedimentation velocity, (ϑ), and corresponding increase of the relative viscosity, (η_r), is not appreciable.

Region II: This region refers to the moderately concentrated suspensions, ($0.2 > \Phi > 0.01$), where the sedimentation is limited due to hydrodynamic interactions between the particles. These suspensions are moderately stabilized that can be evidenced from the corresponding changes appearing in sedimentation velocity and the relative viscosity plots.

Region III: This region refers to the more concentrated suspensions, ($\Phi > 0.2$), where the sedimentation is independent of the particle size. These suspensions are highly unstabilized since a higher packing fraction leads to the system to be flocculated or sedimented. The earlier stated observations regarding changes that occurred in the sedimentation velocity and relative viscosity at a maximum packing fraction, (Φ_p), of the suspension concur with the present discussion.

3.5 Viscosity of Suspension

Viscosity is a rheological property that measures the resistance offered by a fluid or suspension to flow. Figure 3.17 schematically shows the flow of a fluid having viscosity, (η), under laminar conditions in between two imaginary plates, where the bottom plate is stationary, and the top plate moves at a constant velocity "V" in the direction of fluid flow.

Since the flow is laminar, there is a linear change of velocity upward from zero on the bottom plate to V on the top plate that develops a velocity gradient, (dV/dy, s^{-1}), that can be related to the shear stress, (σ, MPa), experienced by the top plate and viscosity of the fluid, (η, Pa.s), using Newton's law of viscosity as:

$$\sigma = \eta \left(\frac{dV}{dy} \right) = \eta(\dot{\gamma}) \tag{3.21}$$

where the velocity gradient (dV/dy, s^{-1}) is equivalent to the shear rate ($\dot{\gamma}$, s^{-1}). So,

$$\sigma = \eta\dot{\gamma} \Rightarrow \eta = \frac{\sigma}{\dot{\gamma}} \tag{3.22}$$

Thus, viscosity is the ratio of shear stress to shear rate. Viscosity of the suspension is a key rheological property in the processing of porous and dense ceramics and their composites. One such typical case is the consolidation of ceramic suspensions by casting techniques (slip casting, tape casting, etc.). Herein, the viscosity is considered as a relative measure of stability of the suspension that controls the structure of the green body in terms of packing density and homogeneity that eventually manipulate pore morphology or sintered microstructure in order to synchronize the targeted engineering properties. However, viscosity of the suspension majorly depends on the various processing parameters like volume fraction of solids, particle size, particle size distribution, and sedimentation velocity. Numerous research works have reported that the consolidated green structures comprising large

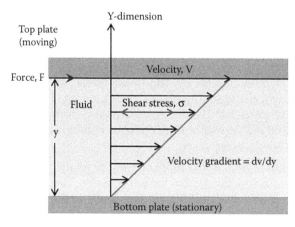

FIGURE 3.17
Two plate model schematically shows the flow behavior of a viscous fluid.

agglomerates lead to the development of heterogeneous packing regions that undergo differential densification in sintering and resulting in poor engineering properties [28–30]. Brief explanations related to rheological behavior of the suspensions are highlighted using viscosity (η) – shear rate ($\dot{\gamma}$) and shear stress (σ) – shear rate ($\dot{\gamma}$) relationships, followed by the effects involved in the interplay of aforesaid processing parameters on suspension viscosity against a range of shear rates have been discussed.

3.5.1 Rheology of Suspensions

Figure 3.18a and b shows the flow behavior of various suspensions including Newtonian and non-Newtonian fluid with respect to change in shear stress, (σ), and viscosity, (η), as a function of the shear rate, ($\dot{\gamma}$). The curve, (i), corresponds to the flow behavior of Newtonian systems like very dilute suspensions, which exhibit a constant viscosity independent of shear rate and governs the Newton's law of viscosity:

$$\sigma = \eta\dot{\gamma} \tag{3.23}$$

The curve, (ii), represents the flow behavior of a Bingham plastic system that displays a yield stress, (σ_β), under low shear rates and a constant viscosity with the further increase of shear rate. The flow behavior of a Bingham plastic system like clay suspensions can be fitted to the following Equation 3.24:

$$\sigma = \sigma_\beta + \eta_{Pl}\dot{\gamma} \tag{3.24}$$

where "η_{Pl}" is the plastic viscosity obtained from the slope of the curve. The curve, (iii), demonstrates the flow behavior of a pseudo plastic or shear thinning system, which exhibits limiting viscosity at low shear rates without showing any yield value. The curve, (iv), signifies the flow behavior of a dilatant or shear thickening system like poly vinyl chloride solution, which displays increasing viscosity with the increase of shear rate due to rearrangement of its microstructure. Shear thinning, (curve 'iii'), and shear thickening, (curve 'iv'), systems obey the power law fluid model, and is given by;

$$\sigma = k\dot{\gamma}^n \tag{3.25}$$

where "k" is the flow consistency index, and "n" is the flow behavior index. Therefore, the flow behavior index for shear thinning systems is, n < 1, and shear thickening systems is, n > 1.

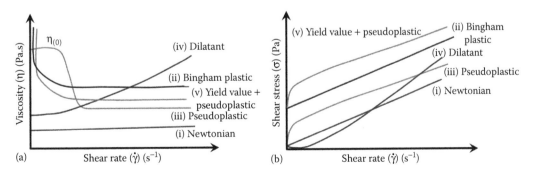

FIGURE 3.18
Schematic representation of flow curves for various systems (a) viscosity (η)-shear rate ($\dot{\gamma}$) relationship and (b) shear stress (σ)-shear rate ($\dot{\gamma}$) relationship.

TABLE 3.7

Reduced Forms of Herschel-Bulkley Equation

Constraint	Reduced Form	Fluid Model	Fluid System
$\sigma_\beta = 0$	$\sigma = k\dot{\gamma}^n$	Power law	Shear thinning and shear thickening
$n = 1$	$\sigma = \sigma_\beta + k\dot{\gamma}$	Bingham plastic	Clay suspensions
$\sigma_\beta = 0; \ n = 1$	$\sigma = k\dot{\gamma}$	Newtonian	Dilute suspensions

Viscosity at any given shear rate for these systems (shear thinning/shear thickening) can be estimated using the following expression:

$$\eta = \frac{\sigma}{\dot{\gamma}} = k\frac{\dot{\gamma}^n}{\dot{\gamma}} = k\dot{\gamma}^{n-1} \tag{3.26}$$

The flow curve, (v), shows a dynamic yield value at low shear rates followed by a shear thinning behavior with further increase of shear rate. Herschel-Bulkley equation can be used to fit and analyze this kind of flow behavior and is given as:

$$\sigma = \sigma_\beta + k\dot{\gamma}^n \tag{3.27}$$

Table 3.7 shows the reduced forms of Herschel-Bulkley equation (Equation 3.27) for different fluid models under a set of constraints.

3.5.2 Particle Size and Volume Fraction on Viscosity

Viscosity enhancement is common incidence with decreasing the particle size and increasing the volume fraction, and it is true for ceramic, metal, and polymeric particles. However, the η and $\dot{\gamma}$ characteristics may alter from material to material because of different degree of surface charge and resultant attraction-repulsion behavior. Herein, a challenging polymeric material, say latex, has been considered to discuss these phenomena [31]. The influence of particle size on the viscosity of a suspension of latex particles under a constant solid loading (particle concentration or volume fraction of particles or packing fraction) is shown in Figure 3.19.

The flow behavior of the suspension is observed to be shear thinning irrespective of the particle size. However, viscosity of the suspension containing smaller sized particles, (~175 μm, curve A), is observed to be high particularly at lower shear rates compared to the suspension containing larger sized particles, (~750 μm, curve B). Under constant solid loading conditions, reduction in particle size increases the number of particles in parallel, increases the interparticle interactions and plausible interlocking phenomena that hinder the free movement of particles, enhances the resistance to flow, and eventually raises the viscosity. However, these competitive interactions are weak in nature, broken down at high shear rates, and progressively resulting in a shear thinning behavior. Figure 3.20 shows the influence of increase in volume fraction of solid particles (particle concentration or solid loading or packing fraction) on the viscosity of a suspension of latex particles across a range of shear rates. The curve 'A' represents the flow behaviour of a suspension at a volume fraction of solids, ($V_f = 45\%$), and is found to be Newtonian with viscosity independent of shear rate [31].

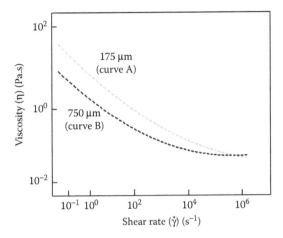

FIGURE 3.19
Change in viscosity behavior of the suspension as a function of particle size against a range of shear rates.

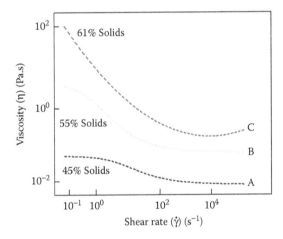

FIGURE 3.20
Change in viscosity behavior of the suspension as a function of particle concentration against a range of shear rates.

The curve "B" shows the flow behavior of a suspension at a slightly higher volume fraction of solids, ($V_f = 55\%$), and the flow behavior appears as shear thinning. The curve "C" shows the flow behavior of suspension with the further increase in volume fraction of solids, ($V_f = 61\%$), and the flow behavior is noticed to be shear thickening at higher shear rates. The increase in viscosity of the suspensions with the increase of solid loading can be mathematically related using Krieger-Dougherty [32] equation, where the viscosity of suspension is proportional to the volume fraction of solids and given as:

$$\eta_r = \frac{\eta}{\eta_{medium}} = \left(1 - \frac{\varnothing}{\varnothing_p}\right)^{-B\varnothing_p} \tag{3.28}$$

where "η" is viscosity of the suspension, "η_{medium}" is viscosity of the base medium, "Φ" is volume fraction of solids in the suspension, "Φ_p" is maximum packing, and "B" is the

Einstein coefficient (also known as intrinsic viscosity of the medium), for example, B is 2.5 for spherical particles. This equation defines the suspension exhibit higher viscosity, (η_r), with smaller particle size compared to coarser ones at the same solid fraction. The increase in viscosity and transition of flow behavior from Newtonian to shear thinning and further to shear thickening with increasing volume fraction of solids over a range of shear rates typically depends on strength of the particle-particle interactions (attractive/repulsive). With the increase of particle concentration from 45% (curve A) to 55% (curve B), the increase in viscosity is primarily attributed to the closer packing of particles, which increases particle-particle interactions, hindering the free particle motion, and eventually enhancing the resistance to flow. However, these interactions are relatively weak in nature, broken down at higher shear rates, resulting in a shear thinning behavior. But these interparticle interactions are observed to be strong when particle concentration further increases from 55% (curve B) to 61% (curve C), cannot be broken down, and significantly hinders the free particle movement. This effect becomes more pronounced with the increase of shear rate and resulting in a shear thickening flow behavior at higher shear rates.

3.6 Concluding Remarks

In fact, ceramic particle surface phenomena are only subject of interest for the manufacturing of traditional ceramics by slip casting. However, in advance, the question including contact angle and hysteresis, work of spreading, hydration energy, water adhesion tension, superhydrophilicity, superhydrophobicity, surface tension, suspension stabilization, zeta potential, isoelectric point, and viscosity play dominant roles in order to use nanoparticle along with coating effectively for functional applications. Contact angle determine the wetting properties of solid surface and thus surface exhibit either hydrophilic or hydrophobic properties. On a flat surface, the same droplet can enhance (advancing) or decrease (receding) the contact angle and its difference hysteresis is more predominant on heterogeneous and rough surface that favors formation of hydrophobic domain. However, an equilibrium contact angle 56° is the result in consideration of water surface energy (72.8 mJ/m^2) and free energy of hydration (–113 mJ/m^2). In consideration of contact angle, work of spreading, free energy of hydration, and water adhesion tension, one can differentiate the superhydrophilic and superhydrophobic surface. High surface roughness and low surface energy favors hydrophobicity, whereas hydrophilicity is the result in high surface energy. There are several methods like surface relaxation, restructuring, adsorption, and composition alteration that are available to reduce the surface energy, and thus a critical surface tension is an important aspect to control the wetting characteristics and can be estimated by Zisman plot. The surface tension of nanofluid prefers to decrease with decreasing nanoparticle size, but high value results of low solid loading and surfactant concentration. Several philosophies, primarily electrostatic, steric, and electrostatic stabilization are responsible to achieve the stable colloidal suspension of nanoparticles that eventually dominated by surfactant, zeta potential and pH of the solution. The resultant viscosity of suspension increases with increasing solid content and decreasing particle size, as more number of particle enhances the more interparticle interactions and hinders the free movement of particles that facilitates the resistance to flow and thus viscosity.

References

1. T. Tadros, *Dispersion of Powders in Liquids and Stabilization of Suspensions*, Wiley-VCH, Weinheim, Germany, 2012.
2. T. Young, An essay on the cohesion of fluids, *Phil Trans R Soc* 95, 65, 1805.
3. K. Holmberg (Ed.), *Handbook of Applied Surface and Colloid Chemistry*, John Wiley & Sons, New York, Vol. 2, 2002.
4. D. J. McClements, *Food Emulsions: Principles, Practices and Techniques*, CRC Press, Boca Raton, FL, 2015.
5. P. Ghosh, *Colloid and Interface Science*, PHI Learning Pvt. Ltd, Delhi, India, 2009.
6. R. E. Johnson, Jr., R. H. Dettre, Contact angle hysteresis: Study of an idealized rough surface, *Adv Chem* 43, 112, 1964.
7. T. Kamegawa, Y. Shimizu, and H. Yamashita, Superhydrophobic surfaces with photocatalytic self-cleaning properties by nanocomposite coating of TiO_2 and polytetrafluoroethylene, *Adv Mater* 24, 3697–3700, 2012.
8. K. Tadanaga, J. Morinaga, A. Matsuda, and T. Minami, Superhydrophobic-superhydrophilic micro patterning on flowerlike alumina coating film by the sol-gel method. *Chem Mater* 12(3), 590, 2000.
9. N. Eustathopoulos, M. G. Nicholas, and B. Drevet, *Wettability at High Temperatures*, Pergamon, Amsterdam, the Netherlands, 1999.
10. C. J. Van Oss, *Interfacial Forces in Aqueous Media*, Marcel Dekker, New York, 1994.
11. M. Callies and D. Quere, On water repellency, *Soft Matter* 1(1), 55–61, 2005.
12. J. Drelich, E. Chibowski, D. D. Meng, and K. Terpilowski, Hydrophilic and superhydrophilic surfaces and materials, *Soft Matter* 7, 9804, 2011.
13. P. L. Du Noüy, An interfacial tensiometer for universal use, *J Gen Physiol* 7(5), 625–633, 1925.
14. W. A. Zisman, Relation of the equilibrium contact angle to liquid and solid constitution, *Adv Chem* 43, 1–51, 1964.
15. Surface tension values of some common test liquids for surface energy analysis, http://www.surface-tension.de/
16. A. J. Kinloch, *Adhesion and Adhesives*, Chapman & Hall, New York, 1987, p. 25.
17. J. Chinnam, D. Das, R. Vajjha, and J. Satti, Measurements of the contact angle of nanofluids and development of a new correlation, *Int J Commun Heat Mass Transf* 62, 1–12, 2015.
18. S. S. Khaleduzzaman, I. M. Mahbubul, I. M. Shahrul, and R. Saidur, Effect of particle concentration, temperature and surfactant on surface tension of nanofluids, *Int J Commun Heat Mass Transf* 49, 110–114, 2013.
19. G. Frens, Controlled nucleation for the regulation of the particle size in monodisperse gold suspensions, *Nat Phys Sci* 241, 20–22, 1973.
20. D. Mayers, *Surfaces, Interfaces, and Colloids: Principles and Applications*, 2nd ed., John Wiley & Sons, Hoboken, NJ, 2002.
21. D. C. Grahame, The electrical double layer and the theory of electrocapillarity, *Chem Rev* 41, 44, 1947.
22. S. G. Malghan, R. S. Premachandran, and P. T. Pei, Mechanistic understanding of silicon nitride dispersion using cationic and anionic polyelectrolytes, *Powder Technol* 79, 43–52, 1994.
23. J. Hierrezuelo, A. Sadeghpour, I. Szilagyi, A. Vaccaro, and M. Borkovec, Electrostatic stabilization of charged colloidal particles with adsorbed polyelectrolytes of opposite charge, *Langmuir* 26(19), 15109–15111, 2010.
24. Horiba Scientific, Zeta potential, http://www.horiba.com/scientific/products/particle-characterization/technology/zeta-potential/
25. C. T. Wamkam, M. K. Opoku, H. Hong, and P. Smith, Effects of pH on heat transfer nanofluids containing ZrO_2 and TiO_2 nanoparticles, *J Appl Phys* 109, 024305, 2011.
26. K. Sellers, *Nanotechnology and the Environment*, CRC Press, Boca Raton, FL, 2008.

27. G. K. Bachelor, Sedimentation in a dilute dispersion of spheres, *J Fluid Mech* 52, 245, 1972.
28. J. S. Reed, *Principles of Ceramics Processing*, 2nd ed., Wiley-Interscience, New York, 1995.
29. Richard J. Mcafee, Ian Nettleship, Effect of slip dispersion on microstructure evolution during isothermal sintering of cast alumina, *J Am Ceram Soc* 89(4), 1273–1279, 2006.
30. O. Lyckfeldt, J. M. F. Ferreira, Processing of porous ceramics by 'starch consolidation', *J Eur Ceram Soc* 18(2), 131–140, 1998.
31. J. Fletcher, *Making the Connection Particle Size, Size Distribution and Rheology*, Malvern Instruments, Worcestershire, UK.
32. I. M. Krieger, T. J. Dougherty, A mechanism for non-newtonian flow in suspensions of rigid spheres, *Trans Soc Rheol* 3, 137–152, 1959.

4

Photocatalytic Phenomena

4.1 Introduction

Noticing the importance in demand of clean environment, catalytic phenomena is being highlighted in the perspective of their use in photo-degradation process and photo-electrochemical (PEC) cells. While encountering this occurrence, removal of water pollutants, synthesis of fuel hydrogen gas without additional pollution, and generation of electrical energy with the help of ceramic semiconductor solid oxide catalyst are prime concern. Water scarcity is exponentially increasing because of excessive contamination of industrial waste, domestic residues, and disposition of agricultural, resulting in depletion of useable and drinking sources. Several remedies are implemented to overcome this issue, among these photo-degradation protocol by using photocatalyst is one of the industrial viable techniques, and thus brief about photocatalyst and their basics have been discussed. Apart from the water purification by photocatalyst, generation of alternate energy is a critical domain to meet the demand without addition of hazardous gas in atmosphere that is an unavoidable incidence during ignition of fossil fuel. Again, effective selection of nanoscale ceramic catalyst can boost up the catalytic conversion in synthesis of alternate fuel or electrical energy. This chapter deals with the essential insight about the characterizations and analysis of nanostrcutred ceramic catalyst in concern of pollutant level reduction to make useable water and generation of alternate energy by abundant cleanest solar energy. Briefing of some module has also been illustrated to understand the use of photocatalyst.

4.2 Photocatalyst for Photo-Degradation

Photocatalysis is derived from two words, photo and catalysis, where "photo" means as light and "catalysis" is the increase in the rate of a chemical reaction in presence of a catalyst, which is never consumed in the main catalytic reaction and can be reused effectively. Hence, photocatalysis is the process that uses light to promote the rate of reaction without being involved itself. It can be defined as the increase in the speed of a reaction by either direct irradiation or by the irradiation of a catalyst that in turn lowers the activation energy for the primary reaction to occur. The photocatalyst may have different forms, however, ceramic photocatalyst has several advantages including band gap tuning to make effective in visible and ultraviolet (UV) light, long stability, and excellent performance in harsh environment. This class of material is very useful for degradation of organic pollutants either dissolved in water or dispersed in air. Thus, significant research focus is given to degrade the different industrial effluents mixed with toxic dyes or making photocatalytic island that is cumulatively made of photocatalytic floor tiles as pedestrian pavement, photocatalytic cobblestone in parking areas, photocatalytic based waterproof tiles, and photocatalytic pavement for road traffic. This novel application opens up wide range of opportunity and permits to move forward to make a pollutant free environment, more smarts, and sustainable. For the last few decades, this class of ceramic oxides are extremely well researched and developed because of practical interest in water remediation, self-cleaning surfaces, and self-sterilizing surfaces. The interaction within semiconductor and photon at proper wavelength splits the electron from its valence band to conduction band, leading to the generation of holes in valence band. These photoassisted species have extensive potential to facilitate different categorized redox reactions, as discussed in following sections.

4.2.1 Band Gap and Valence Band–Conduction Band Position

The energy of a semiconductor's band gap, (E_g), is equal to the difference in energy between the valence band (VB) edge and conduction band (CB) edge, usually determined by UV-DRS method.

   The band gap of most semiconductors is well known, however, it is difficult to measure experimentally of the respective energy levels of valence, (E_v), and conduction bands, (E_c), for many semiconductors. A theoretical approach can be employed to determine the E_c with knowing the value of E_g [1]. In order to calculate the E_c, let's consider the Figure 4.1, where the maximum electron energy is 4.5 eV under vacuum, and, by convention, it is taken to be negative with respect to hydrogen ionization reaction [2]. A flat band configuration represents an identical electron potential that persists within bulk and surface of semiconductor. However, band bending is common phenomena because of surface charge variation, and that concept is further discussed in Section 4.3.1.

   A simple relation within E_c and E_g can be written as Equation 4.1, where X is the Sanderson electronegativity of the solid (the energy difference between the vacuum level and the midgap position of the semiconductor), which is defined as the geometric mean of the Mulliken electronegativities $(\chi_{Mulliken})$ of the atoms [3]:

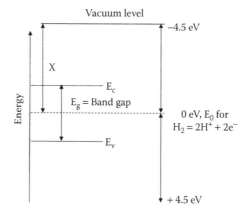

FIGURE 4.1
Relation between band gap (E_g), conduction band (E_c), valence band (E_v), Sanderson electronegativity (X), electron affinity (χ), and H_2 redox potential.

$$-E_c = 4.5\,eV - X + \frac{1}{2}E_g \qquad (4.1)$$

where the $\chi_{Mullikan}$ of an atom is analogous to the negative of the chemical potential, in other words, it is the average of the first ionization energy of the metal and the atomic electron affinity of the anion. As an example, the ionization energies of Ti and O are 6.83 eV and 13.62 eV, respectively, and the electron affinities for Ti and O are 0.083 eV and 1.46 eV, respectively. Thus, the $\chi_{Mullikan}$ for Ti and O are therefore [(6.83 + 0.083)/2] 3.45 eV and [(13.62 + 1.46)/2] 7.54 eV, respectively. It leads to, (X), Sanderson electronegativity of [(3.45 + 7.54)/2] 5.8 eV and, assuming anatase TiO$_2$ band gap of 3.2 eV, the E_c value becomes 0.3 eV higher (more negative) than the hydrogen redox reaction (Figure 7.3a) [4]. In actual, the electron affinity of semiconductor solid also depends on the surface properties, including exposed crystal facets and surface chemistry. This information is useful to select the proper photocatalyst for water splitting, as discussed in Section 4.3.

Quantum effect can alter the energy positions of the valence and conduction band edges, this incidence is prominent if the particle is close to or smaller than the Bohr radius of the first excitation state, causing this smaller size to exhibit both characteristics of bulk and molecular phases. In consideration of previous discussions in Section 1.2.2, as nanoparticle size decreases (preferentially <10 nm), its band gap increases, primarily with an increase in conduction band energy levels and slight decrease in valence band energy levels.

In order to prepare a useful semiconductor photocatalyst, selection of band-position is a basic criterion because it indicates the thermodynamic limitations for the photo-excited reactions essentially related with charge carriers. For example, the valence band level of the semiconductor is preferably lower than the oxidation potential of the surface absorbed H$_2$O [E(H$_2$O/$^\bullet$OH) = 2.29 V] during photocatalytic degradation of pollutants, on the contrary, the conduction band level is at a higher level than the one electron reduction potential of oxygen [E(O$_2$/O$_2^-$) = –0.33 V]. A typical illustration for band gap position for well popular semiconductor TiO$_2$ is represented in view of organic pollutant degradation in Figure 4.2.

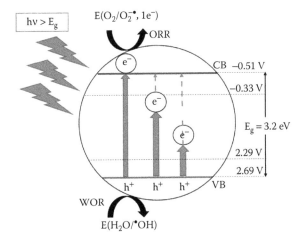

FIGURE 4.2
Generation of e$^-$/h$^+$ in presence of solar energy and promoted to the formation of radicals to degrade the pollutants by the photocatalyst TiO$_2$, where ORR is oxygen reduction reaction and WOR is water oxidation reaction. Electron is only able to shift from VB to CB when hv > E$_g$.

4.2.2 Electron-Hole Charge Transportation Phenomena

Discover and basic privilege of photon-assisted photosynthesis by chlorophyll or organic dye degradation by ceramic semiconductors depends on the generation of electron (e^-) and hole (h^+) under solar irradiation. The probable recombination within electrons and holes reduces the rate of photocatalytic reaction, and trapping of charges can enhance the reaction rate. Hence, a brief discussion on role of electron-hole delivers an insight to understand the reaction mechanism of photocatalyst. Usually, the photo-induced charges have capability to migrate freely to the particle surface that eventually enhances the quantum efficiency and participate in reactions with adsorbed species. But, simultaneously, several incidences on the photocatalyst surface and bulk are happening during photo-excitation process, and these are summarized in Figure 4.3.

Herein, the photo-induced catalytic reaction steps can be illustrated as:

1. light harvesting – path 1
2. electron migration (VB to CB) and charge generation – path 2
3. charge transport in bulk – path 3
4. charge recombination in bulk – path 4
5. photocatalytic surface reduction reactions – path 5
6. photocatalytic surface oxidation reactions – path 6
7. charge recombination on the surface – path 7

Thus, owing to the steps involvement in the photocatalysis, the resultant efficiency is the product of several efficiencies and can be represented by Equation 4.2 [5]:

$$\eta_c = \eta_{abs} \times \eta_{cs} \times \eta_{cmt} \times \eta_{cu} \tag{4.2}$$

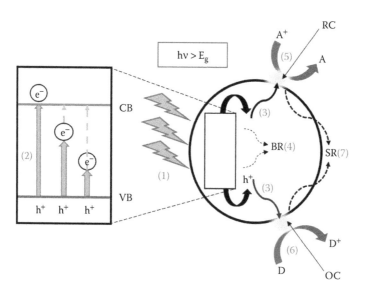

FIGURE 4.3
Schematic representation on the probable charge generation and transport behavior in photocatalyst after irradiation by sunlight, where RC, reduction co-catalyst; OC, oxidation co-catalyst; BR, bulk recombination; and SR, surface recombination.

where η_c is the solar energy conversion efficiency, η_{abs} is the photon absorption efficiency, η_{cs} is the charge separation or excitation efficiency, η_{cmt} is the charge migration and transport efficiency, and η_{cu} is the charge usage efficiency for photocatalytic reactions.

The photo-generated electrons, (Path 1 → Path 2 → Path 3 → Path 5), are primarily used in the reduction of O_2, other reactive species, such as the superoxide radical O_2. or the singlet oxygen excite other radicals and follow chain reactions involving H_2O_2 or O_3. Simultaneously, positively charged photo-generated holes, (path 1 → path 3 → path 6), directly oxidize organic molecules or create highly reactive and short-lived hydroxyl radical •OH once H_2O molecule adsorb on the catalyst surface. This is the probable way to transfer of holes and formation of reactive species (•OH) for the oxidation of other species. However, the charge movement on the particle surface and their involvement in catalytic reaction can be reduced through electron-hole pair (EHP) recombination and entrapment by coordination defects at the surface and by lattice defects in the bulk region of particles.

The main function of electron-hole pair is charge transfer and interaction with adsorbate species. But, a major percentage of charge recombination reduces the population of free electrons or holes and reduces the photocatalytic efficiency, and sometimes the rate of reaction may be reduced up to 90% [6]. Usually, the quantification of the charge dynamics is carried out by ultra-fast time-resolved absorption spectroscopy, identifying all the reactive species, trapped holes and electrons, and quasi-free electrons, besides the determination of the rate of recombination of electron and hole using a decay profile [7]. Time-resolved microwave conductivity technique is preferentially employed for the measurement of charge carrier mobility of nano-structured semiconducting materials.

The transient absorption and emission spectroscopy estimates that the recombination time is shorter, (10^{-15} to 10^{-12} s), than the time it takes for the carrier to diffuse, (10^{-9} s), to the surface. Recombination is an exothermic process and generates additional heat as by-product, and this flash heating can alter the rate of reaction. In general, as the size of the photocatalyst particle size decreases approaching approximately 10–15nm, EHP recombination decreases that attribute as the rapid diffusing charge carriers reach towards the surface (reduction of Path 3 distance) before electron-hole interaction. Factors including substitutional impurities, lattice defects, and vacancies can serve as electron traps near the surface or in the bulk, and thus affect EHP recombination rates, since those defects or inclusions provide possible alternative pathways for the photo-generated electrons to recombine with holes.

During migration toward the surface, the photo-generated charge carriers are sometimes trapped in the shallow trapping sites, thus, have enough potential to successively drop down their position from these shallow sites even energetically dispersed deeper trapping sites permanently by hopping process. In order to reduce the recombination of charge carriers, the trapping of the charge carriers can be feasibly helpful in enhancement of photocatalytic activity if those can be trapped on the surface of the semiconductors, which is generally done by doping of various metal cations into the semiconductor structure or combination of semiconductors. For example, the surface trapping of charge carriers of TiO_2 can be modified by 0.5–1 wt% addition of different dopants V^{4+}, Co^{3+}, Ru^{3+}, Mn^{3+}, Fe^{3+}, Cr^{3+}, Ni^{2+}, or scavengers leading to the decrement in recombination mechanism of electron and hole traps, which enhances the time constant for recombination from 500 ps to 50 ms [8]. This trapping of the electrons and holes by metal cations follows:

$$M^{n+} + e_{cb}^{-} \rightarrow M^{(n-1)+} \, ; e^{-} \text{ trapping} \tag{4.3}$$

$$M^{n+} + h_{vb}^{+} \rightarrow M^{(n+1)+} \, ; h^{+} \text{ trapping} \tag{4.4}$$

where the energy level for conversion of M^{n+} to $M^{(n-1)+}$ should consist below the conduction band edge, (E_{cb}), and the energy level for the conversion of M^{n+} to $M^{(n+1)+}$ should stay above the valence band edge, (E_{vb}), of the photo-catalyst. To restrain this recombination of photo-generated electron-hole pairs and to uplift the charge carrier mobility in the bulk, various tactics are also implemented including reduction in defect sites in the bulk and on the surface, enhancement of interaction surface area by size reduction of photo-catalyst, generation of surface porosity, coupling with nano-carbon materials, fabrication of hetero-junctions, etc. For enhancement of surface reaction kinetics, in addition to the above-mentioned processes, other surface modification strategies are being advocated lately, such as cocatalysts loading, exposing highly reactive facets, and enhancement of interaction surface area [9,10].

4.2.3 Role of Scavenger

Localized surface charge carriers trapping is the desirable factor for the enhanced photocatalytic activity of a semiconductor. To uplift the trapping lifetime of charge carriers, various scavengers are being employed on the surface of the semiconductors. The scavengers for electrons collect the photo-excited electrons, owing to which, the lifetime for trapped holes is enhanced up to 1000-fold with the use of any hole scavenger, thus, $e^- - h^+$ recombination is prevented. Contrary to this situation, the absence of scavenger promotes the recombination of charge carriers, which is a bit less when electron or hole trapping is carried out at the surface. Thus, addition of scavenger to the semiconductor photocatalyst system results in substantial lessening of the recombination rate, which generally follows a second order kinetics. For instance, O_2 and ethanol are used for electron and hole scavenging purposes in TiO_2 during photocatalytic processes to study the hole and electron trapping in the transient absorption spectroscopy. Most likely, electron and hole surface trapping sites are semiconductor –OH groups and semiconductor –O– sites, respectively [11].

4.2.4 Surface Modifications

The photocatalytic efficiency decrement owing to the semiconductor surface, such as, low surface activity, charge carrier recombination, and low absorption due to wide band gap are typical limitations which are essential to modify in target of high degree efficiency. In order to enhance the photocatalytic activities, various strategies are being employed, for instance, surface activation and sensitization, surface plasmon resonance effect, in situ phase formation, etc.

Doping impurities into the semiconductor structure constrict the wide band gap to a narrower one, resulting in a red shift in the absorption spectrum. This doping enhances the absorption efficiency of the semiconductor in the visible range reasonably. Both metal and nonmetal doping shows increase trend in photocatalytic behavior owing to the higher visible light absorption, crystallinity, Three Dimensional (3D) interconnected mesochannels, which in turn give formation of enhanced concentration of hydroxyl radicals, oxidation-reduction potential, or varied lattice parameter [12]. For instance, the metallic doping like chromium doped TiO_2 displays remarkable enhancement in methylene blue decomposition with respect to pure TiO_2 that is attributed to the improved visible light absorption, crystallinity, and 3D interconnected mesochannels.

Nonmetallic surface doping also gives splendid photocatalytic activity, comes from higher porosity, higher interacting surface area, and enhanced active sites present on the semiconductor surface. N-doped TiO_2 nano-materials due to the mesoporous structure and exposed, (001), facet, demonstrate enhanced photocatalytic activity [13]. Metallic and nonmetallic species surface codoped semiconductors exhibit tremendous increment in the activity and stability of nanoscale photocatalyst. The improvement in activity is owed to the combined effect of two doping materials in comparison to the single species surface doped semiconductor [14].

The wide band gap semiconductor coupling with narrow band gap semiconductor reduce the interparticle charge carrier separation by alteration of the energy levels of conduction band and valence band edges, this phenomenon is known as surface sensitization. As an example, quantum size effect of MoS_2 and WS_2 nano-cluster on photocatalyst like TiO_2 particle facilitates the surface sensitized photocatalytic activity enhancement [15]. Construction of surface hetero-junctions has been proven another incitation of photocatalytic activity of wide band gap semiconductors, owing to the improved charge separation and drastic reduction in recombination of charge carriers. Heterojunctions like n–n and p–n junctions are now being conventionally employed broadly for application of degradation mechanism by improved photocatalysis process, for instance, CdS/TiO_2, ZnO/TiO_2, $SrTiO_3/TiO_2$ demonstrates higher photocatalytic activity owing to the restriction of recombination of the photo-generated electron and hole pair [16]. Despite hetero-junction by different semiconductors, a classic approach, namely, hetero-phasic junction of the same photocatalyst is also being used to tune the band gap alignment. This combination reduces the recombination and enhancing the charge carrier separation results in immense photocatalytic activity. For example, a combination of two crystalline phases of mesoporous TiO_2, in preference, anatase, ($E_g = 3.2$ eV), and brookite, ($E_g = 2.2$ eV), exhibits high degree of photo-degradation of n-pentane compared to counterpart anatase only [17].

Along with the phase junction, semiconducting material and nano-structured conducting material junction have also influence to enhance the photocatalytic activity. This is attributed to the promoted charge separation caused by the formation of Schottky junction between the semiconducting photocatalyst and the nano-structured conducting material. For instance, 1 wt% graphene and TiO_2 composite exhibit an excellent upgradation in H_2 production rate up to 41-fold in comparison to pure TiO_2 [18]. Exposure of highly active surface facet is an effective strategy to enhance the photocatalytic activity. In every semiconductor, there exists a facet, which has highest reactivity, and, thus, nano-structured semiconductor with highest reactive facets exposed to the surface is developed by Yang et al. [19]. The facet reactivity can be termed as a function of atomic alignment in distinct directions, which results in variation in optical and electronic properties. The variation of physico-chemical properties of different facets is a harmonious result of photo-adsorption activity and selectivity process of the facets.

Surface fluorination is another tactic used to tackle the surface charge carrier recombination process. It can be clearly demonstrated by taking TiO_2 as a simple example. As conventional pyrolyzed TiO_2 is universally known as the hydrophobic material, this hydrophobicity possibly degrades its photocatalytic activity due to lesser generation of Ti-OH groups at the surface, which can potentially be nullified by surface fluorination leading to significant enhancement in physico-chemical and photocatalytic properties of TiO_2. Surface fluorination traps the conduction band electrons of TiO_2 owing to the high electronegativity of fluorine and formation of free OH radical species by mutating the valence band holes, which results in down turn of the recombination rate of photo-generated charge carriers and enhances the photo-generated OH free radicals. Thus, the availability of the charge carriers for photo-degradation mechanisms enhances degradation of organic pollutants effectively. Fluorite addition not only forms free OH radicals, but also thermodynamically stabilizes the most active anatase phase over rutile and brookite phase along with more reactive, (001), facet over the, (010), facet, which tremendously enhances the photocatalytic activity toward the degradation of organic pollutants. Fluorination of TiO_2 was carried by hydrolyzing titanium tetra-isopropoxide in a mixed NH_4F/H_2O solution with enhanced photocatalytic activity in comparison to pure TiO_2 [20].

Surface decorated porosity potentially enhances the specific surface area, which raises the interaction surface with the photons as well as degrading species owing to which, absorption and degradation activity increases effectively. Not only the size and quantity of porosity, but also the distribution of the porosity, play a critical role in enhancement of photocatalytic activity. In comparison to unimodal porous structure, bimodal porous structure displays high photocatalytic activity. For instance, bimodal porous structure demonstrated better photocatalytic degradation of Rhodamine B (RhB) compared to P25 photocatalyst, only [21].

With consideration of several surface modification protocols, one can expect high rate of photocatalytic activity by exposure of the highly reactive facet with high surface area of the photocatalyst with codoping or hetero junction, in fact, it is a matter of extensive research to find out the suitable condition for particular semiconductor.

4.2.5 Influence of Particle Size

The photocatalytic activity associated with the recombination of charge carriers potentially depends on the surface-to-volume ratio of the semiconductor particles, which is attributed to the bulk and the surface modulate absorptivity and recombination mechanism, respectively. This phenomenon can be linked to the particle size decrement of the semiconductor that enhance the absorbance and eventually photocatalytic activity of optimum size nano-structured materials. For example, nanoscale, (30 nm), TiO_2 particle of anatase phase shows potential increase in photocatalytic activity in comparison to the micron range, (up to 50 μm), TiO_2. A drastic elevation in methylene blue (MB) degradation is obtained when the particle size is decreased to 30 nm. A correlation between initial rate of degradation, (a), and particle diameter, (d), of TiO_2 photocatalyst is given by [22]:

$$a = -0.114 \ln(d) + 0.652 \qquad (4.5)$$

An expression between the half-life period of degradation, $(t_{0.5})$, of MB and particle diameter, (d), of TiO_2 particles is found as:

$$t_{0.5} = 3.905 \ln(d) + 0.707 \qquad (4.6)$$

First order rate constant, (k), of the degradation of MB can be related to the TiO_2 particle size, (d), by the following equation:

$$k = -0.064 \ln(d) + 0.260 \qquad (4.7)$$

where t, k, and d are in minutes, $(minutes)^{-1}$ and μm, respectively. From Equation 4.7, it could be noticed that the k value increases as the particle size decreased. This efficiency further enhanced when anatase particle size reduced to 10 nm. Thus, a conclusion may be outlined that the rate constant is mainly decided by the amount of substrate adsorbed on the photocatalyst surface and their transformation without e–h recombination. However, very small particles enhance the e/h recombination and reduce the degradation efficiency, thus, size optimization for a particular application is essential.

It is worthy to mention that a suspended photocatalytic reaction is industrial viable compared to fixed bed type, as former can overcome the difficulties of the preparation of ultrafine catalyst particles. Furthermore, the equipment for the suspended photocatalyst is simpler than the fixed type, although it facilitates effective separation from the suspension.

4.3 Photocatalyst for Photo-Electrochemical Cell

PECs are the most efficient device for converting solar energy into electrical or chemical energy and have rapidly gained popularity in recent years. It is usually semiconductor liquid-junction based cells. In this cell, the irradiation of a photo-active semiconductor electrode in contact with an appropriate electrolyte (redox couple) changes the electrode potential with respect to reference electrode (under open circuit conditions) or changes the current flowing in the galvanic cell containing the electrode (under short circuit conditions). These devices are quite simple to construct and often consist of a photo-active semiconductor electrode (either n- or p-type) and a metal counter electrode. Irradiation of the semiconductor–electrolyte junction with light of hv $> E_g$ (E_g is the bandgap of semiconductor) results in the generation and separation of charge carriers. Herein, a solid–liquid interface is the prime difference compared to counterpart photo-voltaic (PV) solar cells consisting of solid–solid interface (e.g., Si solar panel).

The majority of carriers are electrons in an n-type semiconductor, which move to counter electrode through an external circuit and take part in a counter reaction. Holes are the minority charge carriers, which in turn migrate to electrolytes and participate in the electrochemical reactions. The commercial use of a PEC solar cell depends on its conversion efficiency and stability. Various efforts have been made to make PECs more efficient, such as electrolyte modification, surface modification of the semiconductors, photo-etching of layered semiconductors, semiconductor septum based PEC solar cells, and so on [23,24]. The prime concern is to capture the energy that is freely available from sunlight and turn it into electric power or chemical conversion. The PECs can be classified in two major categories on the basis of change in Gibbs free energy, (ΔG), as illustrated in Figure 4.4. A schematic of electrochemical photo-voltaic cell, ($\Delta G = 0$), electrochemical photo-electrolytic cells, ($\Delta G > 0$), and electrochemical photocatalytic cells ($\Delta G < 0$) is illustrated in Figure 4.5.

4.3.1 Regenerative Photo-Electrochemical Cell (Solar Cells)

Regenerative PEC solar cells, which are based on a narrow-band gap semiconductor and a redox couple, convert optical energy into electrical energy without bringing about any change in the free energy of the redox electrolyte, ($\Delta G = 0$). Here, the photo-energy is converted into electric energy. The electrochemical reaction occurring at the counter electrode is opposite to the photoassisted reaction occurring at the semiconductor working electrode. Thus, they are also called electrochemical photo-voltaic cells, as depicted in

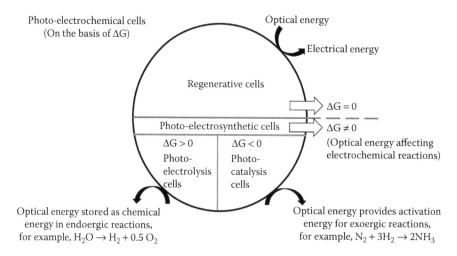

FIGURE 4.4
Classification of photo-electrochemical cells.

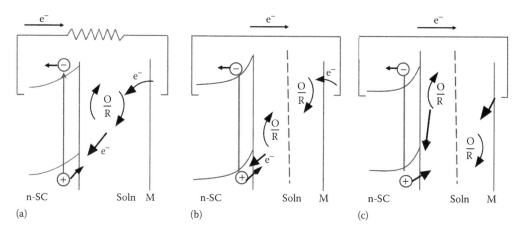

FIGURE 4.5
Schematic representation of (a) electrochemical photo-voltaic cell, $\Delta G = 0$, (solar energy), (b) electrochemical photo-electrolytic cell, $\Delta G > 0$, (water splitting), and (c) electrochemical photocatalytic cell, $\Delta G < 0$ (activation energy for chemical reaction).

Figure 4.5a. A typical example of this cell is a dye-synthesized solar cell (DSSC), and their details are discussed in Section 7.1. In fact, electrochemical photo-voltaic cells made of semiconductors have several advantages over solid-photo-voltaic cells (p–n junction) in the perspective of mode of cell fabrication, conversion mechanism, and performance, listed in the following:

1. In PV, the deposition of grid structure, anti-reflection coating, and high tempera-ture processing of the semiconductor substrate for junction formation by doping, that can be avoided in Photo Electrochemical Cell (PEC).

2. In order to maintain analogous p–n junction of PV cell, a simple electrolyte containing suitable redox species, a counter electrode, and insulation of the back surface of the photo-electrode are required in PEC, this fabrication is relatively cheap compared to PV cell.

3. In solid-state solar cell, the performance severely affected by the defects in semiconductors, this influence is relatively less in semiconductor for PEC.

4. Direct energy transfer from photons to chemical energy, and, thus, PEC device can also be used to store energy in the form of conventional fuels.

5. Unlike conventional solid-state photo-voltaic cells, the potential of the working electrode can be varied with respect to the reference electrode by means of an external voltage source connected between the working electrode and the counter electrode.

4.3.2 Photo-Synthetic Solar Cells

The photo-synthetic cells utilize photon energy input, $(E \geq E_g)$, to produce a net chemical change in the electrolyte solution, $(\Delta G \neq 0)$, when the reaction at the counter electrode is not exactly opposite of the hole transfer reaction at the illuminated semiconductor-liquid interface. Here, the photo-energy is used to affect the chemical reactions, with nonzero free energy change in the electrolyte. These cells can be further classified into two types of cells, namely, photo-electrolytic and photocatalytic.

4.3.2.1 Electrochemical Photo-Electrolytic Cells

In photo-electrolytic cells, the Gibbs free energy, $(\Delta G > 0)$, and the photo-energy is stored as chemical energy in endergonic reactions. Two redox systems are present in photo-electrolytic cells. One redox system reacts with the holes at the surface of the n-semiconductor electrode, while the other reacts with the electrons entering the counter electrode. Anodic and cathodic compartments need to be separated to prevent mixing of the two redox couples. In case of n-SC (Figure 4.5b), water is oxidized to oxygen at the semiconductor photoanode and reduced to hydrogen at the cathode. The overall reaction is the cleavage of water by sunlight: $H_2O \rightarrow H_2 + \frac{1}{2}O_2$. The solar energy conversion efficiency, (η), for photo-electrolytic cells is given by the equation given in the following:

$$\eta = \frac{\text{Energy stored as fuel} - \text{Electrical energy supplied}}{\text{Incident solar energy}} \tag{4.8}$$

The commercial viability of this type of cell (e.g., dye-sensitized photo-electrosynthesis cell) depends on the stability of the surface-bound molecular chromophores and catalysts [25]. Photo-potentials are rarely above 0.6–0.8 V in these cells in presence of single junctions of semiconductors, and, thus, multijunction cells involving multilayer electrodes or a series connection of PECs is required for higher photo-voltage. In section 7.3, a typical example of electrochemical photo-electrolytic cells, referred as "tandem cells", and their details are discussed in order to achieve effective chemical conversion, such as water splitting and generation of H_2 as fuel.

4.3.2.2 Electrochemical Photocatalytic Cells

In electrochemical photocatalytic cells, the rate of reaction increases when $\Delta G < 0$ (Figure 4.5c). Aqueous suspensions composed of irradiated semiconductor particles may be considered to be an assemblage of short-circuited microelectrochemical cells operating in the photocatalytic mode, for instance, the photo-oxidation of organic compounds, or reactions with high activation energy. In photocatalytic cells with $\Delta G < 0$, the photoenergy provides activation energy for exergonic reactions (example: $N_2 + 3H_2 \rightarrow 2NH_3$).

4.3.3 Semiconductor–Electrolyte Interaction and Band Bending

Despite flat band, three probable space charge regions of n-type semiconductor in PECs are illustrated in Figure 4.6. Flat band represents initial condition when there is no net charge transfer from semiconductor bulk to the surface or electrolyte solution, as shown in Figure 4.6a. Now, assume the Fermi energy of solid semiconductor is either lower or higher than the redox potential of the electrolyte solution, in former condition the electron flows from the solution to solid and leads to accumulation, as shown in Figure 4.6b. Excess negative charge facilitates downward band bending and the required potential to be dropped across the space charge region. However, in presence of higher redox potential of the solution, follows reverse phenomenon, as depicted in Figure 4.6c, known as depletion layer.

Here, charge flows out of the solid, leaving ionized donors and positive charge moves in the space charge region below the surface. The mobile charge carriers from the photoelectrode continues to move into the electrolyte until the equilibrium is reached, followed by the formation of a positively charged depletion region of e⁻s, and a diffused layer of negatively charged ionic layer, known as 1st Helmholtz layer. The concentration of the negatively charged ionic species gradually decreases toward the core of the electrolyte, giving rise to a more diffused layer, called as 2nd Helmholtz layer, or simply Gouy layer. When the gradient in chemical potential of the electrode and the electrolyte becomes zero, the Fermi level and the redox level of these two form a line [26]. In fact, n-type oxide based ionized donors experience oxygen vacancies [27]. Thus, it is necessary to extract electrons from the valence band if there are insufficient donors to compensate the charges, this is known as "inversion layer," and the majority charge carrier at the surface changes from

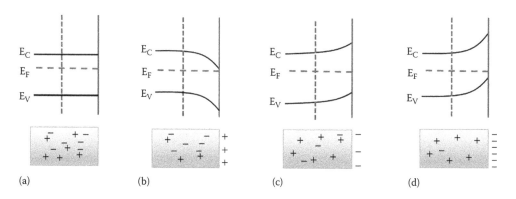

FIGURE 4.6
Schematic diagram of (a) flat, (b) accumulation, (c) depletion, and (d) inversion layer, and their probable surface charge generation.

electrons to holes, as illustrated in Figure 4.6d. An applied potential difference also results in the solid/solution interface potential difference and thus band bending.

4.3.4 Conventional Materials for Photo-Electrochemical Cells

Inorganic materials are commonly used in PECs, but organic materials have certain advantages, and these are [28]:

1. High optical absorption coefficient allows large amount of light absorption, and thus a small amount of material is required compare to inorganic material.
2. Easy adjustability of chemical and physical properties, such as band gap, charge transfer, solubility, etc.
3. Easy synthesis protocol and versatile mode of use including dyes, oligomers, pigments, polymers, dendrimers, and liquid crystals.

Usually, polyacetylene, poly (phenylenevinylene), poly (vinyl carbazole), fullerene, $PC_{61}BM$, Poly (3,4-ethylenedioxythiophene) doped with poly (styrene sulfonate) known as "PEDOT:PSS", metal organic frameworks, and more are used for PECs. However, low efficiency, strength, and stability in harsh environment are the major disadvantages of organic materials. In this contrary, inorganic materials, primarily the metal oxides (TiO_2, Fe_2O_3, WO_3, ZnO), metal chacogenides (MoS_2), doping of metals (Cu, Pt, Ag), and bimetallic nanoparticles, have extensive demand for such application [29,30].

4.3.5 Output and Governing Parameters for Photo-Electrochemical Cells

In PECs, essential components are photo-electrode made of semiconductor-coated transparent conductive glass (e.g. Indium tin oxide, Fluorine doped tin oxide, etc.), an electrolyte and a counter electrode. The photo-excited reaction results in potential difference in the cell, and their electron that flows in the external circuit can be measured by using a potentiometer. An efficient PEC device can be constructed by optimizing basic output including: open-circuit voltage (V_{oc}), short-circuit photo-current (I_{sc}), maximum power (P_{max}), fill factor (FF), incident photon-to-current efficiency (IPCE), and power conversion efficiency (PCE), detailed explanations of each parameter in consideration of DSSC are given in Section 7.2.5. The output characteristics of PECs are governed by several factors, such as band gap of semiconductors, particle morphology, particle porosity, thickness and roughness factor of semiconductor surface, size of dye molecules, transparency of conductive glass electrode, temperature, and light intensity.

1. Bandgap of Semiconductors
   Band gap manipulation is an essential component to utilize photo-excited energy and fabrication of PECs. In order to achieve maximum efficiency of PECs, the band gap, (E_g), of semiconductor electrode material has to be adjusted to absorb maximum solar radiation to utilize for charge separation. This excitation facilitates optimum separation of the photo-generated charge carriers, sufficient amount of electron excitation to the conduction band, and transportation of the charge-carriers through the structure. The resultant band gap energy for the solar cell application is generally preferred in the range of 1.2–1.8 eV, for example, 1.6 eV for TiO_2 based DSSC. On the contrary, if the electrode material band gap deviates

from this range, it leads to insufficient photon absorption and e⁻ – h⁺ separation in the electrode material, and results in the lowered efficiency of the PEC cell [31].

2. Particle Morphology and Porosity

Dye absorption is expected to be more in 1D nanostructure than sintered spherical particles, as it is forming polycrystal and restricting less absorption of dye and visible light, as illustrated in Figure 4.7 [32].

In recent data, it was observed that the e⁻/h⁺ recombination phenomena is also much faster in spherical particles (Figure 4.7a) compare to 1D growth (Figure 4.7b) nanotubes and, thus 1D exhibit higher charge collection and high DSSC efficiency [33,34]. It also has effect on the electron diffusion to reach the conducting glass plate. In fact, maximum obstruction of dye molecule transportation is experienced when particles develop interlocking regions and thus electron transportation path follows random diffuse model. So, a less random diffusion can be expected in 1D nanostructure, but reduction of diameter further enhances the molecule transport rate (Figure 4.7c). So, nanofiber, nanotube, and nanowires deposition on conductive electrode compared to spherical particle is a promising choice to enhance the DSSC efficiency. However, particle size optimization along with morphology can reduce the recombination of e⁻/h⁺, and thus efficiency of cell.

Despite particle morphology, porosity has influence in order to achieve high degree of incident photon-to-current efficiency. In a recent article, the IPCE, as a function of excitation wavelength for single crystal (101) plane and mesoporous anatase TiO_2, has been studied and illustrated in Figure 4.8. In this study, TiO_2 was sensitized by surface-anchored ruthenium complex cis-$RuL_2(SCN)_2$, where L = 2,2′-bipyridyl-4,4′-dicarboxylic acid dye (N3 dye) and electrolyte was a solution of 0.3 M LiI and 0.03 M I_2 in acetonitrile [35].

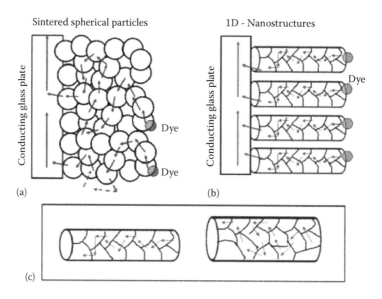

FIGURE 4.7
A probable dye molecule diffusion (a) sintered spherical particles, (b) 1D nanostructures, and (c) diameter minimization can enhance the more unidirectional flow toward electrode. (From Jose, R., *J. Am. Ceram. Soc.*, 92, 289–301, 2009 [32].)

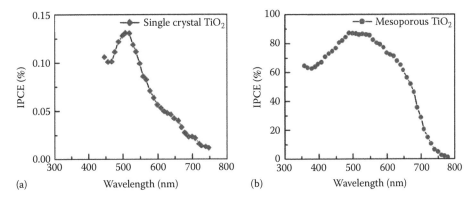

FIGURE 4.8
Difference in IPCE when (a) single crystal and (b) mesoporous TiO_2 was sensitized by the N3 dye. (From Grätzel, M., *Nature*, 414, 339, 2003 [35].)

In PECs, a high fraction of the incident photon is expected to convert into mobile electrons for enhanced performance. Thus, a mesoporous nanostructure metal oxide specifically 1D network with having high specific surface area and consists of pore size more than dye molecule size that eventually facilitate the high IPCE. A distinct difference in between single crystal and mesoporous TiO_2 is attributed to the large surface area that allowed the dye molecule, ($\sim$0.13 mmol/cm^2), anchoring effectively on TiO_2 surface and thereby increasing the absorption cross section. This enhanced cross section boosts the interaction region between the dye anchored-semiconductor layer and photon for the generation of free electron under excitation.

3. Physical Appearance of Semiconductor Electrode

Semiconductor electrode is the prime component in PECs, and thus several important parameters are essential to synchronize including transparency of electrode, thickness of semiconductor coating, and roughness factor of semiconductor surface. A layer of conducting glass electrode is used in the PECs for the purpose of trapping the photons from the incident radiation, owing to purposefully photon absorption by dye-anchored semiconductor surface and generation of charge carriers in the PEC electrode. The optical transparency has to be very high, (>85%), in order to allow a wide and optimum range of solar light wavelength that is suitable for the excitation of the charge carriers on the incidence of light [36].

Thickness of the semiconducting electrode layer should be enough to absorb all the photons involved in the solar radiation spectrum. For the enhanced efficiency of the PEC, the thickness of the semiconducting material has to optimize as the absorption efficiency becomes uniform with respect to thickness for the particular material of interest. In a recent study, the semiconductor thickness made of average $\sim$130 nm diameter, $\sim$1 μm length TiO_2 nanorod was optimized and measured different photo-voltaic features, as illustrated in Figure 4.9 [37]. A slight reduction in V_{oc} is thought to be attributable to the recombination of electron-hole pairs, as well as the high series resistance arising from the weak interconnection of TiO_2 nanorods (Figure 4.9a). When the nanorod layer thickness was raised from 8 to 20 μm, the efficiency was improved, likely due to minimization of electron back transfer and

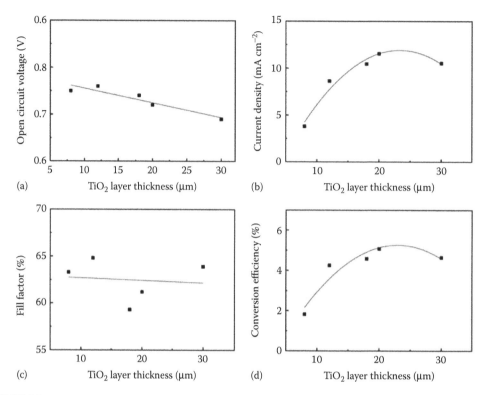

FIGURE 4.9
Photo-voltaic-characteristics relationship with TiO$_2$ layer thickness. (a) V$_{oc}$, (b) J$_{sc}$, (c) FF, and (d) η, respectively.

the increased amount of adsorbed dyes. The efficiency decreased when the coating thickness is beyond 20 μm, likely due to rapid recombination of electron-hole pairs and resistive losses. The best photo-voltaic characteristics of 20 μm were obtained as V$_{oc}$ of 0.72 V, J$_{sc}$ of 11.54 mA·cm^{-2}, FF of 61.2%, and η of 5.07%.

Surface roughness also potentially affects the efficiency of the cell, and author qualitatively analyzed by Transmission Electron Microscope (TEM) study. With increase in surface roughness, the absorption of photons enhances effectively in comparison to smooth surface owing to the increased exposed active surface area for the catalytic activity, thus, an elevated number of photo-generated charge carriers are obtained. This catalytic activity is more if the roughness is structured in nano range dimension owing to the drastic enhancement in the surface energy in comparison to macro range dimension rough surface.

4. Effect of Temperature on V$_{oc}$ and I$_{sc}$

Analogous to PECs, the photo-voltaic Si solar cell performance is also influenced by the temperature and intensity. Some classic results and their analysis are discussed. Under constant light intensity, typical I-V (Current-Voltage) characteristics of Si solar cell as a function of temperature is illustrated in Figure 4.10.

Increase in temperature reduces the band gap of a solar cell, whereby effecting the solar cell output parameters. Elevated temperature increases the energy of the electrons and thus reduces the band gap. The parameter most affected by temperature is V$_{oc}$, as indicated in plot and table. The open-circuit voltage decreases

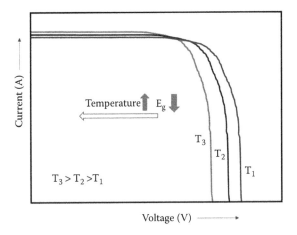

FIGURE 4.10
The characteristic I–V plot shows the effect of temperature at constant light intensity, the V_{oc}, and resultant P_{max} (maximum power) decreasing with temperature.

with temperature as the reverse saturation current, (I_o), depends on the intrinsic carrier concentration, (n_i), and it further depends on the temperature as revealed in Equations 4.9 through 4.12 [38,39]:

$$V_{oc} = \frac{nkT}{q} \ln\left(\frac{I_L}{I_o} + 1\right) \tag{4.9}$$

$$I_o = qA\frac{Dn_i^2}{LN_D} \tag{4.10}$$

$$n_i = N_s \exp\left(-\frac{E_g}{2k_BT}\right) \tag{4.11}$$

$$N_s = \left(\frac{m^*}{m}\right)^{\frac{3}{2}}\left(\frac{T}{300\ K}\right)^{\frac{3}{2}}\frac{2.5\times10^{19}}{cm^3} \tag{4.12}$$

where, V_{oc} = open circuit voltage, n = ideality factor, T= temperature, q = electronic charge, I_o = reverse saturation current (dark saturation current) , I_L = light generated current, A = area, D= diffusivity of minority carrier, L = diffusion length of carrier, N_D = doping concentration, n_i = intrinsic carrier concentration, E_g = energy gap within VB and CB, k_B = Boltzmann constant (1:381 × 10^{-23} Joules/Kelvin), N_s = number per unit volume of effectively available states, and m*/m = 0.543 (for example Si), where m* = effective mass and m = rest mass of electron. Although the resultant value depends on the material characteristics, usually it is of order 10^{19} cm^{-3} at 300K and further increase with temperature. From the earlier equations, it is found that the temperature lowers the band gap, giving a higher intrinsic carrier concentration so higher temperature results in higher n_i, and lowering the V_{oc}. A slight increment can be noticed in I_{sc}, however, both the fill factor and resultant P_{max} are decreasing with temperature in Si solar cell.

5. Effect of Light Intensity on V_{oc} and I_{sc}

The illumination intensity in the range of 160–1000 W/m² is considered to understand their effect on short circuit current (I_{sc}) and open circuit voltage (V_{oc}) of polycrystalline solar cell [40]. Any change of the irradiation causes a proportional change in the short circuit current, while open circuit voltage increases logarithmically with light intensity, as shown in Figure 4.11.

$$I_{sc} = K_E E \tag{4.13}$$

$$V_{oc} = V_{ocn} + \frac{nKT}{q} \ln\left(\frac{E}{E_n}\right) \tag{4.14}$$

where K_E is relative variation of short circuit current as a function of irradiation, here, 0.0051 (A.m²/W), V_{ocn} and E_n are the open circuit voltage and the irradiation under nominal conditions.

Indeed, the extent of variation of the current I_{sc} according to the irradiation is bounded by the values 0.8232 A for 160 W/m² irradiance and 5.1465 A for 1000 W/m². The short circuit current is practically equal to the photo-current. However, the open circuit voltage increases with increasing irradiation, but it is less sensitive to light intensity than the short circuit current. The variation of the open circuit voltage V_{oc} is from 0.565 V for 160 Wm² irradiations to 0.616 V for an irradiation of 1000 W/m². From the previous equation and data analysis, one can observe a doubling light intensity, ($E/E_n = 2$), causes to increment of 18 mV in V_{oc}, but 2 A in I_{sc}. Furthermore, extensive open literature and research data on the influence of temperature and intensity on I-V characteristics of different classes of PEC made of ceramic semiconductor can boost up their utility in harsh environment.

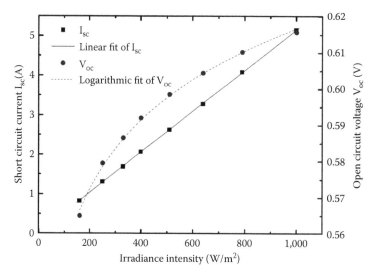

FIGURE 4.11
Open-circuit voltage and short-circuit current as function of light intensity for a polycrystalline silicon solar cell. (From Chegaar, M. et al., *Energy Procedia*, 36, 722–729, 2013 [40].)

4.4 Concluding Remarks

Ceramic semiconductors research is exponentially growing to develop more effective photocatlayst that enable them to participate in solar energy assisted dissociation of pollutant and generation of renewable energy or synthesis of alternate fuel like H_2. Choice of band gap and their position implies the electron transfer process efficacy in visible and UV light, but quantum confinement progressively change the valence band and conduction band edge and increase band gap below critical particle size. Generation of electron-hole and their participation in reaction promote the photocatalytic reaction, and thus several strategies including surface trapping by scavenger, introducing porous structure, composition modification by metallic and nonmetallic dopant, band gap engineering, and coupling with dual semiconductors are common incidence to avoid their (e^-/h^+) own recombination reaction. The recombination happens in picoseconds and this exothermic process increases the localized heat and changes the rate of reaction. In fact, the photocatalytic reaction follows several steps and thus resultant solar energy conversion efficiency depends on photon absorption, charge separation, migration, and usage capability. Dye degradation rate constant increases with decreasing particle size, however, very small particles may promote recombination on surface charges and further decrease the effective photocatalytic conversion. While discussing the importance of photocatalyst in photo-electrochemical cells, the narrow band gap semiconductor and a redox couple converting the optical energy to electrical energy without free energy change, ($\Delta G = 0$), in the cumulative chemical reaction. However, photo-energy is stored as chemical energy in endergonic reactions and resultant $\Delta G > 0$ during synthesis of H_2 from water in photo-electrolytic cells. The band bending and net charge transfer from semiconductor bulk to the surface or electrolyte solution phenomena are discussed as flat band does not respond to charge transfer in initial condition. A successful nanostructured ceramic semiconductor comprehends high optical absorption, easy adjustability of band gap, charge transfer, and synthesis protocol that enable to adsorb wide range of organic molecules. Although high specific surfaces provide high efficiency, yet particle size and shape optimization is required since very smaller size facilitate recombination and reduce the conversion efficiency as like photo-degradation reaction. Effective coating thickness and less random molecular diffusion by confined growth like 1D may be the promising choice for new generation PEC.

References

1. M. A. Butler and D. S. Ginley, Prediction of flatband potentials at semiconductor-electrolyte interfaces from atomic electronegativities, *J Electrochem Soc* 125, 228–232, 1978.
2. R. M. Noyes, Thermodynamics of ion hydration as a measure of effective dielectric properties of water, *J Am Chem Soc* 84, 513–522, 1962.
3. L. Li, P. A. Salvador and G. S. Rohrer, Photocatalysts with internal electric fields, *Nanoscale* 6, 24–42, 2014.
4. S. R. Morrison, *Electrochemistry at Semiconductor and Oxidized Metal Electrodes*, Plenum Press, New York, 1980.
5. J. Wen, X. Li, W. Liu, Y. Fang, J. Xie, Y. Xu, Photocatalysis fundamentals and surface modification of TiO_2 nanomaterials, *Chin J Catal* 36, 2049–2070, 2015.

6. M. R. Hoffmann, S. T. Martin, W. Choi, D. W. Bahnemann, Environmental applications of semiconductor photocatalysis, *Chem Rev* 1995, 95, 69–96.

7. T. Yoshihara, R. Katoh, A. Furube, Y. Tamaki, M. Murai, K. Hara, S. Murata, H. Arakawa, M. Tachiya, Identification of reactive species in photoexcited nanocrystalline TiO_2 films by wide-wavelength-range (400–2500 nm) transient absorption spectroscopy, *J Phys Chem B* 108, 3817, 2004.

8. J. Nešić, D. D. Manojlović, I. Anđelković, B. P. Dojčinović, P. J. Vulić, J. Krstić, and G. M. Roglić, Preparation, characterization and photocatalytic activity of lanthanum and vanadium co-doped mesoporous TiO_2 for azodye degradation, *J Mol Catal A Chem* 378, 67–75, 2013.

9. E. Kowalska, R. Abe, and B. Ohtani, Visible light-induced photocatalytic reaction of gold-modified titanium(IV) oxide particles: Action spectrum analysis, *Chem Commun* 241–243, 2009.

10. W. Q. Fang, X. Q. Gong, H. G. Yang, On the unusual properties of anatase TiO_2 exposed by highly reactive facets, *J Phys Chem Lett* 2(7), 725–734, 2011.

11. A. Credi, *Photoactive Semiconductor Nanocrystal Quantum Dots: Fundamentals and Applications*, Springer, Cham, Switzerland, 2016.

12. W. Choi, A. Termin, M. R. Hoffmann, The role of metal ion dopants in quantum-sized TiO_2: Correlation between photoreactivity and charge carrier recombination dynamics, *J Phys Chem* 98(51), 13669–13679, 1994.

13. Q. Xiang, J. Yu, W. Wang, M. Jaroniec, Nitrogen self-doped nanosized TiO_2 sheets with exposed {001} facets for enhanced visible-light photocatalytic activity, *Chem Commun* 47, 6906–6908, 2011.

14. H. Yamashita, H. Li, *Nanostructured Photocatalysts: Advanced Functional Materials*, Springer, New York, 2016.

15. W. Ho, J. C. Yu, J. Lin, J. Yu, P. Li, Preparation and photocatalytic behavior of MoS_2 and WS_2 nanocluster sensitized TiO_2, *Langmuir* 20(14), 5865–5869, 2004.

16. E. Guo and L. Yin, Tailored $SrTiO_3/TiO_2$ heterostructures for dye-sensitized solar cells with enhanced photoelectric conversion performance, *J Mater Chem A* 3, 13390–13401, 2015.

17. J. Zhang, Q. Xu, Z. Feng, M. Li, C. Li, Importance of the relationship between surface phases and photocatalytic activity of TiO_2, *Angew Chem Int Ed Engl* 47(9), 1766–1769, 2008.

18. Q. Xiang, J. Yu, M. Jaroniec, Enhanced photocatalytic H_2-production activity of graphene-modified titania, nanosheets, *Nanoscale* 3, 3670, 2011.

19. H. G. Yang, C. H. Sun, S. Z. Qiao, J. Zou, G. Liu, S. C. Smith, H. M. Cheng, G. Q. Lu, Anatase TiO_2 single crystals with a large percentage of reactive facets, *Nature* 453(7195), 638–641, 2008.

20. J. C. Yu, J. G. Yu, W. K. Ho, Z. T. Jiang, L. Z. Zhang, Effects of F^- doping on the photocatalytic activity and microstructures of nanocrystalline TiO_2 powders, *Chem Mater* 14, 3808, 2002.

21. X. D. Li, G. Q. Sun, Y. C. Li, J. C. Yu, J. Wu, G. H. Ma, T. Ngai, Porous TiO_2 materials through Pickering high-internal phase emulsion templating, *Langmuir* 30, 2676, 2014.

22. N. Xu, Z. Shi, Y. Fan, J. Dong, J. Shi, M. Z.-C. Hu. Effects of particle size of TiO_2 on photocatalytic degradation of methylene blue in aqueous suspensions, *Ind Eng Chem Res* 38(2), 373–379, 1999.

23. J. Van de Lagemaat, N. G. Park, A. J. Frank, Influence of electrical potential distribution, charge transport, and recombination on the photopotential and photocurrent conversion efficiency of dyesensitized nanocrystalline TiO_2 solar cells: A study by electrical impedance and optical modulation techniques. *J Phys Chem B* 104, 2044–2052, 2000.

24. O. Khaselev and J. A. Turner, A monolithic photovoltaic-photoelectrochemical device for hydrogen production via water splitting. *Science* 280, 425–427, 1998.

25. M. K. Brennaman, R. J. Dillon, L. Alibabaei, M. K. Gish, C. J. Dares, D.L. Ashford, R. L. House, G. J. Meyer, J. M. Papanikolas, T. J. Meyer, Finding the way to solar fuels with dye-sensitized photoelectrosynthesis cells, *J Am Chem Soc* 138(40), 13085–13102, 2016.

26. D. E. Yates, S. Levine, T. W. Healy, Site-binding model of the electrical double layer at the oxide/water interface, *J Chem Soc Faraday Trans 1*, 70, 1807–1818, 1974.

27. T. Bak, J. Nowotny, M. Rekas, C. C. Sorrell, Defect chemistry and semiconducting properties of titanium dioxide: I. Intrinsic electronic equilibrium, *J Phys Chem Solids* 64, 1043–1056, 2003.

28. L. Lu, M. A. Kelly, W. You, L. Yu, Status and prospects for ternary organic photovoltaics, *Nat Photonics*, 9, 2015.

29. S. Giménez and J. Bisquert, *Photoelectrochemical Solar Fuel Production: From Basic Principles to Advanced Devices*, Springer, Cham, Switzerland, 2016.

30. Z. Chen, D. Cummins, B. N. Reinecke, E. Clark, M. K. Sunkara, T. F. Jaramillo, Core shell MoO_3-MoS_2 nanowires for hydrogen evolution: A functional design for electrocatalytic materials, *Nano Lett* 11, 4168–4175, 2011.

31. M. Konstantakou and T. Stergiopoulos, A critical review on tin halide perovskite solar cells, *J Mater Chem A* 5, 11518–11549, 2017.

32. R. Jose, V. Thavasi, S. Ramakrishna, Metal oxides for dye-sensitized solar cells, *J Am Ceram Soc* 92(2), 289–301, 2009.

33. K. Zhu, T. B. Vinzant, N. R. Neale, A. J. Frank, Removing structural disorder from oriented TiO_2 nanotube arrays: Reducing the dimensionality of transport and recombination in dye-sensitized solar cells, *Nano Lett* 7(12), 3739–3746, 2007.

34. K. Zhu, N. R. Neale, A. Miedaner, A. J. Frank, Enhanced charge-collection efficiencies and light scattering in dye-sensitized solar cells using oriented TiO_2 nanotubes arrays, *Nano Lett* 7(1), 69–74, 2007.

35. M. Grätzel, Photoelectrochemical cells, *Nature* 414, 338–344, 2001.

36. S. Zhang and N. Ali, *Nanocomposite Thin Films and Coatings: Processing, Properties, and Performance*, Imperial College Press, London, UK, 2007.

37. Y.-H. Kim, I.-K. Lee, Y.-S. Song, M.-H. Lee, B.-Y. Kim, N.-I. Cho, D. Y. Lee, Influence of TiO_2 coating thickness on energy conversion efficiency of dye-sensitized solar cells, *Electron Mater Lett* 10(2), 445–449, 2014.

38. J. G. Webster, Photoelectrochemical cells. In: *Wiley Encyclopedia of Electrical and Electronics Engineering*, John Wiley & Sons, Hoboken, NJ, 1999.

39. PV Education, Effect of temperature, http://www.pveducation.org/pvcdrom/effect-of-temperature.

40. M. Chegaar, A. Hamzaoui, A. Namoda, P. Petit, M. Aillerie, A. Herguth, Effect of illumination intensity on solar cells parameters, *Energy Procedia* 36, 722–729, 2013.

5

Magnetics and Piezoelectrics

5.1 Introduction

Magnetic and piezoelectric ceramics have been common research interests for the few last decades because of a wide range of material acceptability and their applications from electronic instrumentation, automobiles, energy, and environment to health services. Prior to the focus on the topics, a brief illustration on classification, origin, some important quantities, and basic definitions of magnetic and dielectric phenomena are highlighted to provide a better insight of the subject. Specific properties and influence of particle size and/or grain size are also discussed in relevance. Figure 5.1 describes the brief classification of magnetic materials that depends on permanent dipoles, dipole alignment, dipole direction, and dipole magnitude. This schematic representation enlightens the technical meaning of individual magnetic materials including dia, para, ferro, antiferro, and ferrite including their examples.

Among various classes of magnetic materials, however, ferrites and superparamagnetic have attracted much attention for energy, environment, and health applications [1–3]. Apart from several magnetic behaviors, superparamagnetic is a property occurring principally in *nanoscale* through change in magnetic moment direction and behave like a paramagnet below Curie temperature in absence of magnetic field, although exhibits appreciable susceptibility like ferro or ferrimagnets. Interestingly, this class of magnetic particle develops a strong internal magnetization from exchange coupling of electrons within the domain under external magnetic fields and behaves as superparamagnetic. However, it does not have magnetic memory, relatively less degree of induced magnetization, and single domain activity compared to ferromagnetic or ferrites. For example, Fe_3O_4 nanoparticles, (10 nm), exhibit magnetic saturation (M_s) in the range of 30–50 emu/g, where their bulk (soft ferrite) form experiences near to 90 emu/g. Soft ferrite Fe_3O_4 has high susceptibility compared to nanoscale superparamagnetic Fe_3O_4. Furthermore, the Curie temperature is dramatically reduced from bulk 585°C to 440°C for nanoscale Fe_3O_4. The various hysteresis parameters are not solely intrinsic properties, rather depend on grain size, domain state, stresses, and temperature. Hence, nanoscale ceramics in preference to superparamagnetic and ferrite particles have different features from bulk and can be approached for functional applications. Furthermore, the nanoparticles (NPs) have intense interest to fulfill the demand of miniaturization of devices because their physical properties vary dramatically from their bulk counterparts. In convention, the grain size maintained is near to one tenth of layer thickness during magnetic coating, as the layer thickness preferentially approaches to micrometer and submicrometer thickness, the nanostructured ceramics demand becomes more. Thin magnetic coating and bulk magnets are well known for extensive energy savings and recording devices, however, the utility of magnetic materials and their research

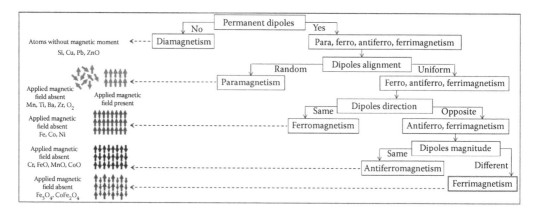

FIGURE 5.1
A brief classification of magnetic material in the perspective of permanent dipoles, dipoles alignment, direction, and magnitude.

in the environment sector is limited apart from, to some extent, cleaning up the oil spilling and mineralogy enrichment.

Under electric field, the ferroelectric has an analogous switchable electric polarization similar to magnetic polarization of ferro or ferrimagnet. Interestingly, it also exhibits a spontaneous electric polarization below the Curie temperature, hysteresis loop, and an additional mechanical strain-electric field butterfly loop. However, a fundamental difference in polarization mechanism eventually promotes different applications. Figure 5.2 demonstrates the basic understanding and classification of dielectric materials, and one can differentiate these among each other and conclude the definition of each group of dielectrics. For an example, the piezoelectrics are noncentro symmetric in which electricity is generated by an applied mechanical stress and vice versa, and, thus, this class of material has extensive academic and technological importance. Among the piezoelectric materials, a subclass of material is referred to as convertible heat-generated electricity resulting from the change of the electric polarization during electrothermal

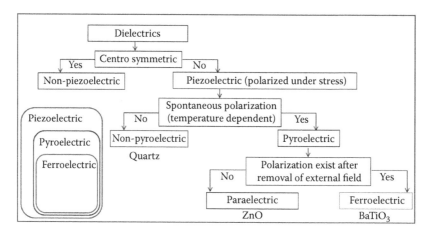

FIGURE 5.2
A brief classification of dielectric materials in the perspective of centro-symmetry, polarization under stress, temperature-dependent polarization, and influence of external field.

and electrocaloric effects. Usually, a small change in dimension is a common feature when a material is subjected to experience external force, such as electric field, mechanical stress, or change in temperature. Depending on the crystal structure, the degree of change in dimension may result in a change in electric polarization and hence result is the incidence of ferroelectric, piezoelectric, and pyroelectric effects. Apart from the definite crystal structure, the particle shape and dimension have influence in order to miniaturize the device.

Interesting to mention that the spherical and tetragonal ferroelectric $BaTiO_3$ promotes better packing efficiency and effective dipole interaction facilitates high dielectric constant after consolidation. In the contrary, 1D $BaTiO_3$ nanowhiskers or nanofibers experience more strain energy under frequency compared to 3D spherical particles and thus high piezoelectric properties. However, it is to be noted that all ferroelectric material is piezoelectric, but vice versa is not true (Figure 5.2). While dielectric materials are classified into several groups, one can encounter the environmentally friendly piezoelectric ceramics is essential to develop for renewable energy, environmental, and health applications.

5.2 Magnetic Phenomena

Nanoscale magnetic materials (metallic or ceramic) in specific magnetic nanoparticles (MNPs) have been the focus as an interdisciplinary topic of interest for several communities, including physics, materials science, chemistry, biology, and working on several research horizons. These MNPs are sought by various synonyms, such as magnetic beads, as solid or ferrofluids, or magnetic fluids consisting of colloidal suspension of magnetic nanoparticles. This is a part of nanotechnology and has potential to develop new prospects of materials and devices for a wide range of applications. A brief of selective and essential magnetic properties is discussed to provide an insight and analyze the MNPs.

5.2.1 Magnetic Moment

This property has grave importance in physical world of science. For an example, a strong magnet consisting of 3.0 T in a magnetic resonance (MR) scanner is an excellent tool to obtain magnetic resonance imaging (MRI) in the hospital. The magnetization of any matter is explained in the terms of dipole moment per unit volume. Some fundamental particles like electrons and protons have also some intrinsic magnetic moment in the form of spin. Thus, the tendency of an object to align with external magnetic field can be defined as magnetic moment. It is a vector quantity and originates through motion of electric charge and spin angular momentum. Quantitatively, the magnetic moment, (m), determines when dipole experience torque, (τ), in presence of magnetic flux density B (Teslas or Newtons per ampere per meter). However, the magnetic field H (ampere per meter) should not be confused with B, and these follow by a simple relation with permeability, (μ). In vacuum, the magnitude of B and H is identical.

$$\tau = m\,\vec{B} \tag{5.1}$$

$$\vec{B} = \mu\,H \tag{5.2}$$

The magnitude of the magnetic moment "m" of a particle is proportional to its volume. However, the magnetization uniformity direction can only be attained either by applying a large enough field or by reducing the particle dimension to prevent domain formation, known as, single domain (SD). When particle is enough small below critical size (r_c), the spherical particle exhibits monodomain and results in high magnetic moment compared to multidomain particles. This phenomenon encompasses superparamagnetism of nanoparticles (Section 5.2.3). Although the critical SD particle size depends on material properties, particularly different anisotropy energy terms [4].

5.2.2 Ferrimagnetic Behavior of Soft and Hard Ferrites

Ferrimagnetic ceramics are like ferromagnets in that they experience a spontaneous magnetization below the Curie temperature, (T_c), and, indeed, no magnetic alignment above this temperature that implies paramagnetism. However, a net zero magnetic moment below Curie temperature, refers as magnetic compensation point for garnets and rare earth-transition metal alloys. Ferrites are preferentially nonconductive ferrimagnetic compounds derived from either iron oxides or a combination of other metal oxides. In the perspective of magnetic properties, it is classified as "soft" and "hard," with respect to low and high coercivity, respectively. However, the ferrites can be further classified based on their crystal structure; spinel [MFe_2O_4, (M = Fe^{2+}, Co^{2+}, Mn^{2+}, Ni^{2+}, Zn^{2+})], hexagonal [$MFe_{12}O_{19}$, (M = Ba^{2+}, Sr^{2+}, Pb^{2+})], and garnet [$M_3Fe_5O_{12}$, (M = Gd^{3+}, Dy^{3+}, Er^{3+}, Y^{3+}, Yb^{3+})]. Low coercivity of soft ferrite facilitate easy reverse magnetization direction without dissipating much energy (hysteresis loss), whereas permanent hard ferrite experience high magnetic flux and store magnetic field that is used for household products, such as refrigerator magnets. Ferromagnetic material has only one type of lattice site, and thus electron spins in one direction within a particular domain, but multilattice site in ferrimagnet prefers to accommodate the combination of spin-up and spin-down with a given domain. This incomplete cancellation leads a net polarization, conventionally weaker than ferromagnetic, but it may be quite strong as well. Herein, intense emphasis has been focused on spinel ferrite compared to other classes of ferrites, as it behaves as both soft and hard ferrites, and superparamagnetic as well, whereas hexagonal and garnet ferrite have hard magnetic properties only. In this aspect, a brief crystallographic concept implies how the opposite direction of electron spin can accommodate under magnetic field. Spinel, (AB_2O_4), lattice has cubic structure without any preferred magnetization direction. It is two types, normal and inverse types. In normal spinel, the divalent A^{2+} ions occupy the tetrahedral voids, whereas the trivalent B^{3+} ions occupy the octahedral voids in the close packed arrangement of oxide ions. However, inverse spinel has extensive research interest and technological importance because of ferrimagnetic properties. Inverse spinel has a [$B(AB)O_4$] arrangement, which implies that the A^{2+} ions occupy the octahedral voids, whereas half of B^{3+} ions occupy the tetrahedral voids, this lattice arrangement and d-orbital electron spin interaction within A and B site encompass the exchange coupling and magnetic properties compared to normal spinel. Most of the common soft and hard ferrites are inverse spinel. A glimpse of the basic magnetic properties of soft and hard ferrite is given in Table 5.1.

TABLE 5.1

Magnetic Properties Different in Soft and Hard Ferrites

Properties	Soft Ferrite	Hard Ferrite
Hysteresis loop	Smaller area	Large area
Remnant magnetization	Low	High
Initial permeability	High	Low
Hysteresis loss	Less	High
Coercivity	Less	High

5.2.3 Superparamagnetic Behavior

In principle, the magnetization is a nanoscale phenomenon, and thus nanocrystalline material carries technological significance in enhancing the performance of existing bulk materials. Around the globe, researchers are exploring size minimization toward nanoscale which consists of different geometries and crystallinity for high performance devices. Magnetism of ferro and ferrimagnetic material depends on the volume of the particles, and thus the bulk magnet always experiences multiple magnetic domain structures. However, below a certain critical radius, nanocrystal behaves as a single-domain state and hence the domain wall resistance can avoid and effectively work at high frequency. A brief illustration on the transition of multidomain to single-domain (SD) with respect to the relation between coercivity and particle size is represented in Figure 5.3. Above Curie temperature, (T_c), the magnetic spins are randomly distributed that lead to zero magnetization, where same magnetic material develops domain walls below T_c.

The strategy toward particle size reduction helps to minimize the formation of domain walls and a sample consists of a single uniformly magnetized domain. The particle size

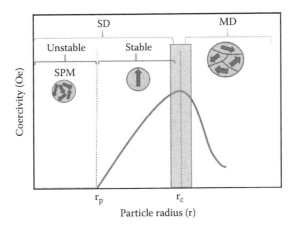

FIGURE 5.3
Alteration of coercivity with particle diameter. Critical size r_c differentiate within single domain and multidomain magnetic particles, but r_p differentiate the stability region of single domain nanoparticles.

below critical radius, (r_c), behaves as a single domain when the energy for creating a domain wall becomes larger than the external magnetostatic energy, and it can be calculated by Equation 5.3:

$$r_c = \frac{9(AK_u)^{1/2}}{\mu_0 M_s} \tag{5.3}$$

where A is the exchange stiffness (constant characteristic of the material related to the critical temperature for magnetic ordering), K_u is the uniaxial anisotropy constant, μ_0 is the vacuum permeability, and M_s is the saturation magnetization. Under another critical dimension, r_p, the amount of coercivity becomes zero due to the predominant thermal effects over the self-magnetization process. Interestingly, the calculated critical radius for ferromagnet is relatively less compared to ferrites, for example, FCC-Co (4 nm), hcp-Co (8 nm), Fe (8 nm), and Fe_2O_3 (45 nm), $CoFe_2O_4$ (50 nm), and Fe_3O_4 (64 nm), respectively [5]. However, below the critical radius, the single domain particle has dual characteristics, stable single domain and unstable single domain, where the unstable region corresponds to the superparamagnetic particles (SPM), as illustrated in Figure 5.3. With a starting particle radius, r_0 (i.e., $<r_p$) experiences high thermal energy and overcomes the anisotropy energy to sustain the magnetic moments along a certain direction. This results in the magnetization array flip easily under external field and does not retain magnetization and coercivity after removal of magnetic field, similar to a conventional magnet. Such a system of particles is referred to as SPM with no hysteresis behavior (both M_r and H_c are zero), and the major difference from the paramagnet is that the magnetic susceptibility, (M/H), is much larger. A typical hysteresis loop of a SPM material is shown in Figure 5.4. This class of superparamagnetic is extensively using for hard disk, MRI imaging, hyperthermia, and other health applications. Fortunately, the superparamagnetism can prevent nanoparticle aggregation or cluster formation in contrast to ferromagnetic particles because the spin relaxes quickly and eases demagnetization at room temperature [6].

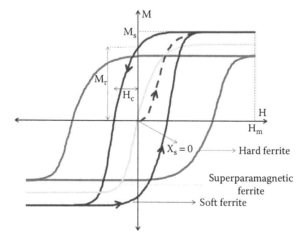

FIGURE 5.4
A typical representation of magnetization and magnetic field for soft ferrite, hard ferrite, and superparamagnetic. The axis and hysteresis loop area are not in scale.

5.2.4 Magnetization and Hysteresis

The ferrimagnetic (soft and hard) and superparamagnetic properties of ceramic magnetic materials can be classified on the basis of saturation magnetization (M_s), remnant magnetization (M_r), coercivity (H_c), and susceptibility (χ), and their typical M-H profile is illustrated in Figure 5.4. Magnetic theory suggests that a magnetization, (M), is the result of more or less alignment of elementary magnets known as magnetic dipoles (electron spins) present or developed under magnetic field (H). Let us consider the blue colored M-H profile for soft ferrite, in which material initially (dotted line) follows a nonlinear magnetization under increment field of H, and in a certain position no further increase, indicates, magnetization saturation, (M_s), with magnetic field is achieved. However, the magnetization drops with considerable degree of magnetization compared to highest magnetization value, when H field drops to zero, this is known as remanence magnetization, (M_r), and this is useful to make a magnetic memory device. In order to get back to zero magnetization, the required magnetic field is known as coercive field, (H_c).

This magnetization characteristic in presence of magnetic field provides a "signature of magnetization history" of a particular material and can be represented as hysteresis loop. In fact, characteristics of a hysteresis loop and detailed analysis of these basic properties provide insight of ceramic magnets. In reality, the magnetization is proportional to applied field ($M = \chi H$) and the constant is defined as susceptibility (χ) per cm^3. The sign and magnitude of "χ" determine the classification of magnetic behavior of materials. For example, diamagnetic material possesses anti-parallel magnetic moment to H, resulting in very small and negative susceptibilities (-10^{-6} to -10^{-3}) and do not retain magnetic properties when the external field is removed. Materials with preferential parallel alignment of magnetic moment to H and susceptibilities in the order of (10^{-6} to 10^{-1}) without persistence of any magnetic memory are called paramagnetic, however, this value dramatically enhances in the order of 10^3 for superparamagnetic (i.e., "super"). While considering the ferro or ferrimagnetic material, the parallel aligned magnetic domains imply large spontaneous magnetization to H which consists of magnetic memory, and their susceptibilities may enhance in the order of 10^5 [7]. Nevertheless, the susceptibilities of these materials depend on their atomic structure, temperature, size of the particles, and the external field H. While discussing the effect of temperature, it is worthy to introduce the concept of 'Weiss domain' that refers ferromagnetic experience quantum mechanical exchange interaction in the crystal lattice, which eventually allows large magnetization or moment per unit volume. The domain of spontaneous vibration disappears in the presence of external temperature and behaves as paramagnetic. This temperature is known as Curie temperature, (T_c).

Ferrimagnetic Curie point encompasses similar characteristics as ferromagnetic Curie point and converts to paramagnetic above T_c. Different range of Curie temperature estimated for ferro and ferrimagnetic materials, for example, ferrimagnetic exhibits 300°C for $MnFe_2O_4$, 520°C for $CoFe_2O_4$, and 440°C for $MgFe_2O_4$, whereas the ferromagnetic has 770°C for Fe, 1115°C for Co, and 358°C for Ni. The variation of susceptibility as a function of temperature is shown in Figure 5.5. Above Neel temperature, (T_N), the antiferromagnetic behaves as paramagnetic. Apart from these common features, additional properties including magnetocrystalline anisotropic constant (K), specific loss powers (SLP), magnetoresistance change (MR%), and exchange spring magnet are discussed in mind of MNPs development for functional applications. The basic features of individual properties and data analysis with respect to nanostructured particles have been emphasized, however, several classic books can refer for the data acquisition techniques and their working principles [8,9].

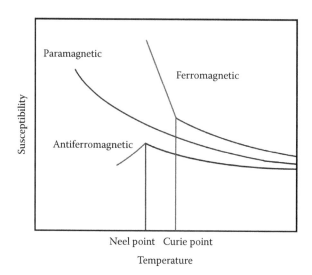

FIGURE 5.5
Schematic representation of susceptibility variation of different class of magnetic materials as function of temperature.

5.2.5 Magnetocrystalline Anisotropic Constant (K)

The dependence of magnetic properties on a preferred direction is called magnetic anisotropy. A magnetically isotropic material has no preferential direction for its magnetic moment, whereas a magnetically anisotropic material prefers to align one of the easiest axes. An easy axis is an energetically favorable direction of spontaneous magnetization that is determined by the crystal structure, grain shape, and stress. This phenomenon strongly affects the hysteresis loops and controls the coercivity and remanence. The material behaves as superparamagnetic without presence of anisotropy. This magnetic anisotropic behavior is represented by a constant, known as magnetocrystalline anisotropic constant, (K). Materials having high K value are difficult to demagnetize, results in permanent magnets. On the other hand, materials with low value of this property are called soft materials, which are used in transformers and inductors. However, K is highly temperature dependent and decreases rapidly as the temperature approaches to the Curie temperature. Let's consider cubic magnetite (Fe_3O_4) in order to discuss the effect of crystal anisotropy, in which <111> is the easiest direction of magnetization, <110> is the intermediate direction of magnetization, and <100> is the hard direction of magnetization, as illustrated in Figure 5.6. The anisotropy encompasses the energy difference within easy to hard direction that arises from the interaction of the spin magnetic moment with the crystal lattice (spin-orbit coupling).

In this contrary, the basal plane of hexagonal hematite (Fe_2O_3) is an easy plane of magnetization, whereas the c-axis is the hard direction. This makes a distinct difference to flip the magnetization out of the basal plane into the direction of c-axis, and, thus, it is difficult to attain saturation under magnetic field 1–2 T, as well [10]. The crystallographic plane directs the resultant magnetic moment and degree of saturation assist to distinguish within crystal structure, like hexagonal hematite and cubic magnetite. Thus, synthesis of nanoparticle having preferred plane is an important research aspect.

Apart from the magnetocrystalline anisotropy, magnetostriction arises due to strain mismatch of the crystal during magnetization, demagnetization process. As an inverse effect,

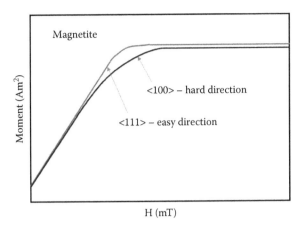

FIGURE 5.6
Magnetic moment for anisotropic material consist of hard and easy direction.

the magnetization changes results under stress. A uniaxial stress can produce a unique easy axis of magnetization if the stress is sufficient to overcome all other anisotropies. The magnitude of stress anisotropy is referred to as magnetostriction constants, (λ). Thus, level of stress is another concerning parameter to describe the change in magnetization behavior of materials. In addition, particle shape has also influence to originate anisotropy. Magnetization develops magnetic charges or poles on the surface and acts as another source of a magnetic field, known as demagnetizing field [11]. This field is less along the long axis compared to short axis in needle shaped particles, however, spherical particle has not shape anisotropy. The magnetic anisotropy behavior facilitates to heat generation in combination of soft and hard ferrites. Precious controlling of d-metal cation ratio eventually implies the interplay between hard and soft phases and maintains behavior of both. This characteristic feature can be achieved either by structural modification or fabrication of core-shell structures.

5.2.6 Specific Loss Powers

Magnetic nanoparticles suspension in a liquid medium, for example, ferrofluids or magnetic nanofluids based cancer treatment is known as magnetic fluid hyperthermia in which selective tumor cells are locally heated and destroyed after injection of biocompatible magnetic nanoparticle suspensions followed by placing the patient under external alternating magnetic field (AMF). This leads to effective heat generation through magnetic losses and kills the tumor cells with minimal damage of the surrounding normal living tissues. The most commonly used physical quantity concerning the calorific power of MNPs under AC magnetic field (i.e., magnetic fluid hyperthermia) is the specific absorption rate, which is also referred as specific loss power (SLP) [12]. The SLP is defined as electromagnetic power lost per magnetic material mass unit and is expressed in watts per gram, (w/g). The estimation of SLP is important for evaluating the heating efficiency of MNPs, for optimizing the parameters of AMF, and for the optimal design of MNPs in an attempt to establish the effectiveness of magnetic hyperthermia. This heat generation mechanism depends on several parameters including external magnetic field strength and frequency, particle size distribution, magnetic anisotropy constants distribution, particle concentration, colloid

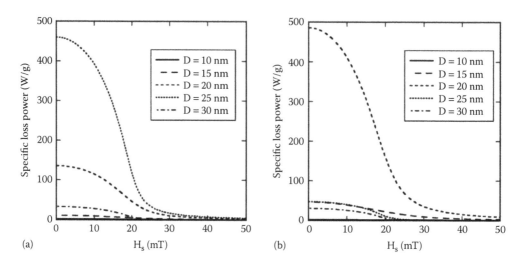

FIGURE 5.7
The calculated SLP with respect to strength of external magnetic field at 20 mT and 300 kHz for (a) maghemite and (b) magnetite. The SLP is too low for 10 nm particles. (From Murase, K., *J. Appl. Sci.*, 6, 839–851, 2016 [13].)

properties, etc. In this backdrop, the SLP can be related with volumetric power dissipation, (P), and density of the suspending fluid, (ρ), as SLP = P/ρ, and P can be defined as:

$$P = \mu_0 \pi f H^2 \chi \qquad (5.4)$$

where μ_0, f, H, and χ are permeability of free space, frequency of magnetic field, magnetic field strength and susceptibility, respectively.

In account of the magnetic nanoparticle size selection for hyperthermia is a critical issue, as an optimum size window provides highest SLP for hyperthermia treatment. In a recent study, the calculated SLP for different nanoparticles of maghemite (γ-Fe$_2$O$_3$) and magnetite (Fe$_3$O$_4$) is studied and illustrated in Figure 5.7 [13]. This result implies a definite particle size 25 nm γ-Fe$_2$O$_3$ and 20 nm Fe$_3$O$_4$ develops the highest SLP in the range of 450–500 w/g under external field of 20 mT and 300 kHz, respectively. However, the SLP was too low for 10 nm particles for both the materials. Very small superparamagnetic nanoparticles experience susceptibility loss, whereas optimum particle size with low anisotropy provide maximum magnetic loss in presence of definite amplitude of the applied field. Thus, particle size selection in account of hyperthermia treatment is a challenging task and depends on several factors as stated earlier. Eventually, a low concentration and nonagglomerated MNPs maximize the heat dissipation under AMF, as narrow particle size distribution (high shape parameter α) is desirable to avoid the heating rate impairing because of polydispersity, and thus, an intelligent selection of biocompatible particles can be considered for this incidence.

5.2.7 Magnetoresistance Change

Magnetoresistance change, (MR%), refers to the magnitude of electrical resistance change in presence of magnetic field. Usually, the MR effect depends on both the strength and direction of magnetic field with respect to current. In the perspective of efficiency level, four distinct magnetoresistances can be classified as ordinary magnetoresistance, anisotropic magnetoresistance (AMR), giant magnetoresistance (GMR), and colossal magnetoresistance

(CMR) [14]. As an example, ordinary magnetoresistance behavior is found in Sn, Cu, Ag, Au, Mg, Cd, Ga, Pt, Pb, etc., AMR in ferro (Fe, NiFe, etc.), and ferrimagnetic (cobalt ferrite, etc.); GMR in zinc ferrite, Fe/Cr multilayer etc., CMR in $La_{1-x}M_xMnO_{3+\delta}$ (M = Ca, Sr; x = 0.33) perovskite structures. In fact, their efficiency level varies with a wide range 2% for AMR, 50% for GMR, and more than 99% for CMR. In fact, these specific features have different modes of utilities, including recording heads, hyperthermia treatment, biosensors, MEMs, magneto transport systems, etc. This can be quantified as:

$$MR\,(\%) = \frac{(R_H - R_O)}{R_H} \times 100 \qquad (5.5)$$

where R_H is the resistance in the presence of external magnetic field and R_O is the resistance in the absence of external magnetic field. Electron spin orbit coupling is the prime resource of resistance behavior. Transverse orientation of field and magnetization with respect to current facilitate the existence of electronic orbits in the same current plane, and thus scattering is minimum within confined cross section, results in low resistance. Conversely, the perpendicular orientation of electronic orbits to the current opens up more cross section and scattering, provides high resistance state. Apart from the electron coupling phenomena, the resistance also depends on the particle aspect ratio and particle size. A high degree of magnetoresistance is observed in high aspect ratio and low particle size, as surface effect plays a leading role to enhancement it. However, other parameters are also needed to encounter during development of magnetic particles for a particular application.

5.2.8 Exchange-Spring Magnet

In presence of external alternating current magnetic field, a reversible magnetization process occurs, and thus thermal energy is the result during magnetic relaxation. The heating effect occurs due to a high specific loss power (SLP) that depends on the magnetic behavior of nanoscale particles and extensively used for medical diagnoses. Recently, the combination of soft (high M_s, low H_c) and hard (low M_s, high H_c) ferrite either in proportionate composite or core-shell results in fairly high M_s and high H_c together, and their characteristic feature is coined as "exchange-spring" magnets. Soft magnet can be readily saturated (high M_s) in a weak field, leading to a very small coercive field (H_c; <1× 10^2 Am^{-1}) to return back to zero magnetization, while hard magnet experiences an obvious hysteresis with a higher coercive field (H_c, >1 × 10^4 Am^{-1}) than the soft magnet (Figure 5.4). However, the magnitude of these typical parameters may differ because of crystal structure, particle size, structural composition, shape, and surface anisotropy. For example, α-Fe_2O_3 nanocrystal possesses saturation magnetization (M_s) less than 10 emu.g^{-1} compared to γ-Fe_2O_3 (>90 emu.g^{-1}) at 300 K. The exchange coupling effect between soft and hard ferrite explores the possibility to construct artificially magnetic structures that can attain high energy product with high magnetization from the soft phase and high magnetic anisotropy from the hard phase. In this aspect, several simulations and finite element methods proposed the characteristic features of soft and hard phases are beneficial to achieve effective exchange coupling within soft-hard interface [15], these are:

1. Enough thin layer of soft phase so that the magnetization direction in both phases could rotate rationally. In preference, this layer thickness is roughly twice than the domain wall width of the hard phase.

2. Soft phase with relatively high anisotropy allows to larger dimension of soft phase and thus construction of composite magnet.

3. Graded intermix soft-hard interface creates intermediate anisotropy, which has more resistance to reversal magnetization than the absolutely soft and sharp interface.

Theoretical studies established a relation to determine the H_c, as illustrated in Equation 5.6:

$$H_c = 2\frac{H_h V_h + H_s V_s}{M_h V_h + M_s V_s} \tag{5.6}$$

where H is the coercivity, V is the volume fraction, and the subscripts "h" and "s" denote the hard and soft phases, respectively. In consideration of equivalent M_s for both phases and high H_c in hard phase, a simplified form of the earlier equation represents the resultant coercivity is propositional to the volume fraction of soft phase:

$$H_c = \frac{2H_h}{M_h}\left(1 - V_s\right) \tag{5.7}$$

In a recent study, Choi et al. developed monodisperesed 15 nm core-shell particles, which consist of ~9 nm hard magnet $CoFe_2O_4$ in the core and ~3 nm soft magnet $MnFe_2O_4$ in the shell ($CoFe_2O_4@MnFe_2O_4$), and extensively studied different magnetic properties in order to realize a hyperthermia treatment for mice [16]. For an identical size range, the SLP for core-shell ($CoFe_2O_4@MnFe_2O_4$) was recorded ~2280 w/g compared to ~450 w/g^{-1} for single component either $CoFe_2O_4$ or $MnFe_2O_4$ phases. The SLP can decrease up to 15% for wider size distribution of particles (i.e., low value of "α" in Weibull statistical analysis for particle size distribution). In the same time, the exchange coupling coercivity, (H_c), of core-shell particle experiences 2530 Oe in between $CoFe_2O_4$ (11,600 Oe) and $MnFe_2O_4$ (0Oe) and demonstrates the feasibility of exchange coupling mechanism. The anisotropy constant K for core-shell particles reduces to 1.5×10^4 J m^{-3} from individual $CoFe_2O_4$, (K = 2.0×10^5 J m^{-3}) and $MnFe_2O_4$ (K = 3.0×10^3 J m^{-3}) single phase. The importance of such research data is discussed in Section 9.6.

5.2.9 Particle Size Effect on Magnetic Saturation

In order to achieve effective performance of nanoparticles for functional applications, one has to concentrate on synthesis protocols to obtain, (a) monodisperse particle size distribution (shape factor, magnitude of distribution, $\alpha \geq 5$); (b) satisfactory high crystallinity with having desired crystal plane; (c) control over the morphology, for example, size and shape of NPs; and (d) stability over long time. Despite these features, finite size effects, surface effects, and interparticle interactions dominate the magnetic properties of MNPs and their ensembles. However, the size reduction leads quantum confinement, and thus surface effect encompasses nonsymmetry from bulk crystal structure at the boundary of each MNP. In this contrary, the magnetic saturation of ferrites with respect to nanoparticle size is a controversial issue.

Representative data are illustrated in Figure 5.8, where saturation magnetization decreases with the high surface area or decreases in the particle size of both γ-Fe$_2$O$_3$ and γ-Co$_{0.06}$Fe$_{1.94}$O$_3$ [17]. The specific saturation magnetizations of both γ-Fe$_2$O$_3$ and γ-Co$_{0.06}$Fe$_{1.94}$O$_3$ particles decreases linearly with increasing specific surface area at both

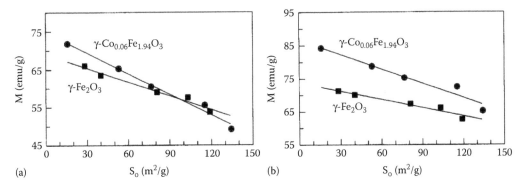

FIGURE 5.8
Magnetization with respect to applied field as a function of surface area (a) 295 K, and (b) 79 K. Saturation magnetization increasing with decreasing specific surface area. (From Han, D.H. et al., *J. Magn. Magn. Mater.*, 136, 176–182, 1994 [17].)

295 K and 79 K. The particle size varies within the range of 20–100 nm. This magnetic behavior reduction can be explained by dominant surface layer spin-canting (angular deviation from a vertical or horizontal plane or surface) phenomenon.

In brief, the fine crystallites consist of two parts, where the first part is the noncollinear surface layer whose magnetic moment cannot be turned completely along the direction of the applied magnetic field, rather encompasses an average canting angle with the field; whereas the inner part can be aligned along the direction of the applied field. Enhancing specific surface area results in smaller particle size and high surface-to-volume ratio, this eventually reduces the saturation magnetization when particle size becomes smaller.

This implies an optimum nanoscale size is prerequisite to achieve maximum efficiency for a definite application. The MNPs obtained under different synthesis protocols may exhibit remarkably different magnetic properties of identical volume of particles of same material because of several factors, including quantum confinement, structural disorder, particle size distribution, particle shape, internal inhomogeneities, surface roughness, porosity, impurities, etc.

5.3 Piezoelectric Phenomena

Piezoelectric effect defines when certain crystal (single-crystal) or crystallites (poly-crystal) became electrically polarized under mechanical stress, and the degree of polarization is proportional to strain. Same material deformed in presence of electric field and is referred to as "converse piezoelectric effect." Specific symmetry of the unit cell determines its degree of polarization and thus dielectric behaviors. Among 32-point groups, 21 classes are noncentrosymmetric, other way, essential requirement for piezoelectricity. However, one out of 21 classes possess other combinations of symmetry, so 20 exhibit polarization induced by mechanical stress. Ten out of 20 can polarize under mechanical stress only (i.e., nonpyroelectric), while other 10 classes comprise spontaneous polarization, so they retain permanent polarization and thus behave with both piezoelectric as well as pyroelectric effects. Furthermore, a subgroup of last 10 classes possess spontaneous and reversible polarization refers to distinct features of ferroelectric from sole pyroelectric and piezoelectric materials [18]. Thus, in brief, piezoelectric (mechanical induced polarization), pyroelectric (mechanical induced as well as spontaneous), and ferroelectric (mechanical induced, spontaneous as well as reversible polarization) are classified in Figure 5.2. Although all probable piezoelectric behaves as paraelectric above the Curie temperature (T_c). As an example of piezoelectric, one can pick up common ferroelectrics (e.g., $BaTiO_3$) in the perspective of functional utility in versatile sectors.

5.3.1 Ferroelectric Behaviors

Spontaneous and reversible polarization is the most fascinating feature of ferroelectrics among all dielectric materials. Ferroelectric single crystal or polycrystalline ceramics have extensive research and technological importance. In current applications, the polycrystalline ceramics offer a more extensive range of easily achievable compositional modifications than single crystals, however, domain wall displacement and extrinsic contribution in preference frequency and field control the properties of polycrystalline ferroelectrics. Consider a common perovskite barium titanate $BaTiO_3$ ($A^{2+}B^{4+}O_3^{2-}$) polycrystalline ceramic that possesses cubic structure above Curie temperature, (T_c), 120°C, while it is tetragonal at room temperature, and further transforms to orthorhombic and rhombohedral structure below room temperature. Below the Curie point, positive and negative charge sites of ferroelectric no longer coincide, and thus elementary unit cells develop electric dipole within the Weiss domain, which may be reversed and also aligned to certain allowed directions by the external electrical field.

Random orientation between neighboring Weiss domains throughout the material neutralize the overall polarization or piezoelectric effect, as illustrated in Figure 5.9a. The ceramic may behave as piezoelectric in preferred direction under a poling treatment that involves exposing the material within a string electric field below Curie temperature. Under this circumstance, the domains experience alignment and expansion in the direction of the field (Figure 5.9b). However, the dipoles remain polarized and locked in near to alignment that results in remnant polarization and a permanent strain or in other ways develop anisotropic (Figure 5.9c).

Under electric field, the charge balance perturbed by the four basic polarization mechanisms: electronic, atomic/ionic, dipolar/orientational, and space charge polarization.

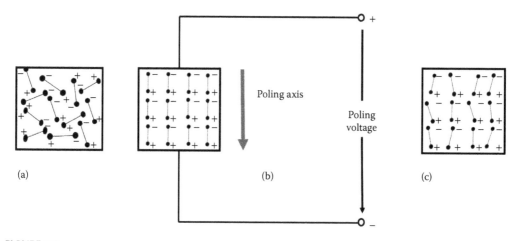

FIGURE 5.9

Electric dipole moments in Weiss domains, (a) before polarization, (b) during polarization, and (c) after polarization.

The ferroelectric component is mandatory to execute poling management to sustain permanent polarization before use in particular application. The electrical behavior of such material has physical analogy with the magnetic behavior of ferromagnetic materials. It also exhibits hysteresis with high dielectric constant resembles with ferromagnetic permeabilities.

Ferroelectrics have reversible spontaneous polarization, as like ferrimagnets that can also be reversed by an applied field in the opposite direction. It exhibits nonlinear relation between the polarization and the applied electric field "E." Conventional and simplest Sawyer-Tower method is the most common to determine the P-E relationship, and a typical schematic representation of hysteresis loop of polycrystalline ferroelectric is shown in Figure 5.10a [19]. Random orientation of crystallites and thus nonuniform axis arrangement of unit cells in polycrystal promotes sluggish reversal, whereas the reverse polarization is quite abrupt in single crystals and form near to square hysteresis loop (Figure 5.10b) [20].

Hysteresis loop (P-E) demonstrates a linear relationship, ($\alpha = 45°$), in the presence of a very small field, and this attribute to domains nonpolarization under this low field. In other words, in presence of $E \gg E_c$, a ferroelectric behaves like an ordinary dielectric ($\varepsilon_r < 100$), but at E_c, the reversal polarization implies a large dielectric nonlinearity. The spontaneous polarization, P_s, likes to increase rapidly beyond the transition point and experience saturation value below Curie temperature (T_c). Beyond saturation polarization, no further increase in polarization is noticed, as all dipoles are aligned with the field. In principle, an extrapolation of saturation polarization toward zero field provides the quantitative data of spontaneous polarization, P_s. If the field, (E), is reduced to zero, the dipoles preferred certain direction within the individual crystallite in absence of external field, this polarization is referred to as remnant polarization, (P_r), and it has lower magnitude than P_s. Further increment of electric field follows the reverse phenomena and thus results in the hysteresis loop, as similar with ferro or ferrimagnetic materials.

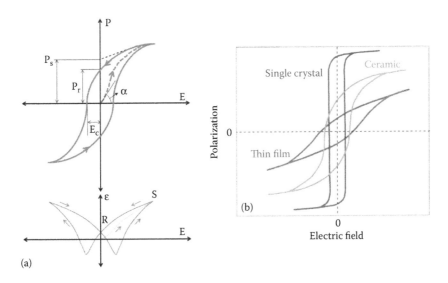

(a)   (b)

FIGURE 5.10
Schematic representation for polycrystalline ferroelectrics (a) P-E characteristics, and bottom part represents the strain-electric field characteristic, (b) a competitive nature of single crystal, polycrystal, and thin coating of ferroelectric. (From Cao, W., *Nat. Mater.*, 4, 727, 2005 [20].)

Noncentro symmetric crystal exhibits piezoelectricity that implies polarization under mechanical stress and hence a voltage across it. In the same time, the reverse polarity and voltage generation are the results of reverse stress direction. In principle, an applied electric field encompasses the mechanical distortion of any geometric shape as matter is condensed of charged nuclei and compensated electron cloud. Thus, polarization affects the charge distribution and hence mechanical distortion. Direct effect of piezoelectricity defines the generation of electric polarization by a mechanical stress (like a generator), while the converse phenomena implies mechanical movement actuated by an electric field (like a motor). The bottom part of Figure 5.10 shows the variation of mechanical strain, (ε), with respect to electric field and referred to as the "butterfly loop" of piezoelectricity behavior of same ferroelectric material. During poling, as described in Figure 5.9, the material experiences tension along the polarization direction (i.e., expansion in the direction) and compression perpendicular to the poling direction (i.e., lateral contraction). Thus, strain along poling direction is positive and lateral strain is negative. A closer look in the strain-electric field demonstrates that in the presence of very low electric field, the converse piezoelectric effect governs positive strain and it follows linear relationship. As the field is increased further, the strain is no longer linear with the field as domain walls start switching and are extended up to saturation state (S) and remnant state (R). In combination with poling and lateral strain-field, the strain-electric field appears as butterfly loop. Both hysteresis (P-E) and butterfly (ε-E) loop area, loop geometry, and ferroelectric parameters depend on crystallinity, crystal structure, defects, particle or grain size, temperature, and frequency. Crystallinity is essential for effective poling, so the ferroelectric behavior depends on the degree of crystallinity. For an example, high crystalline tetragonal $BaTiO_3$ comprises high dielectric and piezoelectric behavior compared to same degree of crystallinity of cubic $BaTiO_3$, although grain size window is an essential factor to achieve effective properties. In addition, the defects such as impurities affect the dielectric properties and their switching phenomenon [21]. It deforms crystalline lattice and thus surrounding volume and modification of the local fields. Eventually, this situation induces additional polarization and may not follow the perfect crystal behavior during depolarization. This difference between perfect crystal and real crystal (consist of defect) results in high coercive field. The distribution of defects has a remarkable effect on the switching properties under external field. If all the defect dipoles appear in the same direction, the hysteresis loop becomes biased and shifted along the axis of electric field. A brief about the influence of particle size, temperature, and frequency effect on the dielectric and piezoelectric properties of $BaTiO_3$ is discussed in Section 5.3.4.

5.3.2 Dielectric Permittivity and Loss

Dielectrics are poor conductors of electricity that exhibit electrical polarization and charge storage capacity under electrostatic field. This eccentric property is quantized or characterized via the parameter complex permittivity, (ε_r^*), which illustrates the material's behavior when subjected to experience an electromagnetic field. This complex parameter, consisting of a real (ε_r') and an imaginary part (ε_r''), can be expressed as:

$$\varepsilon^* \text{ or } \varepsilon_r = \varepsilon_r' - j \cdot \varepsilon_r'' \qquad (5.8)$$

Here, $j = \sqrt{-1}$, ε_r' = real component of ε_r, known as "dielectric permittivity" or "dielectric constant," measures the charge storing capacity of the intended dielectric material, and ε_r'' = imaginary component of ε_r, known as "loss factor," determines the energy dissipated or lost by the material when exposed to an external electric field. The ratio of the imaginary component to the real component of dielectric permittivity counts for another important parameter, $\tan\delta$ = "dissipation factor" or "loss tangent," expressed as:

$$\tan\delta = \frac{\varepsilon_r{''}}{\varepsilon_r{'}} \tag{5.9}$$

This is primarily the electrical energy lost by the material per unit of energy stored during a charging cycle. The dielectric permittivity of any material has been found to be a function of operating temperature, frequency of the applied electric field, structure of the material at the molecular level (polarizability of the dipole varies with the nature of the constituent ions), as well as the microstructural signature. The measurement technique to be used depends on the intended frequency band, as well as nature of the sample prepared. Hence, the selection of an appropriate measurement technique involves careful weighing of significant factors like frequency band, sample dimensions, nature of the material, accuracy in demand, financial compensation, mode of testing (contact/non-H contact mode, destructive/non-H destructive mode), as well as the operation temperature range. The various available dielectric constant measurement techniques are, coaxial probe method, transmission line method, resonant cavity method, free-space method, parallel plate electrode method, and planar transmission line method [22]. Undoubtedly, each technique has its own share of benedictions and blemishes. Depending on the intended area of application, the corresponding technique must be chosen. Hereby, the most commonly used technique known as the parallel plate electrode method has been elucidated in the next paragraph.

Parallel plate capacitor method involves a setup that consists of a thin sheet of the tested material as sandwiched in between the two electrodes exemplary of a capacitor. This measurement is conducted at typically low frequencies (<1 GHz), wherein the measurements are carried out using an LCR meter or impedance analyzer and a dielectric test fixture. Prior to performing the dielectric characterization, the material surface needs to be as flat as possible to avoid air entrapment and misinterpretation of the obtained value. In case of impedance analyzer, the impedance of the sample is converted to complex permittivity using the sample as well as the electrode dimensions. Usually, the $\varepsilon_r{'}$ (dielectric constant) is calculated from Equation 5.10 and loss factor can be estimated from Equation 5.9.

$$\varepsilon_r{'} = \frac{C_P t}{\varepsilon_o A} = \frac{C_P \times t}{\varepsilon_o \times \pi \times (d/2)^2} \tag{5.10}$$

where ε_o is the permittivity of free space; C_P is the measured capacitance; A is the area of electrode; d is the diameter of electrode; and t is the thickness of sample material. Adequate sample preparation and its characterization protocol in relevant frequency and temperature are important criteria to achieve higher accuracy and representative dataset. For an example, the parallel plate (electrode), coaxial probe and free-space methods are effective techniques for the high loss dielectric, while resonant cavity provides higher accuracy for the low permittivity [23].

5.3.3 Piezoelectric Constants

Piezoelectric materials are an all-encompassing class of noncentrosymmetric materials within which consist two other material classes as subsets, namely, pyroelectric materials with a further subset recognized as ferroelectric materials. All materials are known to exhibit diminutive dimensional changes when exposed to an electric field. This strain, (S), caused is directly proportional to the square of the electric field vector, (E). The constant of proportionality is known as "electrostrictive coefficient," (q). This relationship can be shown by the equation:

$$S = qE^2 \qquad (5.11)$$

Out of these generic materials, a class of material can exhibit the reverse of electrostrictive effect, and they are capable of transmuting electrical energy to mechanical energy and vice versa. They are invariably known as "piezoelectric materials." These two transmutations are characterized by two different relationships: direct and converse piezoelectric effect. Direct piezoelectric effect defines the conversion of mechanical energy to electrical energy is accompanied by generation of surface charges on the dielectric due to applied stress and can be defined as:

$$D = d\,T \qquad (5.12)$$

where D = dielectric displacement or surface charge density generated (C/m²), d = piezo-electric charge constant (C/N), and T = applied stress (N/m²). Also, when related to electric field rather than surface charge density, this relationship changes as:

$$E = gT \qquad (5.13)$$

where E = electric field generated (V/m), g = piezoelectric voltage constant (V-m/N), and T = applied stress (N/m²). Converse piezoelectric effect describes the conversion of electrical energy to mechanical energy that is accompanied by generation of strain, and their direction is dependent on the sign of the polarity. A change in the polarizing direction reverses the response direction too and related to S = dE or S = dD, where E − electric field applied (V/m) or D = surface charge density (C/m²), S = generated strain (dimensionless), and d = piezoelectric charge constant (m/V) or (m²/C).

Random orientation of crystal axes in polycrystalline piezoelectric exhibits electrostrictive effect when material belongs to ferroelectric group. Thus, ferroelectric materials are characterized by presence of a spontaneous direction of polarization, which can be switched to certain other selective directions by application of an electric field of certain threshold magnitude. This phenomenon is known as "poling of the material," and this treatment is usually done to use and get the advantage of piezoelectric material, however, this can be depolarized under mechanical, electrical, and thermal agitation [24]. Several parameters including piezoelectric charge constants (d_{33}, d_{31}, d_{15}), piezoelectric voltage constants (g_{31}, g_{31}, g_{15}), elastic constants (S^E_{11}, S^D_{36}), dielectric constant (ϵ^S_{33}, ϵ^T_{11}), coupling factor (k_{33}, k_{31}), and mechanical quality factor (Q_m) can predict the performance of piezoelectric materials. The magnitude of physical constants of anisotropic piezoceramic comprehends both the direction of the mechanical and electrical field and the directions perpendicular to the applied force.

Thus, any constant represents by two subscripts that indicate the direction of the two related quantities, such as stress (force on the ceramic component/surface area of the component) and strain (change in length of component/original length of component) for elasticity. In convention, a superscript index refers to the quantity that is supposed to keep constant. The direction of positive polarization is usually opted to coincide with the Z-axis analogous to the cubic crystallographic system (Figure 5.11). Directions of X, Y, and Z are represented by subscript 1, 2, and 3, respectively, and shear about one of the axes are represented by the subscript 4, 5, and 6, respectively.

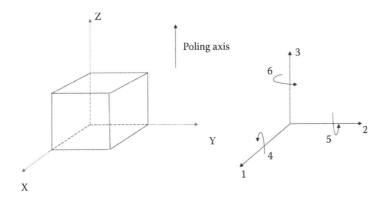

FIGURE 5.11
Schematic representation of cubic crystal and designation of their axes and analogous poling directions.

A brief illustration of individual constants and their physical significance would provide a better insight in order for development and analysis of piezoelectric materials. While discussing about constants, the piezoelectric charge constant, (d), is described as the polarization originated per unit of mechanical stress, (T), applied to a piezoelectric material having density "ρ" and length "l" or in another way, experience mechanical strain, (S), per unit of applied electric field. Herein, the first subscript indicates the direction of polarization (electric field E = 0) or, alternatively, direction of the applied strength. The second subscript refers to the direction of the applied stress or the induced strain, respectively. It is an important indicator to select the material for strain-dependent (e.g., actuator) applications. For example:

- d_{33} is the induced polarization in direction 3 (parallel to direction in which ceramic component is polarized) per unit applied stress in direction 3. In another way, it is the induced strain in direction 3 per unit electric field in direction 3.

- d_{31} is the induced polarization in direction 3 (parallel to direction in which ceramic component is polarized) per unit stress applied in direction 1 (perpendicular to direction in which ceramic element is polarized). Alternatively, induced strain in direction 1 per unit electric field applied in direction 3.

- d_{15} is the induced polarization in direction 1 (perpendicular to direction in which ceramic component is polarized) per unit shear stress applied in direction 2 (perpendicular to direction in which ceramic element is polarized) or induced shear strain about direction 2 per unit electric field applied in direction 1.

As mentioned earlier, the piezoelectric voltage constant, (g), is the electric field generated per unit of mechanical stress, (T), applied, or another way, the experience mechanical strain, (S), per unit of electric displacement applied. The first subscript indicates the direction of the electric field generated in the material or the direction of the applied electric displacement. The second subscript is the direction of the applied stress or the induced strain, respectively. Quantification of such value is essential to develop sensor materials. Some representation "g" values are:

- g_{33} is the induced electric field in direction 3 (parallel to direction in which ceramic component is polarized) per unit stress applied in direction 3. In another way, it is the induced strain in direction 3 per unit electric displacement applied in direction 3.

- g_{31} is the induced electric field in direction 3 (parallel to direction in which ceramic component is polarized) per unit stress applied in direction 1 (perpendicular to direction in which ceramic component is polarized). In another way, it is the induced strain in direction 1 per unit electric displacement applied in direction 3.
- g_{15} is the induced electric field in direction 1 (perpendicular to direction in which ceramic component is polarized) per unit shear stress applied in direction 2 (perpendicular to direction in which ceramic component is polarized). In another way, it is the induced strain in direction 2 per unit electric displacement applied in direction 1.

Elastic compliance, S is the strain produced in a piezoelectric material per unit of stress applied for the 11 and 33 directions. It is the reciprocal of the modulus of elasticity. S^E is the compliance under a constant electric field, and S^D is the compliance under a constant electric displacement. The first subscript indicates the direction of strain, and the second is the direction of stress. So:

- S^E_{11} is the elastic compliance for stress in direction 1 (perpendicular to direction in which ceramic element is polarized) and accompanying strain in direction 1, under constant electric field (short circuit).
- S^D_{33} is the elastic compliance for stress in direction 3 (parallel to direction in which ceramic element is polarized) and accompanying strain in direction 3, under constant electric displacement (open circuit).

The dielectric constant or permittivity, ε defines the dielectric displacement per unit electric field. ε^S is the permittivity at constant strain, and ε^T is the permittivity at constant stress. First subscript indicates the direction of the dielectric displacement, and second is the direction of the electric field. For example:

- ε^S_{33} describes the permittivity for dielectric displacement and electric field in direction 3 (parallel to direction in which ceramic element is polarized), under constant strain.
- ε^T_{11} is the permittivity for dielectric displacement and electric field in direction 1 (perpendicular to direction in which ceramic element is polarized), under constant stress.

The relative dielectric constant, (K^T_3 analogous to ε_r'), is the ratio of the permittivity of the selected materials to permittivity to the free space (8.85×10^{-12} F/m). In another way, it can be calculated from the measured values of capacitance and physical dimensions of the specimen:

$$K^T = \frac{\text{distance between electrodes (meters)} \times C \text{ (pF)}}{\text{area of one electrode (meters}^2) \times 8.85} \tag{5.14}$$

Electromechanical coupling factor, (k), measures the efficiency of the piezoelectric material in converting one form of energy to another, and it is the electrical energy to mechanical energy and vice versa. It is given by the relationship:

$$k^2 = \frac{\text{Electrical (Mechanical) energy converted to Mechanical (Electrical) energy}}{\text{Input Electrical (Mechanical) energy}} \tag{5.15}$$

An essential electrical impedance, (Z_m), with respect to frequency is required in order to estimate the "k." The domains (crystallites with a uniform electronic polarization direction) vibrate during frequency variation. If their vibrational frequency matches with that of the externally applied electric field, resonance may occur causing a sharp decline in

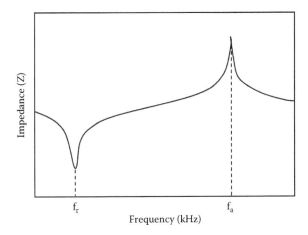

FIGURE 5.12
Electrical impedance versus frequency to measure electromechanical coupling factor.

the electrical impedance for that frequency due to increased conductivity. As shown in Figure 5.12, the trough marked by f_r represents the resonance frequency, which causes a spike in electrical conductivity and represents as f_a, the anti-resonance frequency, which causes an increase in impedance value.

From this plot, we can determine the effective electromechanical coupling coefficient, (k^2), of any shape of ceramic component by the relationship as stated in Equation 5.16:

$$k_{eff}^2 = \frac{f_a^2 - f_r^2}{f_a^2} \tag{5.16}$$

As stated earlier, this factor is generally described by two subscripts, where the first subscript denotes the direction along which the electrodes are applied, and second denotes the direction along which the mechanical energy is applied or developed. Usually, a high "k" value is desirable for efficient conversion and not account any loss or unconvertable energy. Physical meaning demonstrates the ratio of useable energy delivered by the piezoelectric component to the total energy taken up by the component. In accordance with the sample geometry and dimension, a different class of "k" value can be obtained. Some representative parameters are:

- k_{33} is the factor for electric field in direction 3 (parallel to direction in which ceramic element is polarized) and longitudinal vibrations in direction 3.
- k_{31} factor for electric field in direction 3 (parallel to direction in which ceramic element is polarized) and longitudinal vibrations in direction 1 (perpendicular to direction in which ceramic element is polarized).

Mechanical quality factor, (Q_m), illustrates the sharpness of resonance of a piezoelectric component. The reciprocal value of the Q_m is the mechanical loss factor, the ratio of effective resistance to reactance in the equivalent circuit diagram of a piezoelectric resonator at resonance. This can be estimated from the frequencies of minimum and maximum impedance, the magnitude of minimum impedance and capacitance, (C), it is a dimensionless quantity. Some useful relations are given in Table 5.2 to calculate the parameters [25].

TABLE 5.2

Some Useful Relations in Piezoelectric Constants

Constants	Relation
d_{33}	$d_{33} = k_{33}\sqrt{8.85\times10^{-12}\,K_3^T\,S_{33}^E}$
d_{31}	$d_{31} = k_{31}\sqrt{8.85\times10^{-12}\,K_3^T\,S_{11}^E}$
g_{33}	$g_{33} = \dfrac{d_{33}}{K_3^T \cdot 8.85\times10^{-12}}$
g_{31}	$g_{31} = \dfrac{d_{31}}{K_3^T \cdot 8.85\times10^{-12}}$
S^D_{33}	$S^D_{33} = \dfrac{1}{4\rho l^2 f_a^2}$
S^E_{11}	$S^E_{11} = \dfrac{1}{4\rho l^2 f_r^2}$
S^D_{11}	$S^D_{11} = (1 - k_{31}^2)\cdot S^E_{11}$
S^E_{33}	$S^E_{33} = \dfrac{S^D_{33}}{1 - k_{33}^2}$
k_{33}	$k_{33} = \sqrt{\dfrac{\pi}{2}\dfrac{f_r}{f_a}\tan\left[\dfrac{\pi(f_a - f_r)}{2 f_a}\right]}$
k_{31}	$k_{31} = \sqrt{\dfrac{A}{1+A}};\ A = \dfrac{\pi}{2}\dfrac{f_a}{f_r}\tan\left[\dfrac{\pi(f_a - f_r)}{2 f_a}\right]$
Q_m	$Q_m = \dfrac{f_a^2}{2\pi f_r Z_m C\left(f_a^2 - f_r^2\right)}$

5.3.4 Role of Grain Size on Dielectrics and Piezoelectrics

Let's consider BaTiO$_3$, most studied ferroelectric with having perovskite type crystal structure for different functional applications. This class of ferroelectric has also found allegiance as dielectrics in the microelectronics, pyroelectrics in infrared sensors and imaging, and piezo electrics in micromechanical systems [26].

In consideration of the Curie temperature analogous to ferromagnetism, the temperature at which transition from ferroelectric to paraelectric phase takes place, effect of particle size is significantly evident. A sharp decline in the Curie temperature has been observed with decrease in particle size below a critical size. For example, Curie temperature of bulk BaTiO$_3$ has been found to be 120°C for the grain size of 300 nm, whereas it reaches as below as 90°C at the grain size of ~100 nm [27]. In this literature, the latent heat of the orthorhombic to tetragonal transition was noticed only for ceramics with GS ≥ 300 nm. Such grain size dependence on Curie temperature has been demonstrated in Figure 5.13, however, the threshold grain size may vary from material to material.

Nanoscale particle size reduction facilitates to change the surface charge and electronic configuration and results in the distribution change of the electronic energy densities levels (see Section 1.3.2). For polycrystalline ferroelectric materials, the properties are simply evanescent below a certain critical size. It has been invariably argued that the particle size reduction is inevitably accompanied by stability of the underlying

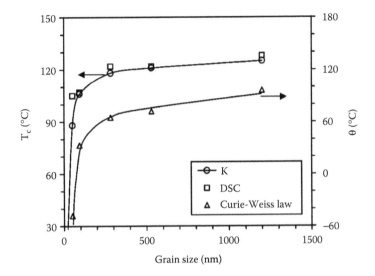

FIGURE 5.13
Variation of Curie temperature of a common ferroelectric BaTiO$_3$ as a function of grains size. Competitive dielectric and differential scanning calorimetry (DSC) measurement is highlighted. Variation of and Curie-Weiss temperature are also given. (From Zhao, Z. et al., *Phys. Rev. B*, 70, 024107, 2004 [27].)

crystal structure at a relatively lower temperature. So, the decrease in Curie temperature or ferroelectric to paraelectric phase transition temperature is highly predictable with particle size reduction. Apart from the particle size-dependent Curie temperature, it is worthy to discuss how dielectric and piezoelectric constants vary with respect to grain size and their reason behind it. Figure 5.14a and b represents the variation of dielectric (poled and unpoled) and piezoelectric constants of BaTiO$_3$ with respect to temperature and grain size, respectively. Herein, the specimens were prepared through conventional uniaxial pressing followed by atmospheric sintering up to 1450°C for 2 h. Prior to measurement, the discs were coated with silver paint on both sides and fired at 575°C for 20 min [27]. Poling was accomplished at 105°C in silicon oil under 5.0 kV mm^{-1} for 30 min. All the dielectric measurements were carried out at room temperature (~25°C) with a frequency of 1 kHz. Both the dielectric and piezoelectric properties were measured with a poling time gap of 24 h. The dielectric constant, (ε_r'), was measured using an Agilent 4294A Precision Impedance Analyzer, whereas the d$_{33}$ value was estimated by a Berlincourt-type d33 meter (YE2730A). An Espec SU-261 chamber was used to meet the accurate temperature and avoid any error percentage during dielectric and piezoelectric constant measurement. Both poled and unpoled BaTiO$_3$ exhibit similar trends with a strong variation with respect to grain size (Figure 5.14a). A particle grain size window near to 1 μm exhibits highest value up to ε = 4045 for unpoled, but dielectric constant decreases near to 15% because of poling in the same grain size window. However, this effect diminishes in case of coarse grains, typically beyond 10 μm. In the same time, the d$_{33}$ shows a strong grain size dependence when grain size is less than 40 μm. Interestingly, the highest d$_{33}$ value of 338pC N^{-1} is observed for fine grain size ~1 μm, also Figure 5.14b. However, a decrement of both constants for grain size ~0.75 μm is attributed to the low degree of density without significant grain growth. Thus, fabrication of nanostructured grains having homogenous microstructure and optimum grain

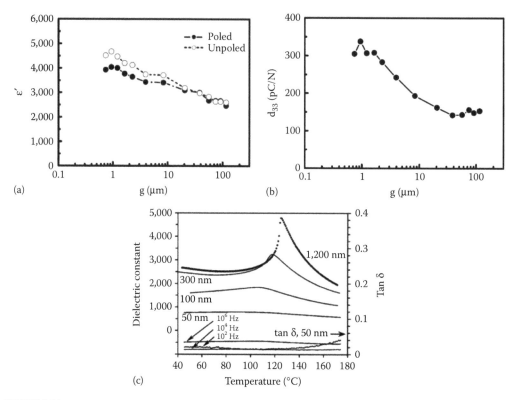

FIGURE 5.14
The dielectric and piezoelectric behavior with respect to grain size, (a) room temperature dielectric constant, (ε_r'), as a function of grain size; after and before poling and (b) room temperature d_{33} of poled BaTiO$_3$ as a function of grain size [28]. (c) Variation of dielectric constant at 10^4 Hz and loss tangent, (tan δ), of BaTiO$_3$ as a function of temperature. T$_c$ is decreasing with lowering the grain size. The loss is reported for the 50 nm grains at different frequencies. (From Zhao, Z. et al., *Phys. Rev. B*, 70, 024107, 2004 [27].)

size are desirable to achieve maximum dielectric and piezoelectric properties. The similarity in the grain size dependence response between d_{33} and ε implies that the extrinsic (domain wall density) contribution is more predominant than intrinsic (ion shift, c/a ratio, and tetragonal-orthorhombic phase transformation) contributions. Reader can consider the classic literature in order to discuss the detail explanations and their mechanisms [28,29]. Another group of researchers fabricated nanoscale BaTiO$_3$ grains through spark plasma sintering 800°C–1000°C for 2–5 min, followed by annealing at 700°C or 800°C depending on the spark plasma sintering temperature. Figure 5.14c depicts the representative dielectric constant and losses, (tan δ), as a function of temperature when measured in air with a frequency range of 10^2–10^6 Hz using an impedance analyzer, temperature variation of 40°C–180°C, and applied voltage of 1 V [27]. The dielectric constant of nanostructured BaTiO$_3$ is strongly depressed up to value of 780 for fine grain size, but less sensitive to temperature in contrast to coarse ceramics at 10^4 Hz. With variation of temperature range 40°C–170°C, a remarkable dielectric constant highest ~4800 at 120°C is noticed for 1.2 μm coarser grain. The reduction of Curie temperature and peak sharpness broadening is noticed with the decreasing BaTiO$_3$ grain size.

Interestingly, the variation of dielectric losses of the fine grain (50–300 nm) ceramics is rather minimum, <5% in the entire frequency range and comprised less than 2% in the range of 3×10^3 to 3×10^5 Hz. It is worthy to note that the dielectric constant is reducing with nanostructured grains, but minimum variation of dielectric loss in any frequency and temperature independent dielectric constant enhances the reliability of the dielectric components. In order to use the piezoelectric material, one has to select the magnitude of piezoelectric constants judiciously and thus grain size, although eventually microstructure depends on the starting particle size, fabrication protocol, and sintering profile.

5.4 Concluding Remarks

This chapter covers a cluster of properties that enable to explain concisely in target to the development of renewable energy, environmental science, and medical diagnosis with the help of nanostructured magnetic and piezoelectric ceramics. In the beginning, both of the materials are classified with relevant examples, highlighted ferrimagnetic and ferroelectric materials, and discussed their properties. Below critical radius, the superparamagnetic does not have any M_r and H_c, although possess high susceptibility compared to diamagnetic and paramagnetic, but lower in value compared to ferro or ferrimagnetic. This particle can avoid aggregation as spin relax quickly and ease demagnetization at room temperature. Magnetically, isotropic material has no preferred magnetic moment direction, without anisotropy, a material behaves as superparamagnetic. However, shape and magnetocrystalline anisotropy eventually facilitate to generate heat in combination of either soft-hard ferrite or core-shell combination. Specific loss power, (watt/gram), determines the heating efficiency and optimum nanoparticle size with having its narrow size distribution is beneficial for high SLP value. High particle aspect ratio with low particle size is an effective management to enhance the magneto resistance change although other parameters are also essential to encounter in target of definite application. In consideration of different magnetic properties, exchange-spring magnet is a new generation choice for cancer treatment. Magnetic saturation is likely to decrease with increasing specific surface area, in other way the reduction of particle size attributes to surface layer spin-canting mechanism.

Crystal symmetry-dependent electrical polarization under mechanical stress is proportional to strain, and reverse mechanical deformations under electric fields are important phenomena of ferroelectrics. Thus, high crystalline and one dimensional particles are supposed to exhibit high polarization under stress, although particles or grain size manipulation is indeed as particle size reduction change the surface charge and electronic configuration. An optimum grain size facilitates the high dielectric constant, but at very low grain size, the minimum variation of dielectric loss enhances the reliability of the dielectric components. A selective synthesis and fabrication protocol can synchronize the particle and grain size for definite application. Different classic examples on the effective utility of such nanostructured ceramics, for example, particle size dependent piezoelectric and magnetic materials are discussed in Sections 7.6, 8.4, and 9.6, respectively.

References

1. O. Gutfleisch, M. A. Willard, E. Brück, C. H. Chen, S. G. Sankar, and J. P. Liu, Magnetic materials and devices for the 21st century: Stronger, lighter, and more energy efficient, *Adv Mater* 23, 821–842, 2011.
2. V. K. Sharma, R. Doong, H. Kim, R. S. Varma, and D. D. Dionysiou, *Ferrites and Ferrates: Chemistry and Applications in Sustainable Energy and Environmental Remediation*, Vol. 1238, American Chemical Society, Washington, DC, 2016.
3. L. Zhang, W. F. Dong, and H. B. Sun, Multifunctional superparamagnetic iron oxide nanoparticles: Design, synthesis and biomedical photonic applications, *Nanoscale* 5, 7664–7684, 2013.
4. K. H. J. Buschow, *Handbook of Magnetic Materials*, Vol. 23, Elsevier, Amsterdam, the Netherlands, 2015.
5. N. T. K. Thanh, *Magnetic Nanoparticles: From Fabrication to Clinical Applications*, 1st ed., CRC Press, Boca Raton, FL, 2012.
6. Q. A. Pankhurst, J. Connolly, S. K. Jones, and J. Dobson, Applications of magnetic nanoparticles in biomedicine, *J Phys D Appl Phys* 36(13), R167–R181, 2003.
7. J. M. D. Coey, *Magnetism and Magnetic Materials*, 1st ed., Cambridge University Press, Cambridge, UK, 2010.
8. N. A. Spaldin, *Magnetic Materials: Fundamentals and Applications*, 2nd ed., Cambridge University Press, Cambridge, UK, 2010.
9. Y. Liu, D. J. Sellmyer, and D. Shindo, *Handbook of Advanced Magnetic Materials*, Springer-Verlag, New York, 2006.
10. B. M. Moskowitz, *Hitchhiker's Guide to Magnetism*, Environmental Magnetism Workshop, Institute for Rock Magnetism, 1991, www.magneticmicrosphere.com.
11. J. A. Brug and W. P. Wolf, Demagnetizing fields in magnetic measurements. I. Thin discs, *J Appl Phys* 57, 4685, 1985.
12. R. Müller, R. Hergt, M. Zeisberger, and W. Gawalek, Preparation of magnetic nanoparticles with large specific loss power for heating applications, *J Magn Magn Mater* 289, 13–16, 2005.
13. K. Murase, A simulation study on the specific loss power in magnetic hyperthermia in the presence of a static magnetic field, *J Appl Sci* 6, 839–851, 2016.
14. J. Nickel, *Magnetoresistance Overview*, Hewlett-Packard, Palo Alto, CA, 1995.
15. F. Liu, Y. Hou, and S. Gao, Exchange-coupled nanocomposites: Chemical synthesis, characterization and applications, *Chem Soc Rev* 43(23), 8098–8113, 2014.
16. J. H. Lee, J. Jang, J. Choi, S. H. Moon, S. Noh, J. Kim, J. G. Kim, I. S. Kim, K. I. Park, and J. Cheon, Exchange-coupled magnetic nanoparticles for efficient heat induction, *Nat Nanotechnol* 6, 418–422, 2011.
17. D. H. Han, J. P. Wang, and H. L. Luo, Crystallite size effect on saturation magnetization of fine ferrimagnetic particles, *J Magn Magn Mater* 136, 176–182, 1994.
18. B. Jaffe, W. R. Cook, and H. Jaffe, *Piezoelectric Ceramics*, Academic Press, London, UK, 1971.
19. C. B. Sawyer and C. H. Tower, Rochelle salt as a dielectric, *Phys Rev* 35, 269, 1930.
20. W. Cao, Ferroelectrics: The strain limits on switching, *Nat Mater* 4(10), 727, 2005.
21. D. Damjanovic, Ferroelectric, dielectric and piezoelectric properties of ferroelectric thin films and ceramics, *Rep Prog Phys* 61, 1267–1324, 1998.
22. *Basics of Measuring the Dielectric Properties of Material*, Agilent-Technologies, Santa Clara, CA, 2006.
23. M. T. Jilani, M. Z. ur Rehman, A. Muhammad Khan, M. Talha Khan, and S. M. Ali, A brief review of measuring techniques for characterization of dielectric materials, *Int J Inf Technol Elect Eng* 1(1), 2012.
24. W. G. Cady, *Piezoelectricity: An Introduction to the Theory and Applications of Electromechanical Phenomena in Crystals*, McGraw-Hill Book, New York, 1946.
25. APC International, Ltd., Piezoelectric ceramics: Principles and applications, 2011.

26. P. Muralt, Ferroelectric thin films for micro-sensors and actuators: A review, *J Micromech Microeng* 10, 136–146, 2000.
27. Z. Zhao, V. Buscaglia, M. Viviani, M. T. Buscaglia, L. Mitoseriu, and A. Testino, Grain-size effects on the ferroelectric behavior of dense nanocrystalline $BaTiO_3$ ceramics, *Phys Rev B* 70, 024107, 2004.
28. P. Zheng, J. L. Zhang, Y. Q. Tan, and C. L. Wang, Grain-size effects on dielectric and piezoelectric properties of poled $BaTiO_3$ ceramics, *Acta Materialia* 60, 5022–5030, 2012.
29. P. Marton, I. Rychetsky, and J. Hlinka, Domain walls of ferroelectric $BaTiO_3$ within the Ginzburg-Landau-Devonshire phenomenological model, *Phys Rev B* 81, 144125, 2010.

6

Mechanical Properties

6.1 Introduction

Individual confined growth particles, for an example piezoelectric nanofiber itself, have substantial potential to generate either useable electrical energy from waste energy (vibration) or to degrade pollutant by nanoelectromechanical system (NEMS) through piezoelectric effect, related to strain [1]. Apart from the direct use of nanostructured particles, the market has a substantial mandate of a consolidated and definite shape of components made from particles (indirect use of nanoparticles) in order to fulfill the load bearing structural applications. Nanostructured bulk ceramics exhibit unique surface characteristics and mechanical properties, such as superplasticity, strength, toughness, hardness, and machinability because of the fine grain size, abundant grain boundaries, and synchronized crystallinity [2]. However, the bulk ceramics are limited in renewable energy or environment issues, rather nanostructured particles and coatings cover multidirectional prospects for renewable energy, water purification, sun light protective coatings etc. Researchers developed classic dense or porous consolidation through oxide and nonoxide ceramics, and, hence, it is worth a brief mechanical response of ceramics to understand developing structural components. This chapter describes elastic modulus, strength, survival probability in the perspective of failure, fracture toughness, hardness, fatigue, superplasticity, friction, and wear behavior of bulk ceramics that are further critically reviewed on the effect of nanoscale grain size and presence of inclusions in wide aspects.

6.2 Elastic Moduli

Elastic modulus (Young's modulus, E) is an intrinsic property, although possibility of variation with respect to grain size, phase assemblage, porosity, and temperature eventually alters the performance of both confined growth particles and heterogeneous ceramics. Despite influence of high temperature effect, the basic information on elastic modulus is very important to scientists and medical professionals, as it provides the information on the possibility of deformation of structural component under load bearing condition. Nanostructured bulk ceramics have large volume fraction of atoms that reside in the grain boundaries and therefore exhibit novel properties compared to conventional coarse-grained ceramics. Thus, plastic deformation behavior particularly in the grain size dependency of the yield stress (Hall-Petch) is most frequently investigated [3,4]. Prior to discussing the estimation and effect of dependent variables, a brief theoretical concept in consideration of grains, grain boundary, and secondary phase, is highlighted. These are classified as "single-phase," consisting of grain and grain boundaries of identical material only (e.g., transparent yttrium aluminium garnet) and "multi-phase," referring to the combination of pores and more than one phase (e.g., hydroxyapatite-gelatin porous scaffold or zirconia toughened alumina dense femoral head), more realistic in order to develop porous or dense composites for different zone of applications.

6.2.1 Single Phase Nanocrystalline Ceramics

Elastic modulus phenomena of single phase nanocrystalline materials can be explained with consideration of Voigt and Reuss models [5]. Both grain and grain boundaries of identical material (sought as apparent composite) are considered to estimate the resultant magnitude of elastic modulus. To describe this incidence, let's assume a unit cubic cell from array of repeating unit cells within nanocrystalline materials and represented in Figure 6.1.

Herein, unit cells consist of grain interior (GI) and grain boundary (GB), but grain boundary is further subdivided in two parts with assumption of different deformation behavior under uniaxial mode of loading [6]. Part-I includes four sides around the cubic grain size of "d" with a definite thickness "δ," whereas part-II grain boundary exists with identical thickness "δ" in upper and lower layers, respectively. If cube is subject to deform uniformly under vertical uniaxial load, the strain of GB_{part-I} is same with that of GI (Voigt model), whereas the $GB_{part-II}$ experience cumulative stress of GI and GB_{part-I} (Reuss model). On the basis of cubical shape, the volume fraction of individual GI (V_G), GB_{part-I} (V_{GB1}), and $GB_{part-II}$ (V_{GB2}) can be represented as:

$$V_G = \frac{(d-\delta)^3}{d^3} \qquad (6.1)$$

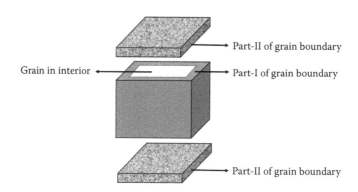

Grain in interior ←

Part-II of grain boundary →

Part-I of grain boundary →

Part-II of grain boundary →

FIGURE 6.1
A unit cubic cell consist of grain interior, two-part grain boundary, Part-I and Part-II; coined as "apparent composite."

$$V_{GB1} = 1 - \frac{\delta}{d} - \frac{(d-\delta)^3}{d^3} \tag{6.2}$$

$$V_{GB2} = \frac{\delta}{d} \tag{6.3}$$

A similar concept can be adopted for regular polyhedron unit cells as well. Thus, a known volume fraction of one phase among grain and grain boundary implies two extreme ends of the resultant E for linear elastic material as theoretical, calculated by both Voigt strain and Reuss stress model. So, the total Young's moduli of single-phase nanocrystalline materials, E, can be represented by:

$$E = \left(\frac{V_G + V_{GB1}}{E_{cb}} + \frac{V_{GB2}}{E_{GB2}} \right)^{-1} \tag{6.4}$$

$$E_{cb} = E_G \frac{V_G}{V_G + V_{GB1}} + E_{GB} \frac{V_{GB1}}{V_G + V_{GB1}} \tag{6.5}$$

where E_G and E_{GB} as the Young's moduli of the constituent grains and grain boundary, respectively, and E_{cb} is the combined Young's modulus of the grain interior and part-I of the grain boundary.

Analogous to the earlier concept, the E for monolithic zirconia matrix with content of two crystallographic phases in an "apparent composite" mixture of tetragonal, (t), and monoclinic, (m), zirconia can be estimated. However, Hill's model proposed to take the algebraic average of the Voigt and Reuss approximations to normalize the extreme end values of elastic modulus, and the resultant value is well matching with experimentally obtained data [7]. If we estimate the volume fraction of tetragonal phase, (V_t), by XRD analysis, and knowing the theoretical elastic modulus of individual tetragonal and monoclinic phase, the E_V, (elastic modulus by Voigt), E_R, (elastic modulus by Reuss, similar to Equation 6.4), E_H, (elastic modulus by Hill), can be estimated by:

$$E_V = V_t \cdot E_t + (1 - V_t) E_m \tag{6.6}$$

$$\frac{1}{E_R} = \frac{V_t}{E_t} + \frac{(1 - V_t)}{E_m} \tag{6.7}$$

$$E_H = \frac{E_V + E_R}{2} \tag{6.8}$$

This approximation is meaningful without content of any pores as secondary phase that resembles with complete dense polycrystalline ceramics. However, a theoretical concept including pores for a multi-phase system is required in the perspective of product development, as the consolidated ceramics can be overall classified based on the relative density (ρ_r = bulk density/theoretical density), transparent ceramics ($\geq 99.99\%$), dense ceramics ($99.99\% \geq \rho_r \geq 70\%$), and porous ceramics ($\rho_r \leq 70\%$).

6.2.2 Multi-Phase Nanocrystalline Ceramics

In order to calculate the Young's moduli of porous, multi-phase nanocrystalline ceramics, Budiansky's self-consistent approach is considered, where "n" number of isotropic constituents are randomly dispersed in composites [8]. Followed by the effect of porosity on Young's modulus, porosity is treated as p^{th} phase in nanocrystalline ceramics with obvious zero strength, and thus resultant bulk modulus and shear modulus are also zero. Now, introduce the volume fraction of the ith phase of nanocrystalline ceramics, and, denoted as f_i, then the overall bulk modulus, (K), and the shear modulus, (G), of the nanocrystalline ceramics are related, by the self-consistent approach, to the moduli K_i and G_i of the ith phase by the following equations:

$$\sum_{i=1}^{n-1} f_i \left[1 - a + a \left(\frac{K_i}{K} \right) \right]^{-1} + \frac{f_p}{1-a} = 1 \tag{6.9}$$

$$\sum_{i=1}^{n-1} f_i \left[1 - b + b \left(\frac{G_i}{G} \right) \right]^{-1} + \frac{f_p}{1-b} = 1 \tag{6.10}$$

where f_p is the fraction of porosity; ν is the overall Poisson's ratio of nanocrystalline ceramics and can be given by

$$a = \frac{1}{3} \left(\frac{1+\nu}{1-\nu} \right), b = \frac{2}{15} \left(\frac{4-5\nu}{1-\nu} \right)$$

$$\nu = \frac{3K - 2G}{6K + 2G}$$

The bulk modulus, (K_i), and shear modulus, (G_i), of ith nanocrystalline phase can be calculated by its Young's modulus E_i and standard relationship $K_i = E_i/3(1 - 2\nu)$ and $G_i = E_i/2(1 + \nu)$, where the E_i can be calculated by eqn. 6.4 or 6.7.

However, experimental evaluation is mandate to support the theoretical concept, and this characterization protocol can be broadly classified into static and dynamic methods. The dynamic method gives adiabatic values, while the static method gives isothermal values. The dynamic methods result in measuring the properties with great accuracy and broadly classified as resonance and ultrasonic wave propagation methods [9].

1. A dynamic elastic property analyzer measures Young's modulus of a test specimen by the impulse excitation of vibration method. For this purpose, the rectangular samples are impacted by a steel rod near the high-frequency sensor of the instrument. The values of Young's modulus is calculated by the following formula:

$$E = 0.9465 \left(\frac{mf_f^2}{b} \right) \left(\frac{L^3}{t^3} \right) T_1 \tag{6.11}$$

where E is Young's modulus, m is the mass, f_f is the natural frequency in the flexure mode, b is the width, t is the thickness, L is the length, and T_1 is a correction factor given by (t/L < 0.05):

$$T_1 = 1 + 6.858 \left(\frac{t}{L} \right) \tag{6.12}$$

2. Another method called "ultrasonic method" measures E from the velocities of longitudinal and transverse waves through the specimen with the help of a piezoelectric transducer (lithium niobate crystals) that transmits and receives signals (waves). Velocities are calculated from the travel time of the waves across the thickness or height of the specimens. The following relationships can be used to determine the modulus:

$$E = \frac{(1+v)(1-2v)}{1-v}(\rho C_L) \tag{6.13}$$

$$v = \frac{1/2(C_L/C_s)^2 - 1}{(C_L/C_s)^2 - 1} \tag{6.14}$$

where v is Poisson's ratio, ρ is the density of the specimen (g/c.c), C_L is the velocity of longitudinal wave (m/s), C_s is the velocity of transverse wave (m/s), and E is the elastic modulus (GPa).

3. Another approach called "flexural/bending strength measurement" provides the Young's Modulus, can be calculated using Equation 6.15:

$$E = \frac{FL^3}{4bd^3\delta} \tag{6.15}$$

where F is the applied force, L, b, and d are length, width, and height of the test specimen, respectively, δ is deflection. However, an indentation method is an effective tool to measure the elastic modulus of coating and particulate within composite matrix and discussed in Section 6.6.2.

6.2.3 Relation in Elastic Modulus and Porosity

Relationship within porosity and deformation of consolidated shape has intense relation, and obviously effective elastic modulus intuitively reduces in presence of certain degree of pore volume content, pore morphology, and pore size distribution. In the perspective of processing protocol, the starting particle morphology and fabrication techniques are the determining factors to synchronize the pore phenomena. Combination of statistical data on porous characteristics and their relationship with modulus is an important research horizon, and it can predict the failure of material under load and deformation. Herein, simple established relation between volume fraction of spherical shaped pore and modulus are encountered and discussed to understand the probable change in elastic modulus of dense to porous ceramics. However, this empirical relation may differ from experimental data because of pore morphology and their distribution, and enormous scope is extended to cellular ceramic research and developing in particular load-bearing applications. Keeping this in view, several relationships are highlighted;

1. Empirical equations for the variation of Young's modulus and the shear modulus with porosity [10]:

$$E = E_0 \exp(-bP) \tag{6.16}$$

$$G = G_0 \exp(-bP) \tag{6.17}$$

Here, "b" is a parameter to be fitted empirically, and its value is 4 for most of the oxides in porosity range of 0%–40%. Herein, uniform size and space pores between sintering spheres of uniform packing are assumed.

2. In more accurate, small values, ($P < 6\%$), of porosity, Hasselman suggested that the relation between porosity and Young modulus is linear and is given by the following equation [11]:

$$E = E_0(1 - b_1 P) \tag{6.18}$$

where b_1 is a constant, and its value is around 2–4.

3. For high volume fraction, ($P \geq 50\%$), of porosity which consist of spherical shaped pores, Rice suggested the relationship as [12]:

$$E = E_0 \left(1 - \exp^{[-b_2(1-P)]}\right) \tag{6.19}$$

where b_2 is about 0.5.

4. Nielsen derived a relation in consideration of porosity shape factor and defined as:

$$E = E_0 \frac{(1-P)^2}{1 + (\rho^{-1} - 1)P} \tag{6.20}$$

where ρ is the Nielsen shape factor. In ceramic system, the lower shape factor, ($\rho \approx 0.25$), indicates the existence of complex and connective pore geometry compared to discrete porous structure ($\rho \approx 0.5$) [13]. For the same porosity level, matrix having small and uniformly distributed pores has higher elastic modulus than those of aggregated pores (effective large critical flaw). Pore location is also important phenomenon when the pores are smaller than grain size, as triple junction pores (more connectivity within pore and grain boundaries) have lower strength compared to lenticular shaped grain boundary (intergranular) pores and intragranular pores [14]. In brief, the mechanical behavior of ceramics strongly depends on the pore content, pore architecture, and nanoscale grain size, some interesting results are highlighted in next section.

6.2.4 Effect of Grain Size and Porosity

In a recent study, the effect of grain size and porosity on the elastic modulus of TiO_2 ceramics have been evaluated and represented in Figure 6.2. Interesting to note that the Young's moduli rapidly decreases below 20 nm grain size, however, gradual increment of E value is noticed with increasing grain size up to 30 nm followed by E value is not changing with further increment of any grain size, and it becomes constant [6]. The characteristic feature of grain size with E value is identical for any percentage of porosity, and in fact, the elastic modulus decreases with increasing porosity content.

In this experiment, a drastic change in E value up to ~40% is reduced with increment of 15% porosity, this experimental observation is well matching with the E and porosity relationship as mentioned in Equation 6.16. Usually, the GI has more elastic modulus than

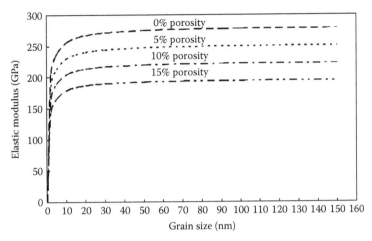

FIGURE 6.2
The calculated grain size and porosity-dependent Young's moduli of nanocrystalline TiO_2.

GB, and thus smaller grain size provides more number of grain boundaries within the same volume of materials. This characteristic feature is significant when grain is too small below 20 nm, and entire microstructure behaves as web rather than individual grains, which dramatically enhances the strain and thus reduction of elastic modulus under load and deformation.

6.3 Strength

Theoretically, strength of a material is the stress at which it fractures by bond rupture and estimated to be in the wide horizon of 10–50 GPa for dense ceramics [15]. Even though these theoretical strength values are very high, the presence of inherent flaws in brittle solids dramatically reduce the stress at which they fail. Different mode of strength values are expressed in consideration of mode of loading, most common are compressive, tensile, shear, and bending, and their brief is discussed to analyze the plausible deformation and fracture at peak load.

6.3.1 Compressive Strength

The ceramics are preferentially high in compression and have exceptional low strength in tension and result in different crack propagation characteristics when loading mode is changing, as described in Figure 6.3a and b. Under tensile mode, the brittle fracture propagate unstably beyond the critical value of crack tip stress intensity, (K_{IC}), however, the cracks in compression tend to propagate stably and twist their original orientation attribute to high degree of intergranular friction and thus slow rate of deflection along the compression axis. During the crack propagation under compression, slow extensive crack links up with multiple cracks and develop a crack enriched zone. Thus, in consideration

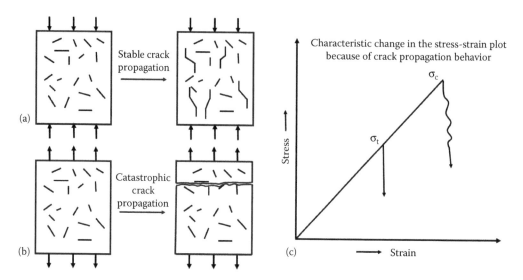

FIGURE 6.3
Schematic representation of existing pre-existing flaws (marked by black line) and their probable propagation behavior (a) cracks compiled, lining up, and deflects under compression, (b) under tension failure attributed to unstable propagation, and (c) competitive stress-strain plot for compressive and tensile mode, tensile strength, (σ_t), is relatively less than compressive strength, (σ_c), for ceramics.

of crack propagation, introduce average crack length, (c_{av}), during compressive mode of failure, (σ_c), and can be represented as Equation 6.21, where Z is constant $\approx$ 15 [16]:

$$\sigma_c \approx Z \frac{K_{IC}}{\sqrt{\pi c_{av}}} \tag{6.21}$$

Compressive strength is usually measured using universal testing machine (UTM) by placing a cylindrical test sample between two parallel plates of the machine and force is applied.

The height-to-diameter ratio of test specimen is normally 1.0 or larger, less ratio may overestimate the value compared to actual strength, however, premature bending and deformation may reduce the resultant compressive strength when this ratio is more than 1.5. During the test, the load-displacement response is like to record by using a computer attached to the UTM. The compressive strength, (σ_c), of the specimen is then calculated by following equation:

$$\sigma_c = \frac{W}{A} \tag{6.22}$$

where W is the maximum load (fracture load) and A is the cross-sectional area of specimen. Combined representation of a typical stress-strain behavior of ceramic under compressive and tensile mode is given in Figure 6.3c. Beyond the liner elastic material up to fracture, ceramics exhibit up to certain extent of nonlinear behavior, and that characteristic is coined as "serration" before complete failure. This behavior is essentially due to some extent of bulging of the specimens along the perpendicular direction of compression and simultaneous piling up the microcracks before ultimate failure. Other way, the tensile strength, (σ_t), follows catastrophic failure and exhibit less value than compressive strength without serration, and few ceramics have near to 6–10 times high σ_c than σ_t. This serration behavior is more prominent in case of porous and polymer-ceramic composites.

6.3.2 Flexural Strength

Ceramic experience bending moment failure, and thus a brief on this property and their relationship with microstructure in preference grain size and porosity is discussed. This bending or flexural strength of ceramics is widely measured by bar geometry sample consisting of either rectangular or circular cross section. Despite the testing advantages, drawbacks including nonuniform stress distribution (top under compression, bottom under tension) and unexpected strain increment due to plastic deformation are critical, and this additional incidence under bending mode of loading mislead the data interpretation.

Usually, the flexural test can be performed in two ways namely: three-point or four-point bending test and the load applied on the specimen can also be either a concentrated load or a distributed load, and the loading surface is placed in compression, while the opposite surface is placed under tension. The flexural strength is calculated on the basis of measured fracture load and the dimension of the test sample, as shown in Figure 6.4. Up to 45° edge chamfering of rectangular bar specimen is mandatory prior to measure the strength in order to avoid unusual sharp stress origin near at the edge and early stage of crack propagation around the stressed zone. For three-point bending test, the relation can be established from bending moment of inertia and eventually used to determine the flexural strength through the following equation:

$$\sigma_f = \frac{3PL}{2bd^2} \tag{6.23}$$

where σ_f is the flexural strength of the material, P is the fracture load, L is the span length, b is the width of the sample, and d is the thickness of the specimen. The deflection, (δ), at the center of the rectangular bar can be calculated as:

$$\delta_{3-pt} = \frac{PL^3}{4Ebd^3} \tag{6.24}$$

In case of three-point bending, a minimum portion of the bar experience stress near to the maximum value, whereas four-point bending experiment is subjected to stress in

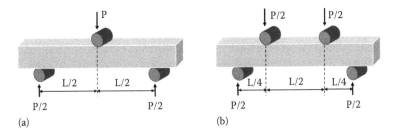

FIGURE 6.4
Schematic illustration of (a) three-point and (b) four-point flexural strength measurement of ceramic bar of rectangular cross section.

wide horizon compared to three-point bending. For four-point bending test, the flexural strength can be obtained from the following equation:

$$\sigma_f = \frac{3PL}{4bd^2} \tag{6.25}$$

The obtained deflection is, where $D = L/4$:

$$\delta_{4-pt} = \frac{PD}{4Ebd^3}(3L^2 - 4D^2) \tag{6.26}$$

$$\delta_{4-pt} = 0.69\,\delta_{3-pt} \tag{6.27}$$

From the previous equations, it is clear that the strength and resultant deflection values obtained in four-point bending test are less compared to three-point bending test. Maximum tensile stress is distributed over a large area during four-point compared to three-point and results lower conservative strength values. In addition, the stress value linearly decrease along the thickness (z direction) of the sample. Thus, four-point bending strength is often considered to estimate the Weibull modulus in the view of designing and failure analysis of ceramics.

6.3.3 Effect of Grain Size on Strength

It is generally known among the researchers that finer grain size ceramics have higher fracture strengths than coarse-grained ceramics. This basic concept can be explained through a relation between stress/strength and grain size, as described by Hall-Petch equation [17]:

$$\sigma_f = \sigma_y + k_y d^{-1/2} \tag{6.28}$$

where σ_y is the yield stress for the easiest slip system of a single crystal and k_y is a constant, and both are independent of grain size. The second part is analogous to Griffith equation: $\sigma_f = K_{IC}(Yc)^{-1/2}$, the critical crack length, (c), is proportionate to grain size, (d), and thus results σ_f versus $d^{-1/2}$. In consideration of Equation 6.28, if the grain size, (d), becomes zero, the resultant stress is equivalent to yield stress σ_y. Rice proposed that higher stress is required for the crack initiation influenced by micro plasticity in fine grain branch compared to fracture initiated by pre-existing flaws and so fracture in the fine-grain branch is preferentially facilitated by pre-existing flaws [18]. In ceramics, the failure through pre-existing flaws is predominate over microplasticity, and that flaws can be generated during fabrication or machining of the specimen, and its size may vary in the range of 20–50 μm, resembles with one grain diameter in size.

Inferior surface finish may be responsible to develop larger flaw and may further reduce the strength of the same specimen. Detailed analysis on the bending strength versus $d^{-1/2}$ for the different processed alumina is shown in the Figure 6.5 [19]. It is not necessary that the continuous strength increment with finer grain size, below a particular grain size, the crack propagation follows via pre-existing flaws developed during processing of the specimen, and thus strength becomes insensitive with very fine grain size as well. In this continuation, it is worth to mention that when grain size is very low,

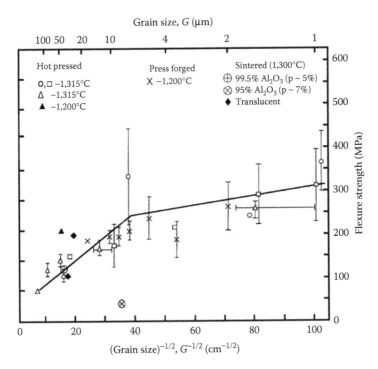

FIGURE 6.5
Relationship between σ_f and $d^{-1/2}$ in sintered alumina fabricated through different techniques [19].

below <30 nm, volume of grain boundary is more compared to grain, and thus the web of grain consist of excessive weak grain boundaries deform at relatively less stress, results in decrement of strength.

6.3.4 Effect of Porosity on Strength

Despite pre-existing flaws, another inclusion known as "porosity" has role to reduce the strength of the ceramics, and can be expressed as similar as elastic modulus, so strength variation follows according to the Equation 6.29:

$$\sigma = \sigma_0 \exp(-bP) \tag{6.29}$$

where σ_0 is strength at zero porosity, "b" is the slope from semi logarithmic plot of strength versus volume percentage porosity P, and it is constant for a given type of porosity and strength

From the earlier equation, it is clear that the strength decreases exponentially with porosity, as the pore volume content and pore size have impact on the resultant strength of the specimens. Different ceramic fabrication protocol ensures the different degree of porosity, and their influence on strength, for example, the conventional and extruded alumina were prepared, and their strength-porosity phenomena is illustrated in the Figure 6.6 [20]. A competitive three-point bending strength and percentage porosity for extrusion and conventional uniaxial pressing followed by sintering are represented in Figure 6.6a. At low porosity regime below 7.5% porosity, the conventional sample has slightly higher value, but the lowering of the strength with increasing porosity is

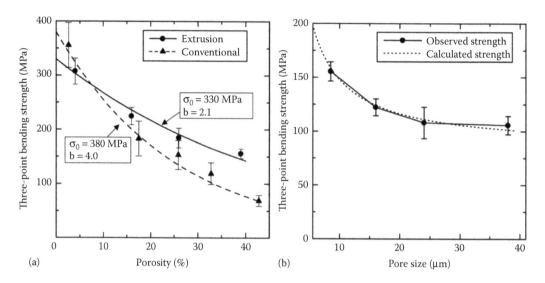

FIGURE 6.6
Variation of three-point bending strength of alumina with respect to change in (a) porosity content and processing conditions and (b) pore size with having constant porosity (~40%) fabricated by extrusion method.

relatively less in the extruded specimen, this is attributed to the more ordered microstructure having unidirectionally aligned cylindrical pores.

Thus, along with the porosity, the pore geometry also plays a major role in determining the strength of a specimen. These pores act as stress concentrators around the pore periphery, and this stress σ_A can be described as:

$$\sigma_A = \sigma\left(1 + \frac{2d_h}{d_w}\right) = K_s\sigma \tag{6.30}$$

where d_h and d_w are the height and width of the crack and, K_s is stress concentration factor and depends only on the geometry of the pore and not its size. Generally, $d_h \le d_w$, however, $d_h = d_w$, and K_s is 3 for spherical pore. Apart from the pore content and geometry, the pore size has also influence on the bending strength, for example 40% porous alumina prepared through extrusion dramatically reduced the strength up to 50% when pore size enhanced from 10 to 20 μm and maintained continuous plateau with respect to pore size, as illustrated in Figure 6.6b. Author reported the good agreement within the experimental data and theoretical calculation [20].

6.4 Weibull Modulus

One should not be misled this statistical parameter with other elastic, shear, and bulk moduli. It is predominately used to design the ceramics comprised of high degree of reliability that in actual depends on the less scattered fracture strength data irrespective of size of the component. In convention, the well-established statistical distributions are used to describe or treat experimental strength data, these are Gaussian or normal distribution, logarithmic distribution, and Weibull distribution [21,22]. Among these, the Weibull distribution is widely used because it is mathematically much more convenient to use (e.g., when incorporating the effect of testing specimen size) and has a more appropriate form for small strengths unlike Gaussian distribution. This may be due to varying sample size resulting in the change in probability of presence of some inclusions including pores, flaw, nonuniform grain size distribution, or crack in the test specimen and measuring technique. This results scatter in strength, although it can be best explained by the statistical strength approach, for example, Weibull modulus. In general, Weibull modulus is represented as "m" and high value represents more reliability during to design load bearing ceramics. In contrary, "m" should not be confused with strength, it is possible to have a weak solid with high "m" and vice versa, in this aspect, ceramic engineer has to select the mechanical properties judiciously for ultimate use. Thus, microstructure synchronization in consideration of grain size and porosity are important aspects to obtain high "m," as well as high reliable ceramics. Weibull distribution is used to generally describe the strength data are on the basis of either two parameter or three parameters (e.g., particle size distribution). In order to estimate the "m," the two parameter is encountered and discussed. The two-parameter distribution equation is given by:

$$P_f(\sigma) = 1 - \exp\left[-\left(\frac{\sigma}{\sigma_0}\right)^m\right]$$

(6.31)

where $P_f(\sigma)$ is the failure probability, σ is failure stress, σ_0 is stress at which probability of failure is 0.63 or survival probability is 1/e or 0.37, and m = Weibull modulus or Weibull shape parameter and is a measure of the distribution width, so a high value of m implies narrow distribution, as highlighted in Figure 6.7a.

The survival probability, S, is given by:

$$S = 1 - P_f(\sigma) = \exp\left[-\left(\frac{\sigma}{\sigma_0}\right)^m\right]$$

(6.32)

$$\ln \ln \frac{1}{S} = m \ln \frac{\sigma}{\sigma_0} = m \ln \sigma - m \ln \sigma_0$$

(6.33)

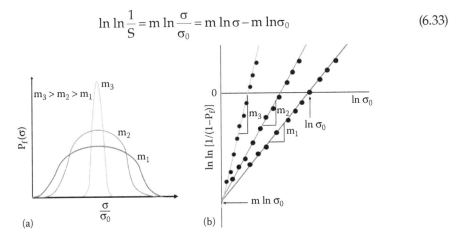

(a)   (b)

FIGURE 6.7
(a) The effect of m on the shape of the Weibull distribution. As "m" increases, the distribution narrows and (b) schematic of Weibull plot showing relationship between linear regression and Weibull parameters.

With the help of Equation 6.33 and least-square data fit method, the slope "m" can be measured, as described in Figure 6.7b. Eventually, knowing the σ_0 and "m," the predictive survival probability at any stress and reliable design stress for ceramics is feasible.

The estimation of Weibull parameters m and σ_0 are required to describe the strength value of "n" number of specimens. The parameters can be determined by two methods: least-square fitting of linearized form of the distribution and the method of maximum likelihood. Keeping the complexity of latter in view, simple least-square method can be considered for ceramics. In least-square fitting method, one can estimate the statistical parameters as following:

1. Measure the flexural strength of ceramic bars (preferentially four point), in preference 25 number of bar specimens, although more number of test specimens provide more reliable data.

2. Rank the specimens in ascending order of obtained strength as σ_n from experimental data, like 1,2,3,...n, n+1,...N, where N is the total number of specimens, herein, N = 25.

3. The survival probability of first specimen is $1 - 1/(N + 1)$...., for n^{th} specimen, it is $1 - n/(N + 1)$, however, estimate the S_n by more accurate statistical expression $S_n = 1 - (n - 0.3)/(N + 0.4)$. Plot $\ln \ln (1/S)$ vs. $\ln \sigma$. The slope obtained from least-square fit to the resulting line is Weibull modulus, 'm'.

4. A more straight forward method for determining σ_0 is to obtain the intercept on the $\ln \sigma$ axis when $\ln \ln (1/S) = 0$, i.e., $\ln \sigma_0$. Hence, S = 0.632 so that σ_0 is the 63rd percentile failure stress, so almost two-thirds of the specimens are weaker than σ_0.

6.5 Fracture Toughness

Fracture toughness indicates the amount of stress required to propagate a pre-exisiting flaw that may exist as cracks, voids or combination of both during processing, fabrication, and machining of sample preparation. In consideration of Griffith energy criteria for the fracture of flaw, the critical condition occurs when energy release rate is greater than consumed energy rate and approached mathematically as [23]:

$$\sigma_f \sqrt{\pi c_{critical}} \geq \sqrt{2\gamma E} \tag{6.34}$$

In order to extend the crack, the system has to consume energy followed by exposing new surfaces, and thus it is related with "γ" intrinsic surface energy and elastic modulus (E). In the same time, a smaller crack length below critical crack length ($c_{critical}$) only consume energy rather than releasing energy and results in stable crack. In contrast, if flaws are greater than the $c_{critical}$, the crack becomes unstable as it releases more energy compared to consumption. Thus, this equation comprises the combination of critical applied fracture stress (σ_f) and critical flaw size to cause failure, and $\sigma_f \sqrt{\pi c}$ is abbreviated to a single symbol K_I with unit MPa.m$^{1/2}$, and referred to as stress intensity factor. However, this factor is comparable with the critical stress intensity factor (K_{IC}) or commonly fracture toughness in mode-I, so the following condition essentially needs to satisfy during fracture:

$$K_I \geq K_{IC} \tag{6.35}$$

In convention, three mode of fracture (K_I - crack opening, K_{II} - crack in-plane sliding, K_{III} - crack anti-plane tearing) appears, however, type mode - I crack opening is predominant in ceramics and referred as K_{IC}. This relation is true only new surfaces are initiated during fracture and no other energy dissipating mechanisms associated like plastic deformation. The relation within K_I and K_{IC} is analogous to relation within "stress" and "tensile stress." In brief, the fracture assisted failure occur if the product of applied stress and the square root of the flaw dimension are equal or greater than the fracture toughness. Some experimental techniques of fracture toughness are discussed and correlated with respect to the grain size.

6.5.1 Competitive Single Edge "V" Notch Beam and Indentation Toughness

Several methods are well established by material scientist in order to determine the fracture toughness of sintered ceramics. In preference, flexural tests consist of predefined crack specimens of rectangular bar are commonly used to determine the fracture toughness. Single edge notch beam (SENB), single edge precrack beam (SEPB), single edge "V" notch beam (SEVNB), and chevron notch test are most common and called "long crack" methods. Sometimes it is difficult to make defect free parallel bar specimens because of unavailability of adequate processing and machining facilities, however, scientists adopted another simple indentation method to determine the fracture toughness. In this method, a single flat surface only needs to generate multiple data point by several indentations on one specimen without complete fracture of the specimen. However, this indentation method perhaps not as accurate as SEVNB method, due to ~25% uncertainty of the magnitude of dimensionless empirical constant, as this constant 0.016 may vary up to (±0.004) [24]. Although indentation technique is a useful research tool, but one should not consider this value for the design and material specification purpose.

In order to overcome the data reproducibility issue, SEVNB method is one of the most effective techniques that needs a small specialized apparatus to introduce predefined flaw shape and geometry. The SENB and SEVNB are almost similar to measure the K_{IC}, the only difference being the shape of the notch, which is of "V" shape in the latter. To estimate the K_{IC}, follow the experimental sequence:

1. Mount the desired number of specimens parallel and side by side (preferentially all in once) by adhesive on a plate, it will help to make uniform notch dimension.

2. With a very thin diamond wheel, cut a straight U-notch to a depth of ~0.5 mm.

3. Further introduce a deeper notch with a razor blade impregnated with diamond paste on rectangular samples into the center of the first U-shaped notch using as a guide.

The notch radius, (S/2), preferentially less than 10 μm. The samples are then fractured using a four-point bending setup on an UTM. A schematic representation of "V" notch with critical and essential dimensions and four-point bending setup is shown in Figure 6.8a.

After introducing definite length of predetermined "V" notch, specimen has to place on to four-point fixture, as illustrated Figure 6.8b, and estimate the fracture load of notched specimen. The fracture toughness is calculated from fracture load by the following equation [25]:

$$K_{IC} = \frac{P_f \left(L_o - L_i \right)}{bd^{3/2}} \frac{3\beta^{1/2}}{2 \left(1 - \beta \right)^{3/2}} f(\beta) \tag{6.36}$$

where, P_f is fracture load, L_i is inner span (L/2), L_o is outer span (L), b is width of specimen and d is thickness of specimen, x is the precrack size, $\beta = x/d$ and:

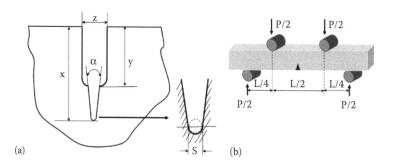

FIGURE 6.8
(a) Schematic showing the V-notch geometry on the specimen, x = 0.8–1.2 mm; y = 0.5–0.6 mm; z ≥ width of blade; x − y ≥ z; α ~ 30°, or as possible; S = V-notch width and (b) SEVNB fracture toughness measurement module.

$$f(\beta) = 1.9887 - 1.326\beta - \frac{\beta(1-\beta)\left(3.49 - 0.68\beta + 1.35\beta^2\right)}{(1+\beta)^2} \tag{6.37}$$

The sample preparation, data acquisition, and their analysis have crucial role to estimate the authentic fracture toughness and designing of ceramics. One can try to estimate the competitive fracture toughness of identical material through Vickers indentation protocol and analyze the reasoning for the deviation within SEVNB and Vickers indentation data.

Fracture toughness measurement by indentation technique is a nontraditional method, it does not break any precracked specimen. Thus, the indentation method (micro hardness method) produces a large variation in the observed values as it depends on location of indents, although it is frequently used to calculate fracture toughness. This technique uses a Vickers diamond indenter to make a hardness impression on a polished specimen surface at particular load and dwell time, say 10 seconds. The indenter creates a plastically deformed region underneath the indenter, as well as cracks that emanate radially outward and downward from the indentation. The cracks are assumed to be semicircular median cracks in most of the analyses. This type of loading produces a stress intensity on the crack that is proportional to $c^{-3/2}$. On the polished surface one sees four cracks that radiate outward from the indentation corners. The final length of the crack, conventionally measured after removal of the indentation load, reported as K_{IC} through the equation 6.38, and represented in Figure 6.9 [26].

$$K_{IC} = \frac{0.016\left(\dfrac{E}{H}\right)^{1/2} P}{c^{3/2}} \tag{6.38}$$

Fracture toughness is computed on the basis of the crack lengths (c), indentation load (P), hardness (H), elastic modulus (E), indentation diagonal size, and an empirical fitting constant ($\eta = 0.016$). It is quite common that the brittle material has different degrees of elastic-plastic damage zone and thus does not deform and fracture underneath an indentation in similar fashion as is assumed in order to derive the Equation 6.38. For example, a covalently bonded, hard ceramic (e.g., silicon nitride) deforms and fractures very differently than an ionic cubic ceramic (e.g., magnesium oxide) or a soda lime silica glass. Hence, consideration of K_{IC} value obtained through indentation method is not apposite for designing of ceramics.

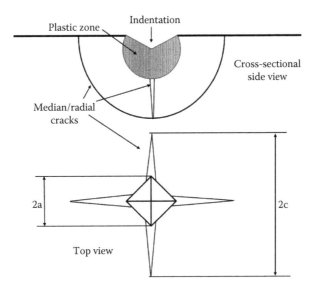

FIGURE 6.9
Formation of cracks around during Vickers indentation, upper part indicates the cross-sectional side view after indentation, and top view after crack initiation, 2a length of indenter and 2c is the cumulative crack length.

6.5.2 Effect of Grain Size on Toughness

A definite grain size, (d), window neither very fine nor very large dominate high fracture toughness (K_{IC}), and it is more noticeable in case of noncubic crystal system. Thus, it is difficult to draw a direct relationship within K_{IC} and "d." Furthermore, the change in grain morphology from equiaxed to high aspect ratio prefers R-curve phenomena, results in high K_{IC} [27]. Prior to discussing the influence of grain size on fracture toughness, it is important to mention why polycrystalline ceramics have higher fracture toughness than single crystal. Grain boundaries matter! (Figure 6.10).

$$K_{tip} = \left(\cos^3 \frac{\theta}{2} \right) K_{app} \tag{6.39}$$

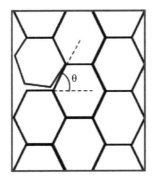

FIGURE 6.10
Schematic of crack deflection in polycrystalline ceramics, reduced the crack tip stress intensity, (K_{tip}), compared to applied stress intensity, (K_{app}), with an angle of θ.

A simple schematic and relation highlights the reasoning of low fracture toughness of alumina single crystal (2.2 MPa.m$^{1/2}$) compared to polycrystalline alumina (4 MPa.m$^{1/2}$), as described in Figure 6.10 and Equation 6.39. In a polycrystalline material, the crack is like to deflect through weak grain boundaries, and a certain deflection angle, (θ), reduce the crack tip stress intensity (K_{tip}), because the applied stress is no more normal to the crack plane. Under certain condition, if $\theta = 45°$, the tip stress intensity factor is being reduced and eventually the fracture toughness may enhance up to 1.25 times. The effect of grain size of the ceramic materials on the fracture toughness can be studied by dividing the materials into two classes [18,28]:

1. The materials having cubic crystal symmetry: for these materials, the fracture toughness preferentially does not seem to vary much with the grain size and remains almost constant, this attributed to the uniform thermal expansion and elastic behavior of the system. Experimental data for Y_2O_3 system are represented in the Figure 6.11.

2. The materials having noncubic crystal symmetry: for these materials, the fracture toughness increases with increasing grain size especially in the low to mid-size range, reaches a maximum and then decreases. The experimental data and their changing behavior with respect to grain size are shown in the Figure 6.11 and phenomenology divided into three regions; fine grain size, fine-to-mid grain size, and large grain size, these increments and further decrement can be explained through three different regions [22,29]:

Region-I: To the right of curve, the maximum grain size at which maximum toughness occurs approximately corresponds to the critical grain size, (d_c), beyond which micro cracking is spontaneous. Considering both thermal expansion anisotropy and elastic anisotropy around triple grain boundaries, the strain energy release rate of cracks of a particular size increases as the grain size increases, results in spontaneous microcracking. As main crack propagates,

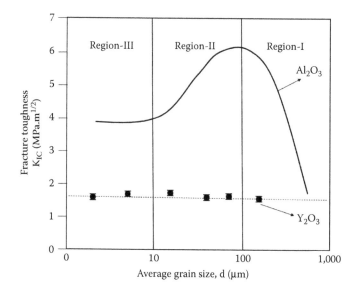

FIGURE 6.11

Variation of fracture toughness, (K_{IC}), with the average grain size for polycrystalline cubic (Y_2O_3) and noncubic (Al_2O_3). (From Rice, R.W., *Mechanical Properties of Ceramics and Composites: Grain and Particle Effects*, Marcel Dekker, New York, 2000; Quinn, G.D., *Ceram. Eng. Sci. Proc.*, 27, 45–62, 2006 [18,24].)

it merely connects the microcracks, compounds the cracking effect, and toughness decreases with larger grain size.

Region-II: Crack deflection takes place in the wake of the crack, and thus toughness increases with increased grain size since larger grains offer a stronger closure force on the crack surfaces and can withstand a higher crack-opening displacement before they fracture. Apart from this grain boundary assisted bridging mechanism, the frictional clamping force on bridging grains by compressive stresses from thermal expansion anisotropy is also responsible to increase the fracture toughness. In this region, grain size range where toughness is increasing, and exhibit crack growth resistance during slow crack growth prior to unstable fracture in the specimen.

Region-III: In presence of fine, say nanostructured, grains, the large volume of weak grain boundary is obvious when grain size is very smaller, and thus probability of bridging effect becomes less and does not have influence on K_{tip} in toughening mechanism, rather deflection exist compared to single crystal.

In this contrary, smaller grain size (above 30 nm) has significant influence on fracture strength and hardness, however, this regime grain size may have low fracture toughness. In this backdrop, one can enhance toughness through introduction of elongated and high aspect ratio grains or martensitic transformed phase, (ZrO_2), as a secondary phase, and thus fabrication of nanocomposite provide high strength, high hardness, and high toughness, together.

Furthermore, the grain boundary triple point is the origin to initiate microcrack, and this occurrence is more severe in presence of intergranular pores generated during consolidation and sintering. Hence, an identical equation that is used to describe the variation of strength and elastic modulus with porosity can also be considered the variation of toughness, (K_{IC}), with porosity, (P), as:

$$K_{IC}(P) = K_0 \exp(-bP) \tag{3.40}$$

where K_0 is the toughness at zero porosity.

6.6 Hardness

Resistance to crack propagation under stress is "fracture toughness," but resistance to plastic deformation or penetration by sharp object is "hardness." While considering the penetration, the sharp indenter assisted localized stress is sufficiently high to initiate local plastic deformation from dislocations, twinning, grain boundary shear, and eventually cracking of the surface. However, this feature varies material to material. One should not be misled and select the material for structural application without enough information about the quantitative magnitude of hardness and toughness together. For example, a ductile non oxide ceramic, titanium silicon carbide (Ti_3SiC_2) is predominantly deformed under applied load without formation of any crack around the indenter and exhibits very low hardness ($H_v = 5$ GPa) and high fracture toughness ($K_{IC} = 9$ MPa.m$^{1/2}$) [30]. Different combination of such properties are preferentially used as structural ceramics, for example, SiC

$(H_v = 27.47$ GPa, $K_{IC} = 3.5$ MPa.m$^{1/2}$), Si_3N_4 ($H_v = 15.38$ GPa, $K_{IC} = 6.89$ MPa.m$^{1/2}$), Al_2O_3 ($H_v = 21.9$ GPa, $K_{IC} = 4.2$ MPa.m$^{1/2}$), ZrO_2 ($H_v = 13.7$ GPa, $K_{IC} = 6.4$ MPa.m$^{1/2}$), $MgAl_2O_4$ ($H_v = 18$ GPa, $K_{IC} = 2.6$ MPa.m$^{1/2}$), etc. [31–34]. In this context, some classic discussion is emphasized to understand the basic philosophy of hardness assessment for bulk, grains, coating, and correlate with grain size and load. Eventually, this brief information may compile to design high wear resistance ceramics for load bearing applications.

6.6.1 Vickers Indentation

This property is mostly measured by indentation on a polished flat surface by an indenter of different shapes (cone, spherical, diamond). Vickers indentation technique (diamond indenter) is mostly used among many other techniques to measure the hardness of a bulk sample because of its simplicity and ease of sample preparation, as described in Section 6.5.1 in order to measure the fracture toughness by indentation method. The other well-known available methods are Brinell hardness, Knoop hardness, Rockwell hardness, etc. The general formula for Vickers hardness measurement is:

$$H_v = 1.854 \frac{P}{(2a)^2} \tag{6.41}$$

where H_v is the Vickers hardness, 1.854 is the geometric factor, (k), obtained from indenter geometry, P is the applied load in Newton, and 2a is the average length of two diagonals (usually 50–200 µm). For Knoop hardness, the geometric factor is 14.229. The primary aspects that are needed to consider to obtain reliable and true hardness values for ceramics are:

1. There should not be any flaws as crack or damage around the indent edge or corner of the specimen. During estimation of multiple data point, the indentation effect should not merge with another indented zone.

2. Care should be taken by applying different indent loads to account for any "indentation size effect" and to obtain true values.

3. Measuring of the indent diagonals should be precise, as it would create excessive difference in the measured hardness values.

In Vickers indentation method, there is possibility to develop two types of crack: Palmqvist cracks with half-ellipse at low loads and median cracks with half-penny at high loads. Most materials exhibit both crack systems based on the applied load, however, very tough material is supposed to follow Palmqvist crack up to an optimum load regime.

After performing the indentation test, the existence of Palmqvist or median crack characteristics can be differentiated after careful polishing up to certain microns, this results in surface crack detachment of Palmqvist from the inverted pyramid intend, while the median crack remains attached, as illustrated in Figure 6.12. In other way, the ratio of c/a >2.5 indicates the material follows median crack.

6.6.2 Hardness by Penetration Depth

Information on the continuous change in indentation depth, (h), with applied load, (P), behavior provides the ability to measure both the plastic and elastic deformation of

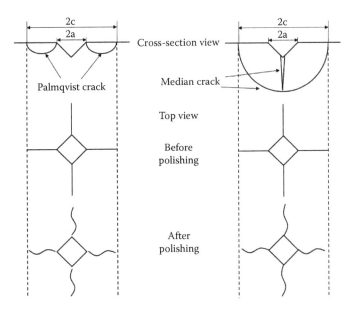

FIGURE 6.12
Different types of cracks observed during indentation test, Palmqvist crack and median crack are the result after indentation, can be distinguished after polishing.

the material. A typical schematic representation of the resultant depth variation and concurrence force displacement hysteresis curve is plotted in Figure 6.13. The characteristic slope (s) during unloading provides the stiffness of material and contact depth h_c can be calculated. The value of resultant contact depth, (h_c), is dependent on the geometry of the indenter, material response and variation of "sinking-in" and "piling-up" that affect equilibrium depth profile and contact area (Figure 6.13a). This precise

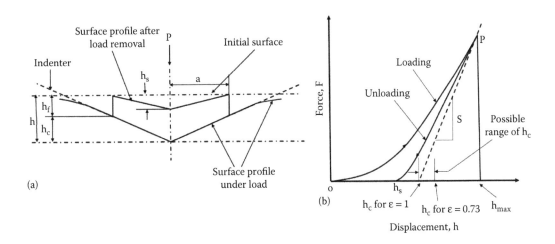

FIGURE 6.13
(a) Schematic of indentation showing the displacements observed during an indentation experiment and (b) typical force-displacement curve to measure the parameters. Parameters are correlated during indentation and corresponding force-displacement plot.

indentation protocol is a very useful tool to measure the hardness and elastic modulus of individual grains, coating, and bulk as well.

Oliver-Pharr proposed a simple model to estimate the hardness through indentation process, and expressed as [35]:

$$H = \frac{P_{max}}{A_r} \tag{6.42}$$

where P_{max} is the maximum applied load and A_r is the real contact area between the indenter and the material, and the simplified equation becomes, $A_r = 3\sqrt{3}h_c^2\tan^2 65.3 = 24.5h_c^2$, so Equation 6.42 becomes, $H = 0.0408P_{max}/h_c^2$. Further, the h_c can be calculated from the load-displacement plot, as represented in Figure 6.13b, the relevant equation is:

$$h_c = h_{max} - \varepsilon\left(\frac{P_{max}}{S}\right) \tag{6.43}$$

where $\varepsilon \approx 0.73$ to 1, and S for a Vickers indenter and S is contact stiffness and can be calculated from the slope (S = dp/dh) of the first third of linear response, recorded during the unloading cycle. The effective elastic modulus, (E*), can be estimated by:

$$E^* = \frac{dp}{dh}\frac{1}{2h_c}\frac{1}{\beta}\sqrt{\frac{\pi}{24.5}} \tag{6.44}$$

where $\beta = 1.034$ (Berckovich indenter) and effective modulus is comprising through both modulus and Poisson ratio of indenter (E_i, v_i) and the sample (E_s, v_s) as given in the following:

$$\frac{1}{E^*} = \frac{\left(1-v_i^2\right)}{E_i} + \frac{\left(1-v_s^2\right)}{E_s} \tag{6.45}$$

Knowing the elastic modulus and Poisson ratio of indenter material, one can estimate the elastic modulus, (E_s), of sample from Equations 6.43 through 6.45. Apart from the estimation of hardness and elastic modulus of nanostructure ceramics, the load-displacement plot implies the elastic and plastic characteristic feature of materials. The area OPh_s represents the regime in between loading, (OP), and unloading, (Ph_s), and comprises the plastic deformation work, (W_p), during indentation. As the load is released, the compressed elastic zone is getting spring back and results the reverse deformation energy released (W_e), and this zone represents by the area of h_sPh_{max}. The material is elastic in nature if W_e is much greater than W_p, and reverse is valid for more plastic material. The work of indentation is high for softer (low hardness) ceramics and can be represented as: work of indentation = OPh_s/h_sPh_{max}; this provides valuable information for the development of new generation of materials including matrix and particulate phase.

6.6.3 Effect of Grain Size

In accordance with Tabor relationship, the measured hardness is related with yield stress and can be written as, $H_v \approx 3\sigma_y$. This implies hardness has simple relationship with strength and it is expected that an identical grain size influence on the hardness, as discussed in

Section 6.3.3. Thus, like strength variation with grain size, hardness also tends to follow the analogous to Hall-Petch equation when it comes to the variation w.r.t grain size, given as follows:

$$H_V = H_0 + kd^{-1/2} \tag{6.46}$$

In order to establish the fact, several researchers considered different materials and estimated the hardness in variation of grain size of consolidated compact. A representative hardness variation with respect to grain size for sintered yttria-stabilized tetragonal zirconia polycrystals (TZ-3YB) and zirconia nanopowder (B261) has been studied and obtained the similar protocol [36]. The hardness of TZ-3YB ceramics decreasing with increasing grain size from H_V = 12.6 GPa at grain size of 0.19 μm to H_V = 10.9 GPa at grain size of 1.79 μm, showing that hardness is function of the inverse square root of grain size. By linear fitting of TZ-3YB data follows Equation 6.46, whereas a deviation is noticed in case of B261, but the linear fit seems to be reasonable.

6.6.4 Effect of Load

Bulk ceramic hardness is conventionally measured either by Vickers or Knoop indentation and critically influenced by the load. Experiment should conduct in an optimum load without any over or underestimate the hardness data prior to develop new class of material. As we know, the hardness is dependent on the volume of material plastically deformed, and thus load dependence is designated as an indentation size effect. The indentation size dependence on the load is often described by Meyer's law and predicts a continuously decreasing load with no plateau value of hardness [37]:

$$P = k(2a)^{\alpha} \tag{6.47}$$

where, k and α are constants and "a" is one-half the diagonal length. If α = 2, hardness is constant with load. The value of α is almost always less than 2, which implies a continuous decreasing hardness with increasing applied load, but α > 2 depicts a reverse behavior (e.g., β-SiAlON). However, experimentally, hardness values often reach a plateau at some critical value of load, P_c, where cracking around the indentation takes place. In this context, a modified equation is proposed with consideration of indentation size effect:

$$P = A_1(2a) + A_2(2a)^2 \tag{6.48}$$

where, constant A_1 is related to surface work that is associated with both friction energy and elastic resistance energy and A_2 for plastic work in creating the indentation volume, respectively. Multiplying the equation 6.48 by $k/(2a)^2$, we can obtain equation 6.49:

$$\frac{kP}{(2a)^2} = \frac{kA_1}{2a} + kA_2 = H \tag{6.49}$$

where k is geometric constant. The first term $kA_1/(2a)$ is proportional to the suface or contact area of the indentation, and the second term kA_2 is related to the volume of the deformed indentation. Thus, hardness consists of both surface and volume energy contributions. Again from the Equation 6.49, the first term controls the early part of the size effect curve and second part is dominative at large loads, where hardness is independent of load and considered as true hardness.

Analogous to this concept, a characteristic Vickers hardness behavior of polycrystalline Al_2O_3 predicts in the existence of two halves, a continuous hardness decrement from 40 GPa at 0.5 N up to 14.1 GPa at 20 N is estimated, that is related to the surface term, eventually follows a continuous plateau associated with volume deformation irrespective of load up to 140 N. The diagonal to depth ratio for Vickers indenter is 7:1, that is much lower than 30:1 for Knoop hardness and results in 15% higher Vickers hardness compared to Knoop hardness [38]. In consideration of the earlier data, it imparts that the optimum load 20 N (2 kgf) is required to produce a representative hardness data to design structural component made of bulk alumina ceramics.

6.7 Fatigue of Ceramics

Generally, fatigue failure of a material comes into limelight when a specimen is subjected to experience cyclic loading. This is very common type of failure in metals, as it results early stage of plastic deformation and rapid crack propagation even at very low stress values compared to the threshold strength limit. Recently, with the development of high fracture toughness ceramic material, this kind of failure is also necessary to encounter in preference when it exhibit R-curve behavior [39]. Usually, the fatigue test is performed on a specimen by subjecting it to an alternating stress of a given amplitude and frequency. The cyclic stress amplitude is defined as:

$$\sigma_{amp} = \frac{\sigma_{max} - \sigma_{min}}{2} \qquad (6.50)$$

where σ_{min} and σ_{max} are minimum and maximum stress experienced by the specimen. A periodical measurement in presence of either short crack (up to 250 μm) or long crack (up to 3 mm) specimen determines the number of cycles, (N), required for the failure of specimen as a function of either maximum stress, (S), or amplitude of stress, results in S-N curve. In metals, the crack propagation results from dislocation activity at the crack tip, whereas, ceramics with existence of R-curve appear to be the most susceptible to fatigue. Many ceramics experience fatigue behavior at room temperature, and mechanism appears to involve degradation of the toughening elements of the microstructure rather than plastic deformation. However, the crack growth behavior depends on the grain size and morphology as analogous to crack growth resistance behavior. In order to understand the effect of grain size on the cyclic fatigue, a typical S-N curve for two different grain sizes, say 19 and 5 μm polycrystalline alumina, is represented in Figure 6.14 [40].

The experimental data are plotted and normalized by a straight line, follows Equation 6.51:

$$\sigma^n N = constant \qquad (6.51)$$

where σ is the stress amplitude, N is the number of cycles to failure, and n is the experimental constant. The values of n are 13.8 and 22.1 for grain sizes 19 and 5 μm, respectively, so, higher for finer grained material. Thus, grain size has competitiveness and considered as a primary factor in enhancing the fatigue strength, and fatigue limits can be assumed to

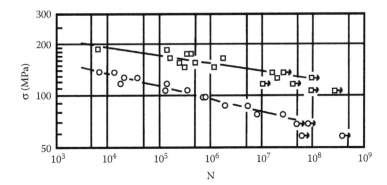

FIGURE 6.14
Influence of grain size on the S-N behavior of polycrystalline alumina, where (o) 19 μm and (□) 5 μm are average grain size of alumina. (From Ko, H.N., *J. Mater. Sci. Lett.*, 14, 56–59, 1995 [40].)

be about 70 and 120 MPa for grain sizes 19 and 5 μm, respectively. However, the determination of grain size window consisting of high aspect ratio for fatigue strength is opened up enormous scope of research. As an importance of such study, the obtained fatigue strength data can simulate with the failure analysis of knee prosthesis [41].

6.8 Superplasticity

Fine grain size (up to micrometer or less) and equiaxed ceramics experience substantial deformation at high temperature without changing any grain shape, known as superplasticity. Eventually, this is different from room temperature plastic deformation of ductile polycrystalline ceramics where grains elongated under tension before failure. In a recent article, high strain-rate superplasticity of silica incorporated zirconia, alumina, and spinel composite experience extensive tensile elongation up to 1050% at the strain rate of 0.4 s^{-1} [42]. However, low temperature assisted (<500°C) superplastic ceramic may have research interest and is awaiting. Zirconia is a common and promising ceramics that have functional, structural, and high temperature applications, and thus, herein, the variation in strain rate with stress for specimens with initial grain size of 65, 110, and 140 nm has been highlighted in Figure 6.15 [43].

The monoclinic zirconia polycrystal (MZP) dense compact was prepared from TZ-0Y Tosoh zirconia. The high purity powder was compacted in die pressing at 20 MPa followed by cold isostatic pressing at 200 MPa. The green specimen was coated with boron nitride powder and then encapsulated by borosilicate glass at 1063 K, finally, hot isostatic pressing at 1273 K with applied pressure 196 MPa for 30 minutes. The grain size was measured and desired size of specimen was machined and performed high temperature compression with a constant crosshead speed and estimated the strain rate by the experimental data and creep Equation 6.52:

$$\dot{\varepsilon} = A \frac{Gb}{RT} \left(\frac{\sigma}{G}\right)^n \left(\frac{b}{d}\right)^p D_o \exp\left(\frac{-Q}{RT}\right) \tag{6.52}$$

where, strain rate, ($\dot{\varepsilon}$), is conventionally expressed as a function of the applied stress (σ), the absolute temperature (T), grain size (d), D_o diffusion constant, Q activation energy,

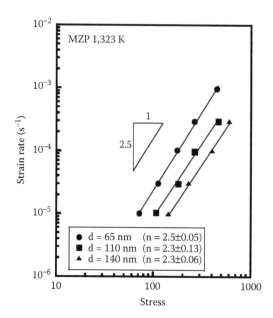

FIGURE 6.15
High temperature 1050°C strain rate variation with stress for monoclinic for three different grain size of 65, 110, and 140 nm.

R gas constant, G shear modulus, b Burgers vector, p the exponent of the inverse grain size, n stress component, and A is a dimensionless constant. The stress exponent, (n), for nanocrystalline 65 nm MZP was 2.5 over a wide stress range. However, the stress exponent n for MZP with grain sizes of 110 and 140 nm was 2.3. The strain rate was dependent on grain size, however, the stress exponents were almost the same regardless of the difference in grain size. Grain boundary sliding and migration is certainly evident to exhibit superplastic deformation of finer grain size ceramics. In the same work, author reported MZP exhibits higher strain rate at low temperature than that of nanocrystalline 3Y-TZP.

Apart from plastic deformation phenomena, most of the common ceramics lowering the elastic modulus, strength, and fracture toughness with enhancing temperature, however, some exceptional materials and relevant discussion is focused in an excellent review article by Rice [19]. Grain size increment and enhancing microcrack population results reduction of the thermal expansion of the polycrystalline ceramics [44]. However, thermal conductivity of SiC ceramics is increasing with growth of grain size attributed to the rapid scattering of the phonon and free electrons when grain boundaries are minimum compared to nanostructured ceramics [45].

6.9 Friction and Wear

Prior to brief the subject, it is remembered that tribology (tribos = rubbing) deals with friction, wear, and lubrication, and it is a system-dependent property. A stationary body never experience friction, but vibration does. Relative motion is obligatory to encompass the friction and subsequent wear, so lubrication is required to reduce the friction and wear, in contrast of life expectancy improvement of the device. Friction can be illustrated as a resistance to the motion of two bodies in contact with relative motion (sliding, rotating, to and fro fretting, etc.) between them. This resistance is opposite to the direction of motion and is called frictional force, (F). If W is the normal load of a body under friction over a surface, then frictional force is given by:

$$F = \mu W \tag{6.53}$$

where μ is the coefficient of friction. From this equation, it is clear that the more value of coefficient of friction is attributed to the more frictional force and resistance of motion that results high wear of the surface. Among several mechanisms, adhesion, deformation, ploughing, and cutting are most predominate to explain the wear mechanism that eventually originated during friction within counter bodies. A detailed explanation of these mechanisms can be found in Basu and Kalin [46]. In consideration of these basic wear behaviors, the wear mechanism can be classified as (a) adhesive wear, (b) abrasive wear, (c) tribochemical wear, (d) erosive wear, (e) fretting wear, and (f) fatigue wear. Several mechanisms can act simultaneously, but most likely one mechanism become predominant for a particular zone of interest and testing condition or application.

For components that have contact surfaces and relative motion including vibration, ball bearings, pump seals, prosthetic devices, and more should have high wear-resistance to avoid premature failure that is related with several mechanical properties, however, most predominant is hardness. The extent to which the grinding grits or asperities on one of the two rubbing surfaces penetrate into each other represents the wear rate of materials by abrasion (analogous to the definition of hardness). Thus, according to Archard, the wear rate is directly related to the hardness by the following equation [47]:

$$w' = k\frac{W}{H} \tag{6.54}$$

where w' is the wear rate (volume removed per unit sliding distance from a surface), k is the wear coefficient, W is the normal load, H is the hardness, and $1/w'$ is the wear resistance. Wear rates can be divided into (a) mild wear, (b) severe wear, and (c) very severe wear. In maintaining the mild wear regime of commercial component, the performance and life can be extended. The transition from mild to severe occurs suddenly as the load, velocity, or travel length is increased beyond critical value. Severe wear is important in applications like grinding and is characterized by cracking along the wear track, a much increased surface roughness, a higher coefficient of friction, and a large increase in wear rates. The transition to severe wear is related to the onset of cracking, which is similar to the cracking at a high load under a micro hardness indentation and leads to ceramic particles ejecting out of the surface, causing very rapid wear rates.

One can argue that the friction and wear is a tribological phenomenon rather than mechanical properties only. Partially true, but research community and industrial appliances cannot avoid the concurrence effect of the vibration of machineries, articulating surfaces, etc., and their early stage of failure when tribocouple are under load and relative motion. In this context, excellent surface finish consists of low average roughness ($R_a < 0.1$ μm) and high degree of mechanical properties are responsible to minimize the friction and thus wear. However, it is difficult to draw a direct relation with friction and wear, as it is system-dependent properties. Friction may vary with time under particular condition and exhibit different plateau. For example, polymer-metal counter bodies exhibit low coefficient of friction, (μ), but high wear rate of polymer under load and relative motion, however, a material consist of high hardness and high fracture toughness comprises moderate μ, but low wear volume, and enhance the life expectancy of components [48]. In the same time, high compressive strength is desirable to avoid premature failure under relative motion. From previous discussion, it can be noticed that the targeted mechanical properties depend on grain size, for example, hardness and strength demand nanostructured grains, whereas a particular window of grain size is advantageous for toughness. Thus, synchronization on grain size in monolithic ceramics or composites is advantageous to improve the wear resistance of components.

6.10 Concluding Remarks

Functional ceramics are in forefront to develop nanotechnology compared to structural ceramics, as it facilitates device miniaturization in electronics, optics, biotechnology, and so on. However, there is always need to a wide extent to develop structural materials for artificial implant, armor, high temperature refractories, construction concrete and glass, high wear resistance seals, cutting tools, household appliances, etc. Thus, grain size engineering is a new horizon of research to synchronize the microstructure that eventually control the properties of ceramics. Controlling over grain size and inclusions are prime interest to achieve desired mechanical properties of structural ceramics. Reduction of elastic modulus with increment of porosity volume is a common phenomenon, but same time the pore shape, size, and location can alter the E value for same pore content. However, the modulus becomes constant below a critical grain size, where entire matrix behaves as web and deforms under low stress. This is also true for strength behavior as well. Compressive strength is always higher compared to tensile strength of ceramics as former experience intergranular friction and thus slow rate of flaw propagation, whereas latter facilitates rapid crack opening up and early stage of failure. Processing protocol, porosity content, and size are also strength determining factors. A definite grain size window exhibits maximum fracture toughness for noncubic ceramics, whereas cubic system does not significantly vary with respect to grain size. Hardness measurement at optimum load is essential to provide representative data, as these information are described how to resist elastic-plastic deformation and eventually fracture under load, leads to high wear resistance performance. Although smaller grain size has better fatigue resistance than larger grain size, still several research studies are required to understand the influence of grain size distribution, aspect ratio, and inclusion content for wide range of ceramics. When grain size in nanoscale, more grain boundaries are expected and result in migration and superplastic behavior at high temperature. Finally, yet importantly, only token specimen preparation and material characterizations are not enough good to design a particular ceramic, rather extensive data generation and their interpretation are in mainstream to develop reliable nanostructured ceramics for structural applications.

References

1. A. Koka, Z. Zhoub, and H. A. Sodano, Vertically aligned $BaTiO_3$ nanowire arrays for energy harvesting, *Energy Environ Sci* 7, 288–296, 2014.
2. A. Mukhapadgyay and B. Basu, Consolidation-microstructure-property relationships in bulk nanoceramics and ceramic composites: A Review. *Int Mater Rev* 52(5), 257–288, 2007.

3. E. O. Hall, The deformation and ageing of mild steel: III discussion of results. *Proc Phys Soc London* 64, 747–753, 1951.

4. N. J. Petch, The cleavage strength of polycrystals. *J Iron Steel Inst London* 173, 25–28, 1953.

5. R. S. Lakes, Extreme damping in compliant composites with a negative-stiffness phase, *Philos Mag Lett* 81(2), 95–100, 2001.

6. J. Zhou, Y. Li, R. Zhu, and Z. Zhang, The grain size and porosity dependent elastic moduli and yield strength of nanocrystalline ceramics, *Mater Sci Eng A* 445, 717–724, 2007.

7. C.-S. Man and M. Huang, A simple explicit formula for the Voigt-Reuss-Hill average of elastic polycrystals with arbitrary crystal and texture symmetries, *J Elast* 105(1–2), 29–48, 2011.

8. B. Budiansky, Thermal and thermoelastic properties of isotropic composites, *JComposMater* 4, 286–295, 1970.

9. B. Basu and K. Balani, *Advanced Structural Ceramics*, Wiley, Hoboken, NJ, 2011.

10. R. M. Spriggs, L. A. Brissette, and T. Vasilos, Effect of porosity of elastic and shear modulii of polycrystalline magnesium oxide, *J Am Ceram Soc* 45, 400, 1962.

11. D. P. H. Hasselman, On the porosity dependence of the elastic moduli of polycrystalline refractory materials, *J Am Ceram Soc* 45, 452, 1962.

12. R. W. Rice, Evaluating porosity parameters for porosity-property relations, *J Am Ceram Soc* 76, 1801–1808, 1993.

13. L. F. Neilsen, Elasticity and damping of porous materials and impregnated materials, *J Am Ceram Soc* 67(2), 93–98, 1983.

14. W. E. Lee and M. Rainforth, *Ceramic Microstructures: Property Control by Processing*, Springer, Dordrecht, the Netherlands.

15. A. Kelly and N. H. MacMillan, *String Solids*, 3rd ed., Oxford University Press, Oxford, UK, 1986.

16. M. Barsoum, *Fundamentals of Ceramics*, McGraw-Hill, New York, 1997.

17. J. R. Rice, C. C. Wu, and F. Borchelt, Hardness-grain size relations in ceramics, *J Am Ceram Soc* 77, 2539–2554, 1994.

18. R. W. Rice, *Mechanical Properties of Ceramics and Composites: Grain and Particle Effects*, Marcel Dekker, New York, 2000.

19. R. W. Rice, Effects of environment and temperature on ceramic tensile strength–grain size relations, *J Mater Sci* 32, 3071–3087, 1997.

20. T. Isobe, Y. Kameshima, A. Nakajima, K. Okada, and Y. Hotta. Gas permeability and mechanical properties of porous alumina ceramics with unidirectionally aligned pores, *J Eur Ceram Soc* 27, 53–59, 2007.

21. W. Weibull, *A Statistical Theory of the Strength of Materials*, Ingeniörsvetenskapsakademiens Handlingar, Vol. 151, pp. 1–45, Generalstabens litografiska anstalts förlag, Stockholm, Sweden, 1939.

22. J. B. Wachtman, W. R. Cannon, and M. J. Matthewson, *Mechanical Properties of Ceramics*, 2nd ed., Wiley, New York, 2009.

23. A. A. Griffith, The phenomena of rupture and flow in solids, *Philos Trans R Soc Lond A* 221, 163–198, 1921.

24. G. D. Quinn, Fracture toughness of ceramics by the Vickers indentation crack length method: A critical review, *Ceram Eng Sci Proc* 27, 45–62, 2006.

25. T. Nishida, Y. Hanaki, and G. Pezzoti, Effect of Notch-Root Radius on the fracture toughness of a fine-grained alumina, *J Am Ceram Soc* 77, 606–608, 1994.

26. G. R. Anstis, P. Chantikul, B. R. Lawn, and D. B. Marshall, A critical evaluation of indentation techniques for measuring fracture toughness: I, direct crack measurements, *J Am Ceram Soc* 64(9), 533–538, 1981.

27. J. R. Rice, A path independent integral and the approximate analysis of strain concentration by notches and cracks, ARPA SD-86, Report E39, 1967.

28. R. W. Rice, S. W. Freiman, and P. F. Becher, Grain size dependence of fracture energy in ceramics: I, experiment, *J Am Ceram Soc* 64, 345–350, 1981.

29. R. W. Rice, Mechanisms of toughening in ceramic matrix composites, *Ceram Eng Sci Proc* 2, 661–701, 1981.

30. D. Sarkar, B. Basu, M. J. Chu, and S. J. Cho, R-curve behavior of Ti_3SiC_2, *Ceram Int* 33(5), 789–793, 2007.

31. M. Khodaei, O. Yaghobizadeh, H. R. Baharvandi, and A. Dashti, Effects of different sintering methods on the properties of SiC-TiC, SiC-TiB_2 Composites, *Int J Refract Metals Hard Mater* 70, 19–31, 2018.

32. A. Nevarez-Rascon, A. Aguilar-Elguezabal, E. Orrantia, and M. H. Bocanegra-Bernal, Compressive strength, hardness and fracture toughness of Al_2O_3 whiskers reinforced ZTA and ATZ nanocomposites: Weibull analysis, *Int J Refract Metals Hard Mater* 29, 333–340, 2011.

33. S. Benaissa, M. Hamidouche, M. Kolli, G. Bonnefont, and G. Fantozzi Characterization of nanostructured $MgAl_2O_4$ ceramics fabricated by spark plasma sintering, *Ceram Int* 42, 8839–8846, 2016.

34. L. Kvetková, A. Duszová, P. Hvizdoš, J. Dusza, P. Kun, and C. Balázsi, Fracture toughness and toughening mechanisms in graphene platelet reinforced Si_3N_4 composites, *Scr Mater* 66, 793–796, 2012.

35. W. C. Oliver and G. M. Pharr, An improved technique for determining hardness and elastic modulus using load and displacement sening indentation experiments, *J Mater Res* 7(6), 1564–1583, 1992.

36. M. Trunec, Effect of grain size on mechanical properties of 3Y-TZP ceramics, *Ceramics—Silikáty* 52(3), 165–171, 2008.

37. E. Jimen and J. Terraza, The Meyer law for hardness tests, *Nature* 16, 359, 1950.

38. T. Wilantewicz, R. Cannon , G. Quinn, The indentation size effect (ISE) for Knoop hardness in five ceramic materials, *Ceramic Engineering and Science Proceedings*, 2008.

39. J. J. Kruzic, R. M. Cannon, J. W. Ager, and R. O. Ritchie, Fatigue threshold R-curves for predicting reliability of ceramics under cyclic loading, *Acta Mater* 53, 2595–2605, 2005.

40. H. N. Ko, Effect of grain size on cyclic fatigue and static fatigue behaviour of sintered Al_2O_3, *J Mater Sci Lett* 14, 56–59, 1995.

41. B. Gervais, A. Vadean, M. Raison, and M. Brochu, Failure analysis of a 316L stainless steel femoral orthopedic implant, *Case Stud Eng Fail Anal* 5–6, 30–38, 2016.

42. B. N. Kim, K. Hiraga, K. Morita, and Y. Sakka, A high-strain-rate superplastic ceramic, *Nature (London)* 413(20), 288–291, 2001.

43. M. Yoshida, Y. Shinoda, T. Akatsu, and F. Wakai, Superplasticity-like deformation of nanocrystalline monoclinic zirconia at elevated temperatures, *J Am Ceram Soc* 87(6), 1122–1125, 2004.

44. F. J. Parjer and R. W. Rice, Correlation between grain size and thermal expansion for aluminum titanate materials, *J Am Ceram Soc* 72(12), 2364–2366, 1989.

45. Y.-J. Lee, Y.-H. Park, and T. Hinoki, Influence of grain size on thermal conductivity of SiC ceramics, *IOP Conf Ser: Mater Sci Eng* 18, 162014, 2011.

46. B. Basu and M. Kalin, *Tribology of Ceramics and Composites*, Wiley, Hoboken, NJ, 2011.

47. J. F. Archard, Contact and rubbing of flat surface, *J Appl Phys*, 24(8), 981–988, 1953.

48. D. Sarkar, B. Basu, M. C. Chu, and S. J. Cho, Is glass infiltration beneficial to improve fretting wear properties for alumina? *J Am Ceram Soc* 90(2), 523–532, 2007.

7

Nanostructured Ceramics for Renewable Energy

7.1 Introduction

Nanostructured ceramic aided renewable energy modules have extensive potential to fulfill the energy demand for several sectors, and thus it is interesting to emphasis on different horizons including:

1. Solar energy conversion to electrical energy by dye-sensitized solar cell (DSSC)
2. Synthesis of hydrogen gas through water cleavage by tandem cell
3. Solar energy storage and electric energy by photovoltaically self-charging battery (PSB)
4. Gaseous fuel to electrical energy generation by solid oxide fuel cell (SOFC)
5. Vibration to electrical energy by nanoelectro mechanical system (NEMS)

Hence, a combination of solar energy and vibration assisted waste mechanical energy have had much focus to develop renewable energy. In order to elucidate this fact, four important coupon ceramics titanium dioxide (TiO_2), tungsten tri-oxide (WO_3), zirconia (ZrO_2), and barium titanate ($BaTiO_3$) are considered to demonstrate five case studies. The advance research of such group of materials is in advance progress either in synthesis of different morphologies or fabrication and characterization of device modules [1–5]. A brief illustration on the probable alternative energy and utility by these selective nanoparticles is addressed in Figure 7.1.

While addressing these materials, several aspects are essential to look after in order to design material and device together. Despite stable thermodynamic phase, development of different crystal structures and preferential morphologies are in well advance state to fulfill the research objective. For example, tetragonal anatase TiO_2 ($E_g \sim 3.2$ eV) is preferable for photoassisted conversion in both regenerative and photovoltaic cell compared to thermodynamic stable rutile ($E_g \sim 3$ eV) [6]. In this context, an important semiconductor WO_3 is established as a prospective candidate when is in series connection with DSSC to make photoanode for tandem cell and water splitting [7]. However, a specific crystal structure with having more intercalation space may eventually facilitate solar energy harvesting [8]. Interestingly, less than 100 nm thin nanostructured yttria stabilized zirconia (YSZ)

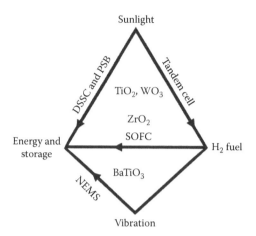

FIGURE 7.1
Nanostructured TiO_2, WO_3, ZrO_2, and $BaTiO_3$ are effective ceramics to generate electrical energy, synthesis of hydrogen gas, store solar energy, generate electricity by H_2, and electrical energy by vibrational energy. Probable devices made of these materials and their applications are represented in this tetragonal. Interlink in between effective H_2 fuel generation by waste vibrational energy is missing in current research practice!

electrolyte consists of nanograin attains high power density because of low ohmic, as well as activation loss, and rapid O_2 dissociation facilitate O^{2-} diffusion compared to micron grains [9]. Thus, it is obvious the performance of targeted device depend on the material and microstructure. For example, high degree of electro-mechanical coupling coefficient of poled piezoelectric nanowire $BaTiO_3$ harvesting vibration assisted mechanical energy and producing effective power density compared to conventional ZnO nanowire [10]. Herein, aforesaid five different case studies including synthesis, characterization, analysis, and future scopes are discussed to highlight and establish the significance of nanostructured ceramic research and development.

7.2 Nanostructured TiO_2 for Solar to Electrical Energy, Dye-Sensitized Solar Cell

7.2.1 Background

Discovery of photoelectric effect is the benchmark to achieve electric power or synthesis of chemical fuels in presence of light. Basic philosophy on the way to transformation of solar light to electric energy is now a broad approach for the generation of clean energy. In this context, commercial grade crystalline Si (silicon)-semiconductor solar panel is proficient to 25% conversion in order to generate 15 watt per square foot of panel as an alternative energy. Consecutive stacking of doped phosphorous in silicon (one more electron; N-type) and doped boron in silicon (one less electron; P-type) capture photon from the sunlight and excite electrons to migrate and produce direct current (DC) electrical power. Hence, semiconductor in presence of electron transport media is the prime resource to generate electricity through photovoltaic cell. In this continuation, nanostructured oxides so-called ceramic semiconductors TiO_2, ZnO, In_2O_3, Nb_2O_5, SnO_2, $SrTiO_3$, and more are established as potential carrier for the absorbance of sensitizer through their high surface area and electron transfer medium to the conducting surface. In the beginning, a large band gap semiconductor (3.2 ev) ZnO was sensitized by chlorophyll and achieved low degree of energy conversion efficiency in the order of 1%–2.5% [11]. A breakthrough efficiency up to 10% is reported in combination of few microns film made of mesoporous titanium dioxide nanoparticles sensitized with ruthenium complex on organic electrolyte under solar illumination, and the National Renewable Energy Laboratory in USA certified this potential. However, several approaches are coming across by several group of researchers to enhance the efficiency of DSSC. A brief illustration including basic philosophy, working principle, fabrication of a typical low cost testing module, characterization and analysis of ceramic particles, and DSSC are discussed to envisage research prospectus. However, recently, very limited number and small size of such devices are available in the market and demands extensive progress for commercialization.

7.2.2 Importance of Nanostructured TiO$_2$, Anatase

Titanium dioxide offers some unique properties compared to other semiconductors in consideration of DSSC. It is worthwhile to mention that the wide band gap and nanoparticles are not only enough good for effective performance. Starting from synthesis to performance are needed to monitor through several steps and in this background few aspects on the material features are summarized in the following:

1. Nanoscale oxide particles are enough small to build-in electric field and thereby the rapid charge transport occurs via diffusion compared to bulk, however, below optimum size, the rapid recombination of e^-/h^+ reduce the photocatalytic activity.

2. Agglomerate free, high surface area, pure phase of tetragonal anatase (band gap 3.2 eV) facilitate rate of reaction that can be prepared through low temperature synthesis protocol.

3. Conduction band of TiO_2 zone appears just below the excited state energy level of reasonable number of dyes that promote the effective electron injection from dye to semiconductor.

4. Relatively high dielectric constant, ($\varepsilon = 80$), compared to ZnO, ($\varepsilon = 10$), encompass excellent electrostatic shielding of the injected electron from the oxidized dye molecule attached to the TiO_2 (anatase) surface and inhibit their recombination before reduction of dye by the redox electrolyte.

5. The wide band gap, ($E_g = 3.2$ ev), anatase compared to WO_3 ($E_g = 2.6$–3.1 ev) preferentially absorb ultraviolet light comprised to an absorption edge of $\lambda_g \sim 390$ [λ_g (nm) $= 1240/E_g$ (eV)], whereas, remaining visible light further promoted by anchored dye on semiconductor surface.

7.2.3 Materials and Methods

Semiconductor nanostructured TiO_2 particles were prepared through hydrolysis of titanium tetraisopropoxide followed by autoclaving for 12 h at 200°C [6]. Prior to fabricate the film, the nanoparticle-dispersed sol was concentrated by evaporation of water at 25°C (without drying) until formation of the viscous liquid. Carbowax M-20,000 (40% by weight of TiO_2) was added, and the viscous dispersion (TiO_2 content 20% by weight) was spread on the 85% transparent and conducting fluorine-doped SnO_2 (FTO) that finally heated at 450°C for 30 min to make 10 μm thick coating. Continuous three-dimensional network formation without TiO_2 particle growth was observed after heat treatment. Before dye absorption or coating, additional few monolayers of TiO_2 were electrodeposited onto the TiO_2 film from a Ti (III) solution in order to enhance the roughness factor. A definite quantity of trimeric ruthenium complex RuL_2 (μ-(CN)Ru(CN)L_2')2, where L is 2,2' bipyridine-4,4'-dicarboxylic acid and L' is 2,2'-bipyridine, dye adsorbed (up to 15 min) on the rough surface of TiO_2 to finalize the working electrode with an effective area of 0.5 cm^2.

The counter electrode was prepared from few layer deposition of platinum (Pt) on conducting FTO glass and placed directly on the top of the fabricated working anode. Prior to light expose, a thin layer of redox electrolyte (iodide/triiodide) in organic solvent was added to complete the DSSC circuit and dispersed along three-dimensional TiO_2 scaffold through capillary action. Absorption spectra for both nanostructured TiO_2 and dye adsorbed TiO_2 film was measured, whereas photocurrent spectra was measured for dye coated TiO_2 film and plotted together to understand the effective current formation due to electron injection from dye to the conduction band of TiO_2. Yield was corrected for 15% loss of incident photons through light absorption and scattering by the FTO conducting glass support. Photovoltaic measurements were employed through Air Mass (AM) 1.5 solar simulator equipped with a 400 W xenon lamp. More details on such experimental work is described in another pioneer work [12].

7.2.4 Working Principle of Dye-Sensitized Solar Cell

A simple representation on conversion of solar light into electricity in DSSC is shown in Figure 7.2a. A connotation PERP describes the working principle and commensurate electron transfer mechanism, where P—photoexcitation, E—electron injection, R—regeneration, and P—percolation. In brief, the anchored dye-sensitizer, (S), molecule becomes photoexcited (*step-1*) during irradiation of sunlight followed by electronically excited dye-sensitizer, (S*), inject an electron into the slight below conduction band of the TiO_2 semiconductor electrode (*step-2*). The electron deficient oxidized dye sensitizer, (S+), is subsequently reduced and regenerated in original state by electron donation from the ionic liquid redox electrolyte (*step-3*). The available donated electron by dye-sensitizer in photoanode follows percolation across the nanocrystalline TiO_2 scaffold through diffusion process that further capture by redox electrolyte at counter electrode (*step-4*) and the circuit being completed through the current collector. Herein, it is important to remember that the intramolecular relaxation of S* through lattice collision and phonon emission may restrict the electron injection process. The maximum photovoltage, (V_{oc}), generated by the cell corresponds to the difference between the Fermi level of the electron in the semiconductor electrode and the Nernst potential of the redox electrolyte. Thus, electric power is the result under sunlight illumination without permanent chemical transformation of any electrode, dye-sensitizer, and redox electrolyte.

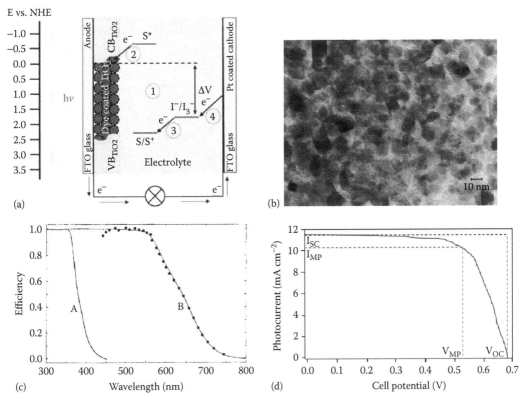

FIGURE 7.2
(a) Schematic representation of the working principle of DSSC, (b) TEM of TiO_2 particles used to make thin film, (c) absorption efficiency of the bare (A) TiO_2 and (B) dye-coated TiO_2; full circles (•) monochromatic current yield at short circuit as a function of excitation wavelength, and (d) photocurrent-voltage characteristics of dye adsorbed TiO_2 under simulated sunlight AM 1.5 spectral distribution.

7.2.5 Characterization and Analysis

This section describes fundamental aspects in order to assess the performance of DSSC made of nanostructured TiO_2 particles and their coating on indium tin oxide (ITO) working electrode. Essential experimental data and propositions are published in *Nature* that encompasses how to utilize nano TiO_2 to generate alternative energy, henceforth, promoting research in DSSC [6]. Transmission electron microscopic image of TiO_2 particles used in fabrication of electrode is shown in Figure 7.2b. Synthesized particles are uniform in size of 15 nm without any agglomeration, and maintained their continuity with remarkable pore content in 10 μm film on FTO during low temperature sintering at 450°C that eventually allow the redox electrolyte diffusion to obtain efficient light harvesting efficiency. The absorption spectra for transparent and colorless porous anatase film, (A), and dye (trimeric ruthenium complex) adsorbed brownish-red film, (B), is shown in Figure 7.2c. The light absorption efficiency is almost 100% for both bare TiO_2 and dye-adsorbed film within visible region below 550 nm and drastically reduced for dye-TiO_2 film at 750 nm. Dye-coated film harvested the light in presence of a definite intensity of solar emission AM 1.5, where, AM = 1/sin α, α is the angle of incidence of the solar rays at the Earth's surface. The light harvesting efficiency (LHE) at 478 nm is found 99% by Equation 7.1:

$$LHE = 1 - 10^{-\Gamma\sigma} \tag{7.1}$$

where Γ is the number of moles of sensitizer per square centimeter of projected surface area of the film (1.3×10^{-7} mol cm^{-2}) and σ is the surface absorption cross section (780 cm^2/mol). Herein, the smaller roughness factor 780 (measured from inner surface of film) is reported because of necking between TiO$_2$ particles and partially restricts the bigger dye molecules (area, 1 nm^2) movement through very small pores. The photocurrent spectrum is depicted similar to absorption spectrum that reveals the generation of effective current, and it is attributed to the electron injection from dye to conduction band (CB)-TiO$_2$. However, photocurrent yield measured at 520 nm varies with respect to the source of counter ion of the iodide/triiodide (I$^-$/I$_3^-$) redox electrolyte. After correction for the light absorption by the conducting glass, however, the highest up to 97% can be achieved in presence of Li$^+$ electrolyte. Photocurrent (J$_{sc}$; mA/cm^2) – voltage, (V), characteristics for thin layer cell under illumination by simulated AM 1.5 solar light are represented in Figure 7.2d. Important estimation of fill factor describes the ratio of maximum power to the electric power, power conversion efficiency, incident photon-to-current conversion efficiency (IPCE), and absorbed photon to current conversion efficiency (APCE) provide the better perception to judge the overall performance and behavior of DSSC. One can calculate these characteristics from the current set of data through Equations 7.2 through 7.5:

$$FF = \frac{I_{MP}\,V_{MP}}{I_{sc}\,V_{oc}} \tag{7.2}$$

$$PCE = \eta = \frac{I_{MP}\,V_{MP}}{P_{in}} \times 100 \tag{7.3}$$

$$IPCE\,(\%) = \frac{1240}{\lambda}\frac{I_{sc}}{P_{in}} \times 100 \tag{7.4}$$

$$APCE\,(\%) = \frac{IPCE\,(\%)}{LHE} \tag{7.5}$$

where $I_{MP} \times V_{MP}$ = maximum power, I_{sc} = short circuit current, and V_{oc} = open circuit voltage (refer to Figure 7.2d). P_{in} = input power, 83 W/m^2, λ = 520 nm. The surface area of the dye-coated TiO$_2$ photoanode was 0.5 cm^2. Under one tenth of sunlight, the observed fill factor and power conversion efficiency are 0.685% and 7.12%, respectively. Furthermore, enhanced performance up to 12% reported under direct sunlight during June afternoon, that eventually supports the application feasibility of nanostructured TiO$_2$ based DSSC. In this observation, incident photon-to-current conversion efficiency at 520 nm was calculated ~35%. The high light harvesting efficiency, high fill factor and conversion efficiency, long term cell stability, very fast electron injection, high chemical stability in presence of nanostructured TiO$_2$ particles coating, trimeric ruthenium complex dye and 0.5 M tetrapropylammonium iodide, and 0.02 M KI/0.04 M I$_2$ redox electrolyte comprise a potential photovoltaic cell for alternative energy. In addition, the cell exhibits long term stability under visible light illumination ($\lambda > 400$ nm) with a minor photocurrent reduction up to 10% that implies high stability, renders very attractive for practical development.

7.2.6 Research and Development Scope

This photo electrochemical device follows several and rapid (less than microsecond) electron transfer processes in parallel and competition, and thus selection of nanocrystalline semiconductor oxide particle morphology, high particle roughness factor, high porous network hierarchy of film and their optimum thickness, dye sensitizer absorption efficiency, and redox electrolytes are critical issues to achieve high degree of conversion efficiency. One can attempt to deposit mesoporous TiO_2 nanofibers on FTO to enhance the surface area, and thus enhanced roughness factor boosted up the dye adsorption and efficiency of DSSC. In variation of electron-doped (n-type) II/VI or III/V semiconductors along with different class of electrolytes sulphide/polysulphide, vanadium(II)/vanadium(III), or I_2/I^- redox couples are also attempted to enhance the conversion efficiency of regenerative cells. Conversion efficiencies up to 19.6% have also been reported for multijunction regenerative cells [13]. It is essential to consider that a sustainable dye molecule up to 10^8 redox cycles during photoexcitation, electron injection, and regeneration, establish a commercial grade device service life of 20 years.

7.3 Nanostructured WO_3/TiO_2 for H_2 Gas through H_2O Splitting, Tandem Cell

7.3.1 Background

Renewable energy by hydrogen, (H_2), gas is a challenging and upcoming technology. Industrial process utilizes large amount of fossil fuel to produce H_2, resulting in emission of carbon dioxide. Thus, an alternative solar radiated photon driven conversion of abundant water to gaseous hydrogen attracted much interest as a recyclable clean energy carrier. There are two fundamental approaches implying: (a) one-step splitting by single visible-light-responsive photocatalyst and (b) two-step photoexcitation mechanism between oxide semiconductor based O_2 evaluation photocatalyst (WO_3, Fe_2O_3, etc.), and H_2 evaluation photocatalyst (TiO_2, ZnO, SnO_2, etc.), known as "Z-scheme." The optimal absorption threshold for a single photon converter is theoretically calculated at an energy of ideal 1.6 eV that eventually absorb all solar photons below 770 nm in order to boost up the efficiency of the tandem cell. Apart from regenerative cell, one-step photo electrochemical cell (PEC) attempted by TiO_2 (rutile, $E_g \sim 3$ eV) anode and a Pt cathode to generate H_2 fuel through water splitting; unfortunately, this semiconductor absorbs only ultraviolet part, accordingly low conversion efficiency [14]. In photocatalytic process, light energy is converted to chemical energy and Gibbs free energy increases through water splitting reaction, as indicated in Equation 7.6:

$$H_2O + h\nu \rightarrow H_2 + \frac{1}{2}O_2; \quad \Delta G_o = 238 \text{ kJ mol}^{-1} \tag{7.6}$$

A conversion efficiency (solar to hydrogen; STH) of PEC at least 10% is essential for practical application, and can be estimated by:

$$STH = \frac{100\% \times J \times 1.229}{I_{light}} \tag{7.7}$$

where J is the photocurrent density and I_{light} is the light intensity. A classic theoretical analysis illustrates on the hydrogen gas production rate of 0.7 mg/s or 7.8 mL (STP)/s per m^2 of collector surface at maximum solar peak intensity at 1 kW/m^2 [15]. In this circumstance, this section discusses a basic strategy to synthesis hydrogen gas from water in presence of nanostructured dual semiconductors and photon by means of sunlight.

7.3.2 Importance of Dual Semiconductors

Reader should pay attention to select suitable photocatalyst for high efficiency conversion and hence ideal materials possess: (1) band edge potentials suitable for overall water splitting, (2) band gap energy smaller than 3 eV, and (3) stability in the PEC process under photo irradiation. Selection of band gap and CB-valence band position is an important criteria to initiate solar hydrogen production through photo electrochemical approach (see Section 4.2.1). Electricity for applied potential is often required for some semiconductors whose CB level is more positive than hydrogen-evolution potential [16]. Although, bottom level of the conduction band must be located at a more negative potential than the reduction potential of H^+/H_2, or the top of the valence band is more positive than the oxidation potential of H_2O/O_2 during water splitting (Figure 7.3a). Thus, visible-light-responsive photocatalyst,

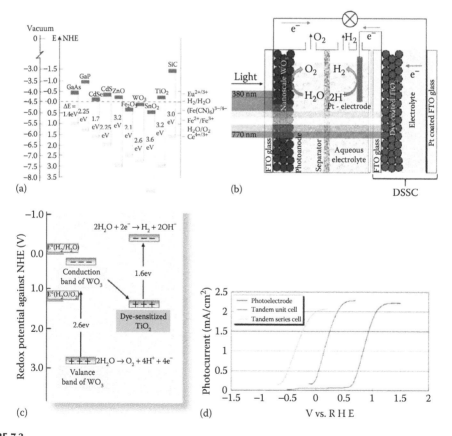

FIGURE 7.3
(a) Band gap of different oxide, nonoxide ceramics, (b) structure of a tandem cell system composed of WO_3 photoelectrode and DSSC, (c) Z-scheme of photocatalytic water splitting by a tandem cell, and (d) comparative current-potential plots for only WO_3 electrode, single tandem cell, and two tandem cell in series.

such as WO_3 oxide is not enough good to split water to hydrogen because of their low conduction band for H_2O reduction, and thus employed z-scheme (i.e., dual semiconductors, e.g., WO_3 and TiO_2) to generate H_2 from water in presence of solar irradiation.

7.3.3 Materials and Methods

In this incidence, a couple of semiconductor, say WO_3 and TiO_2 coated electrodes are the prime component to build the tandem cell, that promotes two-step O_2 and H_2 generation in different chambers, respectively. Apart synthesis of TiO_2, several groups are attempted to nanoscale WO_3 particles research, either to synchronize different morphologies or crystallinity. One-dimensional (1D) nanostructures of WO_3 in preference of nanowires, nanotubes, nanofibers, nanorods, and nanoribbons exclusively offers attractive properties for most of the electro-functional devices. Many different approaches have been disclosed for the synthesis of WO_3 nanostructures like spherical, cuboid, fiber, ribbon, etc., via different wet chemical methods and thereafter predefined processing protocols. Wet chemical method specifically includes sol-gel, acid precipitation, hydrothermal, solvothermal, and combustion method are more common [2,3,8]. These methods offer a better control of the material morphology through hydrolysis, condensation, etching, and oxidation during the reaction. Detailed discussion on synthesis of different semiconductor particles are out of scope in this book. A similar electrode fabrication protocol can be adopted, as described in Section 7.2.3.

7.3.4 Working Principle

Photocatalysis by semiconductor particles comprehends three major steps:

1. Absorption of photons with higher energies (λ_g (nm) = $1240/E_g$ (eV)) than the semiconductor band gap, leading to the generation of electron (e^-) – hole (h^+) pairs in the semiconductor particles.
2. Charge separation followed by migration of these photogenerated carriers in the semiconductor particles.
3. Surface chemical reactions between these carriers with various compounds (e.g., $H_2O \rightarrow H_2 + \frac{1}{2}O_2$); however, probability of electron (e^-) – hole (h^+) recombination with each other without participating in any chemical reactions decreases the overall efficiency.

In convention, it is difficult to separate two different gases of H_2 and O_2 during simultaneous formation and evaluation in a conventional single step water splitting system. As a remedy, a low cost tandem device consists of two photosystems: WO_3 (O_2 evaluation) based on its outstanding stability over a wide range of acidity and dye-sensitized TiO_2 (H_2 evaluation) connected in series to split the water is proposed (Figure 7.3b). Thin film of nanocrystalline mesoporous WO_3 ($E_g \sim 2.6$ ev) coated on transparent ITO serves as photo anode that absorbing blue region ($\lambda = 475$ nm) of irradiated solar spectrum. Excited valance band create holes (h^+) in film and oxidize H_2O to O_2:

$$2H_2O + 4h^+ \rightarrow O_2\uparrow + 4H^+ \tag{7.8}$$

This cell does not generate enough voltage to split the water effectively, so the available electron (e^-) in conduction band fed into the second photosystem consisting of the dye-sensitized mesoporous TiO_2 cell (see Section 7.1). The latter is positioned just adjacent of water reservoir connected through Pt wire to transmit the electron. The longer wavelength green (510 nm) and red (622–770 nm) immediately pass through the front cell and capture by the attached back DSSC (band gap 1.6 eV) that increasing the potential of the electrons, and eventually CB electron flow to the hydrogen cathode to form H_2 gas:

$$4H^+ + 4e^- \rightarrow 2H_2\uparrow \tag{7.9}$$

For a fixed water volume, dimensional variation of cavity may reduce penetration path and promote effective solar spectrum transmission through water. A transparent membrane is required in the water cell to separate the hydrogen and oxygen generate chamber. The overall reaction, $(2H_2O \rightarrow 2H_2 + O_2)$, corresponds to the splitting of water by visible light through a close analogy to the "Z-scheme," named in the perspective of electron flow diagram shape and represented in Figure 7.3c. This splitting mechanism by ceramic semiconductor follows natural photosynthesis comprising effectively charge carries without unwanted recombination redox processes. Herein, no electricity is required, no emission of CO_2 or other gases during synthesis of H_2. In the practical scenario, the additional sacrificial reagents (e.g., methanol, sulphide ions, Ag ions) are often also employed as hole scavengers (i.e., as electron donors) to overcome the unwanted photoexcited e^--h^+ recombination during H_2 generation in tandem cell.

7.3.5 Characterization and Analysis

The quantitative estimation of photocurrent density provide a better comprehension the performance of PEC cell. A comparative study on current-potential for three separate systems: (a) only mesoporous WO_3 photoelectrode (red), (b) tandem cell consists of mesoporous WO_3 photoelectrode and a single DSSC (blue), and (c) tandem cell consists of mesoporous WO_3 photoelectrode and twin DSSC connected in series (green) are shown in Figure 7.3d. Each experiment, (I–V), was conducted under simulated solar-light irradiation (AM 1.5, 100 mW cm^{-2}, 1 sun) and added 0.1 M H_2SO_4 electrolyte in aqueous solution. A uniform mesoporous WO_3 film thickness of 1.5 μm was used for each system along with typical DSSC unit (as discussed in previous section). The onset of the photocurrent of single electrode was about 0.7 V versus RHE (reversible hydrogen electrode), and steady-state photocurrent of 2.25 mA cm^{-2} attained at 1.2 V versus RHE. Blue curve depicts the characteristics I–V behavior of a tandem cell composed of a mesoporous WO_3 photoelectrode and a single DSSC made of TiO_2. A minimal photocurrent increment is noticed without any applied potential and onset photovoltage shifted to −0.7 V in this system. Thus, an attempt was taken to enhance the output by series assembly of two unit of DSSC in a tandem cell. The green I–V curve shows the other tandem cell system composed of a mesoporous WO_3 photoelectrode and a two-series connected cell of DSSC, which has a photovoltage of 1.4 V, as onset of the photocurrent of this system shifted about 0.7 V again negatively down to −0.7 V. In this case, the photocurrent at 0 V applied potential was 2.1 mA cm^2. The STH efficiency of this tandem cell system using a two-series connected DSSC cell was 2.58%, which was about six times higher than that of the only mesoporous WO_3 photoelectrode (0.44%) under identical experimental condition. Thus, a significant improvement of STH efficiency was observed by a tandem cell system composed of a mesoporous WO_3 and a two-series connected cell of DSSC.

In consideration of their optical and electronic characteristics, it is essential to recall that the reduced light harvesting caused by scattering and reflection strongly influences

TABLE 7.1

Overall Water Splitting Using Z-scheme Photocatalysis Systems Composed of Various Photocatalysts without Electron Mediators

Run	H_2-Photocatalyst	O_2-Photocatalyst	Weight Ratio[a]	Incident Light (nm)	Activity ($\mu mol\ h^{-1}$)	
					H_2	O_2
1	Ru/SrTiO$_3$:Rh	BiVO$_4$	0.1/0.1	>420	40	19
2	Pt/SrTiO$_3$:Rh	BiVO$_4$	0.1/0.1	>420	1.3	0
3	Ru/SrTiO$_3$:Cr, Ta	BiVO$_4$	0.1/0.1	>420	D.3	0.1
4	Ru/(Cu Ag)$_{0.22}$In$_{0.5}$ZnS	BiVO$_4$	0.1/0.1	>420	4.4[b]	0[b]
5	Ru/SrTiO$_3$	BiVO$_4$	0.1/0.1	>420	0.05	0
6	Ru/TiO$_2$(anatase)	BiVO$_4$	0.1/0.1	>420	Trace	0
7	Ru/SrTiO$_3$:Rh	AgNbO$_2$	0.1/0.1	>420	1.9	0.7
8	Ru/SrTiO$_3$:Rh	Bi$_2$MoO$_6$	0.1/0.1	>420	12	5.2
9	Ru/SrTiO$_3$:Rh	TiO$_2$:Cr, Sb	0.1/0.1	>420	6.7	3.3
10	Ru/SrTiO$_3$:Rh	TiO$_2$:Rh, Sb	0.1/0.1	>420	5.1	2.2
11	Ru/SrTiO$_3$:Rh	WO$_3$	0.1/0.1	>420	5.7	2.4
12	Ru/SrTiO$_3$:Rh	SrTiO$_3$	0.1/0.1	>300	19	8.2
13	Ru/SrTiO$_3$:Rh	TiO$_2$(rutile)	0.1/0.1	>300	67	31
14	Ru/SrTiO$_3$:Rh	BiVO$_4$	0.04/0.1	>420	13	6.0
15	Ru/SrTiO$_3$:Rh	BiVO$_4$	0.2/0.1	>420	22	9.6
16	Ru/SrTiO$_3$:Rh	BiVO$_4$	0.4/0.1	>420	14	5.8
17	Ru/SrTiO$_3$:Rh	–	–	>420	0.6	0
18	–	BiVO$_4$	–	>420	0	0

Source: Abe, R., *J. Photochem. Photobiol. C*, 11, 179–209, 2010; Sasaki, Y. et al., *J. Phys. Chem. C*, 113, 17536–17542, 2009.

Note: Reaction conditions: reactant solution; aqueous H$_2$SO$_4$ solution, pH 3.5, 120 mL, light source; 300-W Xe-arc lamp, cell; top-irradiation cell with a Pyrex glass window; ph = 7.

[a] Weight ratio of H$_2$-photo-catalyst/O$_2$-photocatalyst (g/g).
[b] pH 7.

the overall conversion efficiency, moreover, they attenuated the light transmitted to the middle and back cells and caused the DSSC photocurrent to limit the overall performance of the tandem cell. Applying Z-scheme, a potential H$_2$ and O$_2$ evaluation up to 67 $\mu mol\ h^{-1}$ and 31 $\mu mol\ h^{-1}$ is successfully synthesized in presence of SrTiO$_3$ (H$_2$) and rutile TiO$_2$ (O$_2$), and a list of current ongoing research findings are listed in Table 7.1.

Future device engineering is like to focus on improving the performance of water splitting behavior by reducing reflection, scattering, and resistive losses.

7.3.6 Research and Development Scope

A narrow band gap semiconductor coated photoanode effectively harvest visible light in the longer wavelength regions and enhance the photogenerated charge separation in photocatalysis. Electronic structure of a photoelectrode material is the most important factor in determining its PEC properties. The band-gap engineering for the modification of band structure of semiconductor photocatalysts using ion doping, semiconductor sensitization, or solid solutions implies significant opportunities to strengthen the activity of such materials in the visible light region. Despite band-gap phenomenon, several important factors like electronic properties, chemical composition, structure and crystallinity, surface states, and morphology eventually determine the photocatalytic activity of such materials, and they need to be further elucidated in detail during synthesis of new class of materials. Recent data exploring on enormous scope on tandem cell research for socio-economic benefit that are already been adapted for refineries, vehicle refueling, and in supplemental energy for industrial and domestic buildings. Now-a-days, an interesting development describes about a prospective module, in which 7×7 m array of tandem cells exhibits near to 10% efficiency and producing enough hydrogen for 11,000 miles of driving per year. Furthermore, placing of such arrays on car rooftops or incorporated onto industrial buildings can facilitate the reduction of transportation costs for the hydrogen that could be used as alternative energy. The GE Global Research, University of CA-Santa Barbara, M V Systems, Inc. and Midwest Optoelectronics are performing extensive research on the tandem cell to generate cheap H_2; funded by US Department of Energy.

7.4 Nanostructured WO₃/TiO₂ for Solar to Electrical Energy and Harvesting, Photovoltaically Self-Charging Battery

7.4.1 Background

The first solar battery concept was proposed by Hodes et al. in 1976, where the efficiency was measured in an assemble of polycrystalline CdSe photoanode and sulphide catholytes [19]. Upon illumination, photogenerated electron (e⁻)—hole (h⁺) pairs are separated at the photoelectrode (PE)—catholyte interface, followed by the holes are injected into the catholyte, thus the electrons are transported through the external circuit. In this circumstance, the device simultaneously stores harvested solar energy and outputs electric power. In the dark, reduction of electron Fermi level of PE, and the connected Ag anode, is spontaneously getting oxidized followed by supplying electrons and comprises the previously stored electrochemical energy as electricity. In this cell design, the catholyte was in direct contact with the anode and exhibits relatively low storage efficiency. Therefore, researchers introduced a membrane separator and an anolyte to protect the anode and avoid the severe anode-catholyte recombination that eventually enhance the efficiency. Depending on the specific PE-electrolyte combination, solar batteries are preferentially categorized into three different types: (a) the photoanode-catholyte design, (b) the photocathode-anolyte design, and (c) the photoanode-catholyte|anolyte-photocathode tandem design. Because thus far most solar batteries, herein, the prime focus will be given on photoanode-catholyte design and working mechanism of a solar battery. It is worth to mention that the DSSC is an electrochemical solar cell, in fact, it has similar to most of the battery components; one redox couple (I^-/I_3^-), an electrolyte (mostly organic solvent), and two electrodes (TiO₂ and Pt on FTO), however, a second redox potential demands to make a complete battery, known as photovoltaically self-charging battery (PSB). In this background, the high redox potential difference in between TiO₂ and Li⁺ intercalation restricts the charge storage process and thus an additional redox couple in the form of an additional WO₃ layer has been introduced and measured the solar energy harvesting behavior in a recent article [8]. In order to assess and justify the importance of nanostructured ceramics in energy sector, combination of consecutive WO₃-TiO₂-coated transparent conductive oxide (TCO) has been considered as a photoanode. The design of such PSB is shown in Figure 7.4a, and their characterization and analysis are discussed in following sections.

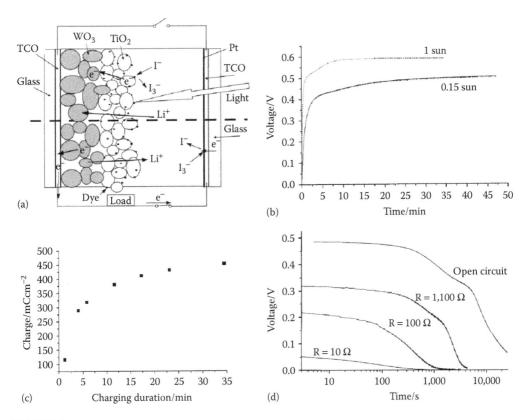

FIGURE 7.4
(a) Schematic representation of self-charging battery, (b) influence on illumination intensity on open circuit voltage with respect to time, (c) charge storage capacity with respect to time after charging under 1 sun illumination, and (d) change in voltage during discharging in the dark.

7.4.2 Materials and Methods

In order to demonstrate the working principle, in the beginning Hauch et al. [8] used sol-gel paste and prepared a 25 μm thick WO_3 layer followed by 5 μm TiO_2 layer onto the WO_3 layer through a doctor-blading technique. A relatively thick layer of WO_3 was developed in order to increase the charge storage capacity. Details of similar TiO_2 paste preparation has already been discussed in Section 7.2.3. In brief, the WO_3 paste was made from peroxopolytungstic acid solution followed by evaporating and heating at 450°C to make nanocrystalline WO_3 powders. Out of this powder, a paste was made using a similar method as for TiO_2. The obtained WO_3 layer shows highly porous, adhesive, and micrometer thick. Precursor deposition followed by sintering at 450°C for 30 min, both the TiO_2 and WO_3 layer develop a continuous porous layers. The top TiO_2 layers were sensitized with *cis*-bis (isothiocyanato) bis (2,28-bipyridyl-4,48-dicarboxylato)-ruthenium(II) dye. Average 25 nm spherical particle size and 30% porosity with having porosity size in the range of 25–50 nm was observed by scanning electron microscopy (SEM) technique. The opposing Pt-electrode layers were sputtered with a thickness of 2 nm on TCO (F-doped SnO_2 glass) glass substrate approaching to resistance of 8 Ω sq^{-1} and maintained same degree of transparency. The two electrodes are filled with an electrolyte 0.5 M LiI + 0.05 M I_2 in propylene carbonate with 10% vol of 4-*tert*-butyl pyridine.

The resultant sample area was 2 cm^2. Electrodes (WO$_3$ Photoanode – Pt cathode) were keep apart 1 mm thick electrolyte by silicone frame and entire management was sealed in a polymeric epoxy adhesive. Different characterization and their analysis were performed and discussed in Section 7.4.4.

7.4.3 Working Principle

A two-electrode design concept has been introduced in recent literature, where dye-sensitized TiO$_2$ as photoanode (PE), LiI/LiI$_3$/propylenecarbonate catholyte, Pt-coated TCO cathode, and layer of WO$_3$ underneath the PE to act as the anode [8,20]. Meanwhile, the mesoporous PE (TiO$_2$ layer) acted as an inbuilt separator to allow Li$^+$ transport and prevent the self-discharging reaction. The assembled entire cell has been divided into two hypothetical halves, upper and lower, to understand the details working principle of PSB, for example, charging and discharging phenomenon in solar light and dark, respectively. Under illumination (upper part of Figure 7.4a), the chemisorbed dye molecules are photoexcited and injected electrons to the CB of TiO$_2$, as described in DSSC. The oxidized dye molecules are then regenerated by I$^-$ anions from electrolyte and transformed to I$_3^-$. Meanwhile, the redox-active species on the, (WO$_3$), anode side is reduced by electrons from the nanocrystalline TiO$_2$ semiconductor separator. To keep the charge balanced, Li$^+$ cations are intercalated into the WO$_3$ through the membrane separator and balance the internal charge (Equations 7.10–7.14). In open circuit, this leads to the charging of the WO$_3$|LiWO$_3$|| LiI|LiI$_3$ redox system. dye + hv → dye*:

$$dye* + TiO_2 \rightarrow dye^+ + TiO_2 - e^- \tag{7.10}$$

$$2dye^+ + 3I^- \rightarrow 2dye + I_3^- \tag{7.11}$$

$$TiO_2 - e + A^{ox} \rightarrow TiO_2 + A^{red} \text{ (through external circuit)} \tag{7.12}$$

$$I_3^- + 2e^- \rightarrow 3I^- \tag{7.13}$$

$$A^{red} \rightarrow A^{ox} + e^- \tag{7.14}$$

Under discharging conditions (lower part of Figure 7.4a), the electricity output is obtained through electrons transfer from WO$_3$ via an external load to the Pt electrode encompasses electrochemical oxidization of the A$_{red}$ to A$_{ox}$ on the anode side. The Pt catalyzes the electron transfer to I$_3^-$ which is reduced to I$^-$ on the cathode side and releases the intercalated Li$^+$ ions. Both the electrochemical process is described in Equations 7.13 and 7.14. Thus, when connected to a load, the device acts as a power supply. During illumination it works as a solar cell, after illumination as a battery. The self-discharging process can manipulate through synchronization of crystal structure, morphology, porosity, thickness, and roughness factor of both semiconductor layers.

7.4.4 Characterization and Analysis

The developed PSB was characterized in several aspects, however, a cluster of properties with the variation of following parameters are highlighted in order to understand the charging and discharging efficiency:

1. Open circuit voltage during charging under two different illumination intensity by halogen lamp; 1000 W/m² for a solar spectral distribution (1 sun) and 150 W/m² for a solar spectral distribution (0.15 sun)

2. Charge density with respect to time variation under highest illumination intensity and external load of 10 Ω

3. Voltage change during discharging as a function of time in dark

4. Discharging phenomenon with respect to electrolyte concentrations

The charging behavior of the device in open circuit is shown in Figure 7.4b. An initial rapid increment open circuit voltage is detected after immediate illumination and achieved maximum up to 0.6 V after 4 min, however, not changed with time under 1 sun, whether the initial charging was slower and reached up to constant 0.5 V under 0.15 sun intensity. Rapid voltage change is attributed to change in reversible transformation of higher symmetry WO_3 crystal structure. Stored charge was measured by connecting the device to a load resistance of 10 Ω in the dark, and represented in Figure 7.4c. As maximum as 0.45 C/cm² charge was measured after 30 min at 1 sun, but 0.27 C/cm² is the result under 0.15 sun. Herein, the overall efficiency of this device still is rather low. For example, after 300 s a charge of 300 mC/cm² is stored by illumination with 0.1 W/cm² (Figure 7.4c), and the device achieves a potential of about 0.55 V (Figure 7.4b). Therefore, the irradiated energy is 30 J/cm², and the stored energy not higher than 0.55 V × 0.3 C/cm² = 0.165 J/cm². This corresponds to a ratio of about only 0.6%, however, utilizing the basic principal, one can enhance the efficiency through optimization of different parameters including selection of intercalation materials other than WO_3 or other electrolytes may be more suitable. The device was connected to different load resistances, and the voltage change during discharging was measured as a function of time (Figure 7.4d). Self-discharging under open circuit condition is attributed to the electron injection from TiO_2 to WO_3 that eventually forms Li_xWO_3, and this process favors because reduction potential of Li^+ intercalation into WO_3 is close to 0 V and TiO_2 conduction band minimum is about –0.5 V, followed by the Equation 7.15:

$$xLi^+ + WO_3 + xe^- \leftrightarrow Li_xWO_3 \tag{7.15}$$

The rate of this loss reaction can be optimized up to several days by optimizing the semiconductor layer thickness. A high concentration of 2 MLiI develops as high as charge capacity 1.8 C/cm² compared to 0.66 C/cm² in 0.5 MLiI, whereas open circuit voltage and charge retention time dramatically reduces under 1 sun. It is emphasized that excessive coverage of Pt-surface by LiI hinders the reduction of I_3^- to I^- on the Pt surface and thus reduction potential and overall voltage, whereas excessive WO_3 surface coverage suppresses the loss reactions and decreased self-discharging of the device with the high LiI concentration.

7.4.5 Research and Development Scope

Electrolyte solvent controls the electrolyte solubility, electrolyte-solvent molecular interactions, and thus determine the opening voltage window. It is an important constituent, as it is used in larger amount and determines the solar battery lifetime, fabrication cost, and environmental friendliness. While photocharging, potential that is more negative necessary for the catholyte to have sufficient electron-injection driving force to deliver a large current to the PE. However, during discharging, a potential that is as positive as possible maximize the obtainable output voltage. Therefore, there must be a trade-off on the catholyte potential to optimize both the operating current and the voltage [20]. With regard to possible applications, the proposed device made of nanostructured WO_3 and TiO_2 is especially suitable for low-cost and low-power applications, and that it can, in many cases, replace primary batteries or combinations of conventional solar cells with storage systems; examples included microelectronics, watches, smart cards, or transponders. In principle, it is possible to use this setup for applications with a higher demand for charge capacity, like radios or energy-saving lamps.

7.5 Nanostructured ZrO_2 for Gaseous Fuel to Electric Energy

7.5.1 Background

Fuel cells convert wide range of fuels into electrical energy, and thus polymer electrolyte membrane fuel cells and SOFCs are the subject of intense study in recent era. Usually, SOFC is a fully solid-state electrochemical generator consist of two electrodes and an electrolyte that promote the reaction between a fuel (H_2, natural gas, biomass, diesel) and an oxidant at an elevated temperature and simultaneously producing heat. Electrodes are either electronic or mixed ionic-electronic, and thus most common lanthanum strontium manganite cathode and Ni-YSZ cermet anode has porous structure to permeable the gases (fuel or effluent) during operation. In contrast, the solid electrolyte YSZ is a popular oxygen ionic conductor being necessarily dense to avoid any recombination between gaseous oxygen and hydrogen. A typical solid oxide fuel cell arrangement comprised of simple material combination of porous Pt (cathode)—nanostructured YSZ (electrolyte)—porous Pt (anode) is shown in Figure 7.5a. High temperature (700°C–1000°C) comprise effective ionic conductivity of solid electrolyte, but require specific material for sealing of the cell, as well as plausible reaction between material component ultimately result in degradation of fuel cell performance.

   Thus, the largest disadvantage is relatively prolonged start-up times, as well as mechanical and chemical compatibility issues. In this context, several strategy including different fuel, cathode and electrolyte materials, cell design, and utilization of waste heat are attempted to reduce the operating temperature (400°C–500°C) with effective cell yielding electrical efficiencies up to 85% and peak power density of 440 mW/cm². Apart different upcoming technologies, herein, a discussion is focused on how nanostructured ceramic like YSZ as an effective electrolyte for SOFC to generate high power density [9].

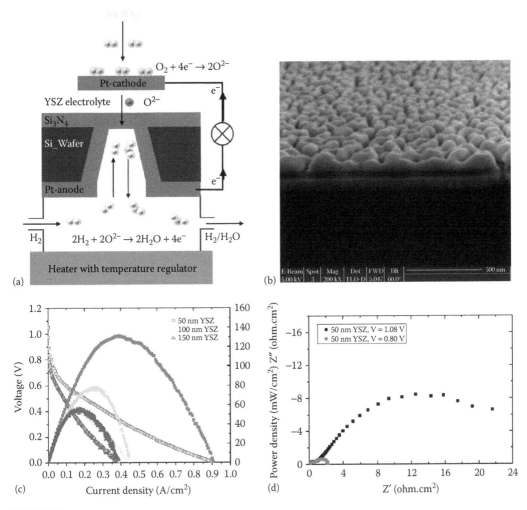

FIGURE 7.5
(a) Schematic representation of SOFC including testing arrangement; (b) SEM image of cross section of an ultra-thin SOFC, dark YSZ sandwiched within two Pt electrode (grey); (c) I–V plots measured at 350°C for difference thickness of YSZ electrolyte film sandwiched identical electrodes; and (d) electrochemical impedance spectra for 50 nm YSZ electrolyte with respect to voltage.

7.5.2 Importance of Nanostructured YSZ

A prerequisite dopant like yttrium dioxide (Y_2O_3) alters the thermodynamically stable monoclinic zirconia (m-ZrO_2) crystal structure to either tetragonal zirconia (t-ZrO_2) or cubic zirconia (c-ZrO_2) or combination of both, and in fact, this synthetic phase is apposite in the perspective of different mode of applications. For example, m-ZrO_2 is an effective for advance high temperature resistance, 3 mol% Y_2O_3-tetragonal zirconia polycrystal particulate prefers for bioimplants due to high fracture toughness through martensitic transformation and cubic YSZ as solid electrolyte because of high ionic conductivity. In due of functional application as SOFC, it is essential to understand how a solid material is capable to diffuse oxygen ions through their matrix and used as solid electrolyte. Dopant Y_2O_3 addition to pure zirconia (i.e., yttria stabilized zirconia) comprises Zr^{4+} cationic sublattice replacement by Y^{3+} ions, as a result the oxygen vacancies are generated to maintain charge neutrality as described in Equation 7.16:

$$Y_2O_3 \rightarrow 2Y'_{zr} + 3O_o^x + V_o^{\gamma\gamma} \tag{7.16}$$

It depicts two Y^{3+} ions generate one vacancy on the anionic sublattice and thus facilitates high conductivity of yttrium stabilized zirconia for O^{2-} ions at elevated temperature. This phenomenon promotes O^{2-} ions transportation in yttria-stabilized zirconia and considered for application as solid electrolyte in solid oxide fuel cells. Despite crystal structure, grain size has also impact to achieve effective ionic conductivity when operating temperature is $\leq 500°C$. Let the ionic conductivity of an electrolyte material is given as "σ," then the total resistance of cell, $R_{int} = \delta/(\sigma\,A_{cell})$, where δ is the thickness of the electrolyte layer and A_{cell} is cell area. Thus, nanostructured dense YSZ electrolyte facilitate to enhance the power density by reducing electrolyte thickness or increasing the ionic conductivity with help of high operating temperature.

7.5.3 Materials and Method

Sputtering is a well-developed technology to make fully dense or highly porous films through adjustment of gas pressure, deposition power, temperature, and time. In recent, Huang et al. fabricated porous Pt films for the use of both cathode and anode by DC sputtering at 10 Pa of Ar pressure, 100 W, and room temperature. Dense YSZ electrolyte films were deposited at 200°C by radio-frequency sputtering. Both the targets were 99.9% pure, DC sputtering target was Pt and the radio-frequency sputtering target was $Y_{0.16}\,Zr_{0.84}O_{1.92}$. Ultra-thin 50 nm YSZ was coated on Si wafer through an intermediate 500 nm Si_3N_4 to prevent the electrical current from leaking and avoiding reaction between Si and YSZ. By using physical masks, 80 nm porous Pt films cathode and anode were patterned on both sides of YSZ. Details fabrication procedure of such ultra-thin SOFC is found elsewhere [9]. The cross-sectional view of ultrathin Pt and YSZ coating were examined with the help of SEM and represented in Figure 7.5b. Both the top cathode and bottom anode made of Pt represents grey in color, whereas the yellow layer represents the YSZ electrolyte. The pores in Pt electrode are in the range 100 nm and the porosity was estimated to 30%–40%. The grain size of YSZ film was observed to be 20–30 nm in average. X-ray Photon Spectroscopy (XPS) surface and depth analysis depicts the presence of Y to Zr ratio close to 1:5 through the whole thickness. A testing chip (1 cm²) containing 16 individual thin SOFCs was used for the performance analysis. The area of the tested SOFCs was confirmed to be 100×100 sq µm based on SEM images. Before fuel cell characterization, the fuel cell structures were annealed at 400°C for 1 h in air (Figure 7.5b). Thin SOFCs containing 100 nm and 150 nm thick YSZ were also fabricated by the same processes in order to compare and understand the influence of nanostructured electrolyte on the resultant power density and open circuit voltage.

7.5.4 Working Principle

A planar SOFC consists of three consecutive layers in arrangement of high dense electrolyte like YSZ sandwiched and separated by two porous Pt-electrodes, as illustrated in Figure 7.5a. A conventional fuel gas such as hydrogen (can be obtained through water splitting) is introduced into the anode and heated up to certain extent and in another end, the oxygen usually carried by air, enters the cell via the cathode. The anode disperses the hydrogen gas uniformly over the entire surface and conducts the electrons that are free from hydrogen molecule, to be used as power in the external circuit. The cathode disburses the oxygen fed to it onto its surface and conducts the electrons back from the external circuit, where they can recombine and interact with oxygen ions, passed across the YSZ electrolyte, and hydrogen to form water. Part of the unreacted hydrogen may go off with the effluent water and reduce the efficiency of SOFC. In this context, the anodic and cathodic reactions are represented in Equations 7.17 through 7.19:

$$\text{Anode: } 2H_2 + O^{2-} \rightarrow 2H_2O + 4e^- \tag{7.17}$$

$$\text{Cathode: } 2O_2 + 4e^- \rightarrow 2O^{2-} \tag{7.18}$$

$$\text{Resultant: } H_2 + \frac{1}{2}O_2 \rightarrow H_2O \tag{7.19}$$

The ideal electromotive force (EMF) of a SOFC is determined by the Nernst equation and represented in Equation 7.20. In a solid oxide fuel cell, the Nernst equation is:

$$E = E_o + \frac{RT}{2F}\ln\left(\frac{P_{H_2}P_{O_2}^{1/2}}{P_{H_2O}}\right) \tag{7.20}$$

where E is the electromotive force or reversible open circuit voltage (V), E_o is the EMF at standard pressure (V) and expressed as $E_o = -\Delta G^\circ_{rxn}/2F$, where ΔG°_{rxn} is the Gibbs free energy of the reaction $H_2 + 1/2\,O_2 \rightarrow H_2O$ (evaluated at standard pressures and the operating temperature of the cell) and P_{H_2} or P_{O_2} are the partial pressures of the reactant H_2 and O_2, respectively. High concentration of reactant gases enhance the partial pressure that eventually increase the reaction rate, E and current. Similarly, an increase in product pressure reduces the reaction rates. In common, when the fuel cell is connected to a load in a closed circuit, a current is produced through the electrochemical reaction. However, the cell potential is reduced by predominately three nonreversible voltage losses (ohmic, activation, concentration), and one can enhance the cell efficiency by decreasing the losses through different techniques. In brief, the plausible strategies to minimize:

1. "ohmic loss" through decrease electrical resistance of the electrodes (area specific resistance, ASR_c), low resistance of ion transport in the electrolyte membrane (ASR_e), appropriate interconnect materials and design, and nanoscale electrolyte
2. "activation loss" through high rate of electrochemical charge-transfer reactions at cathode and anode pore architecture and hierarchy
3. "concentration loss" at the electrochemically active reaction zone (triple phase boundaries) by increment of partial pressure of hydrogen and oxygen through increasing the electrode porosity

The operating cell potential thus can be calculated as a subtraction of the earlier different losses from the reversible potential. In assumption, if the concentration loss is small and the cathode activation loss is dominant, the resultant voltage of a fuel cell, (V), is obtained by Equation 7.21:

$$V_{SOFC} = EMF - J \cdot ASR_e - \frac{RT}{\alpha nF} \ln \frac{J}{J_o} \qquad (7.21)$$

where EMF is the theoretical electromotive force, J is the current density, ASR_e is the area specific resistance of the electrolyte, R is the gas constant, F is the Faraday constant ($9.6 \times 10^4 c.mol^{-1}$), n is the number of electrons participating in the reaction, J_o is the exchange current density, an indicator of the electrochemical charge-transfer reaction rate, and "α" is the corresponding transfer coefficient, which is typically between 0.2 and 0.3. Thus, both electrode and electrolyte materials, size and density of electrolyte, and porosity of electrodes are essential parameters in the perspective of material selection for high degree efficiency of SOFC.

7.5.5 Characterization and Analysis

While developing renewable energy by such SOFC, the most important properties including open circuit voltage, power density, electrode interfacial resistance (ASR_c), and electrolyte resistance (ASR_e) are essential to evaluate and understand the influence of nanostructured ultra-thin YSZ. In order to fulfill the objective, the I–V data under constant operating temperature 350°C were measured with Solartron 1287 using both the galvano dynamic (scanning current) mode and the galvano static constant current mode. Electrochemical impedance spectra were obtained with Solartron 1260/1287 in the frequency range of 50 kHz to 0.1 Hz with an ac amplitude of 20 mV. At 350°C, open-circuit voltages in the range 1.05 V to 1.10 V were measured across the ultra-thin SOFC, and the obtained value was slightly lower than 1.16 V as predicted by Nernst equation due to probable gas leakages. Figure 7.5c presents the I–V curves of the thin-film SOFCs recorded at 350°C. The maximum power density was achieved 130 mW/cm² at 350°C out of the ultra-thin SOFCs consisting of 50 nm YSZ electrolyte. In addition, SOFCs consist of 100 nm and 150 nm thin YSZ were also fabricated separately by the same processes, and the obtained maximum power densities at 350°C were 85 mW/cm² and 60 mW/cm², respectively. Herein, no characteristic concentration loss was detected in the I–V curves, that suggesting sufficient gas diffusion through the thin nanoporous Pt electrode. However, 50 μm thick YSZ consist of 600 nm grain enveloped with 80 nm thin Pt electrodes encompasses power density less than 1 mW/cm² at 350°C, and it reflects the necessity of nanoscale electrolyte in SOFC. The high performance of the ultra-thin SOFCs consisting of 50 nm thick nanocrystalline YSZ was not only beneficial due to the very low ohmic resistance (decrease of ohmic loss), but also high electrochemical charge-transfer reaction rate (reducing the activation loss). Figure 7.5d shows the electrochemical impedance spectra of an SOFC consisting of 50 nm YSZ recorded at 350°C. It is well known that the impedance arc behavior is the result of frequency based two important phenomena: high-frequency region corresponds to ionic conduction in the electrolyte and low-frequency region corresponds to the reaction processes at the electrode/electrolyte interface. Accordingly, the ionic conductivity at 350°C of 50 nm YSZ is calculated to be 2×10^{-5} S/cm, consistent with 50 μm YSZ (2.5×10^{-5} S/cm). It is obvious from Figure 7.5d that the area specific electrode interfacial resistance, (ASR_c), is around one order of magnitude higher than electrolyte resistance, (ASR_e). This indicate major loss is originated from electrode reaction process. The same behavior can be observed from the I–V curves, in which voltage drops dramatically at low current densities. On the other hand, ASR_c is decreased significantly at 0.8 V as well. The electrode kinetics of the dominant process accelerated upon increasing the over potential, where the cathodic reaction kinetics are much slower than that of anode reaction in SOFCs [21]. These observations suggest that the ASR_c is mainly related to cathodic reaction. In other words, the cathode reaction dominates the activation loss. Cathodic reactions were found much faster in highly ion conductive electrolytes when the charge-transfer reaction at the triple phase boundaries is the rate-determining step. In order to enhance the efficiency of SOFC, an additional 50 nm gadolinia-doped ceria, $(Gd_{0.1}Ce_{0.9}O_{2-\delta})$, as an active interlayer between the YSZ electrolyte and cathode was incorporated to minimize the cathodic activation loss and achieved as high as powder density 400 mW/cm^2 at 400°C. This promising result promotes to develop SOFC through nanostructured ceramics for high power output in the future.

7.5.6 Research and Development Scope

Development of different anode, cathode, and electrolyte with consideration of (chemical, phase, morphological, and dimensional) stability, dopant concentration, electrolyte density, thickness and grain size, pore content and their size distribution in electrodes, choice of fuel, design of cell, and stacking sequence, operating temperature are recent practice to enhance the power density and fulfill the technological challenge. Apart, from the physical properties, thermal and mechanical properties have great interest to enhance the functional performance of SOFC. A competitive coefficient of thermal expansion between electrodes, electrolyte, and intermediate material may reduce the thermal stress followed by cracking and change in diffusion of gaseous product and resultant performance of fuel cell in operating temperature. In addition, the high strength and toughness properties, easiness to fabrication, amenable to particular fabrication conditions, compatibility at higher temperatures of ceramic structures, and relatively low cost are needed to encounter. 3D printing is also explored to make SOFC from slurry made of nanopowder, binder, solvents for complicated, flexible, and efficient design. As an interesting example, one can envisage Department of Energy (DOE), USA target requirements are 40,000 h of service for stationary fuel cell applications and greater than 5000 h for transportation systems (fuel cell vehicles) at a factory cost of $40/kW for a 10 kW coal-based system without additional requirements [22]. Rolls-Royce has also developed SOFC for turbine hybrid fueled by natural gas for power generation applications in the order of a megawatt (e.g., Futuregen).

7.6 Nanostructured BaTiO$_3$ for Mechanical to Electrical Energy

7.6.1 Background

Despite conversion of solar to electrical energy or harvesting through nanostructured ceramics, utilization of waste vibrational mechanical energy to electrical energy transformation in railway platform, runways, and beneath the tiles of temples, road is a new area of research and economic interest. For example, one kilometer path of a six-lane highway may be capable to produce 1 MWh electricity through suitable set of vibrational assisted (1 hz–1 khz) piezoelectric device for thousands of households. In this context, it is meaningful to install such a device in automobile in order to produce electricity during vibration or continuous impact on road. The heart of such system is a piezoelectric material that describe the stress-voltage relation; most popular quartz has common use in electrical oscillators, clocks, and microbalances. In practice, a specific crystal consists of nonsymmetrical atom position, but perfectly balanced electrical charge experience upsetting the balance of positive and negative charge, in another way the charge separation under mechanical stress. Thus, squeezing the crystal, a voltage or potential energy difference appears in opposite faces in piezoelectric crystals. Recently, NEMS has gained enormous interest because of the ability to generate sufficient electric power by harnessing several sources of mechanical energy such as sound waves, ultrasonic waves, vibrational energy, atomic force microscope tip induced stimuli, and biomechanical energy to replace a battery and power for small electronic devices [10]. In this context, a classic study has been demonstrated by a lab scale module made of nanoscale piezoelectric $BaTiO_3$ fibers.

7.6.2 Importance of Nanostructured $BaTiO_3$

Device miniaturization with effective performance of nontoxic material, nanoscale $BaTiO_3$ is an excellent choice over semiconductor ZnO and lead (Pb) based ferroelectric ceramics for higher energy conversion, in specific renewable energy. The higher power generating capability is attributed to the higher piezoelectric coefficient (d_{33}) of $BaTiO_3$ compared to same size of ZnO. Furthermore, the self-standing 1D nanoscale enable to exhibit higher strain compare to bulk of same material and thus enhance the piezoelectric energy conversion. A brief of such nanowire $BaTiO_3$ has enormous potential for new generation alternative and resource of cheap energy that is discussed in subsequent sections. Usually, the phase transition of $BaTiO_3$ depends on the temperature and crystallite size, and thus selection of adequate phase and morphology is a technological challenge. Ferroelectricity in tetragonal $BaTiO_3$ is due to an average relative displacement along the c-axis of Ti-ion from its centrosymmetric position in the unit cell and consequently the creation of a permanent dipole moment. The dielectric and ferroelectric properties of $BaTiO_3$ are known to correlate with size, and the device miniaturization demands decreasing dimension with high poling capacity [5]. Hence, nanoscale particles preferentially fibers with high tetragonality is expected for the fabrication of such device.

7.6.3 Materials and Method

Vertically aligned $BaTiO_3$ nanowire (NW) were directly grown on a conductive fluorine doped tin oxide (FTO, 2 mm thick, 7 Ω sq^{-1}) glass substrate with the help of two-step hydrothermal process. In first step, vertically aligned TiO_2 nanowire array was grown on ~10 × ~10 mm FTO glass from precursor titanium isopropoxide at 200°C for 3 h. Following the first hydrothermal process, the resultant FTO glass substrate with an array of vertically aligned TiO_2 NW was rinsed with deionized water and dried in ambient air. In presence of Ba^{2+} ions at 150°C–240°C, the $BaTiO_3$ nanowire was grown on the presynthesized TiO_2 NW, finally cleaned, dried, and heat treated at 600°C for 30 min to remove any hydroxyl defects before their use as NEMS energy harvesters. The resulting $BaTiO_3$ NWs exhibits a length of ~1 μm and a diameter of ~90 nm. The details processing conditions, crystallographic and microstructure analyses are illustrated in a recent literature [10]. The conductive FTO glass acts as the bottom electrode with the $BaTiO_3$ NW arrays sandwiched between the two (indium beam as top and FTO as bottom) electrodes as shown in Figure 7.6a. The fabricated $BaTiO_3$ NW NEMS energy harvester was poled at room temperature by supplying a high DC voltage of ~120 KV cm^{-1} across the two electrodes for 24 h to ensure that the dipoles align in the electric field direction. This configuration allows the NEMS energy harvesting device to achieve electrical energy at low resonant frequency effectively. The details poling phenomena have been described in Section 5.3.

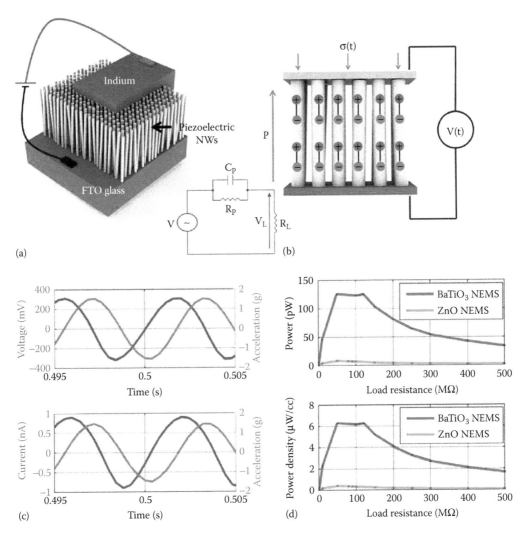

FIGURE 7.6
(a) Schematic diagram of the NEMS energy harvester fabricated using BatiO$_3$ NW arrays, (b) voltage generation protocol (where P is polarization direction, σ denotes stress, and V voltage generated by nanowires), (c) V_{oc} and I_{sc} from 1 g RMS sinusoidal acceleration input near resonant frequency (~160 Hz), and (d) peak power and peak power density ~6.27 μW cm^{-3} at 1 g RMS acceleration.

7.6.4 Working Principle

Vibration-influenced stress alters the inertial force of the vibrating indium beam on the BaTiO$_3$ NW arrays that results in charge generation through the direct piezoelectric effect and developing an alternating potential difference across the two electrodes, as shown schematically in Figure 7.6b. In this piezoelectric NEMS energy harvester, the common denotation P refers to polarization direction; σ, stress, and V, piezo-voltage generated by the nanowires. A representative electrical equivalent circuit for the NEMS energy harvester is shown as an inset in Figure 7.6b with the piezoelectric voltage, V, induced from the vibration acceleration in series with the inherent capacitance of the source, C$_p$, and piezoelectric leakage resistance, R$_p$, connected in parallel.

The voltage, V_L, is measured across the load resistor, R_L, to calculate the AC power dissipation. The details could be found elsewhere [10]. The resonance frequency promotes maximum strain on the NW arrays and results the maximum piezoelectric open circuit voltage, (V_{oc}), and short circuit current, (I_{sc}), of the NEMS energy harvester. In this resonance frequency, the root mean square (RMS) voltage, (V_L), is usually determined across the external resistive load, (R_L), and thus resultant AC power, (P_L), experimentally from the NEMS energy harvester, as shown in Equation 7.22.

$$P_L = I^2_{L(RMS)} \, R_L = \left[\frac{V_{RMS}}{(Z_S + R_L)} \right]^2 R_L = \frac{V^2_{L(RMS)}}{R_L} \tag{7.22}$$

The peak AC power is dissipated when the external resistive load, (R_L), is matched with the source impedance ($Z_S = 1/(\omega_n C_p)$), where resonant frequency, (ω_n), as per maximum power transfer theorem. In this concept, the effect of leakage resistance on the overall impedance is considered as negligible.

7.6.5 Characterization and Analysis

The NEMS energy harvester is fabricated to harvesting the ambient mechanical vibrations up to maximum range of 1 kHz. The capacitance and impedance measurements from the NEMS energy harvesters were conducted using a high precision LCR (inductance, capacitance, resistance) meter. Mechanical vibration was generated from a miniature permanent magnet shaker, and the voltage measurements from the NEMS energy harvester under vibration excitation were performed using a voltage follower/buffer amplifier. The capacitance of the $BaTiO_3$ NW energy harvester at 1 kHz was measured 8.21 pF at 1 kHz. The sinusoidal excitation at resonant frequency yielded a high peak to peak voltage V_{MP} of ~623 mV from 1 g RMS (acceleration g = 9.8 m/s²; root mean square) base acceleration input, as shown in Figure 7.6c. The high voltage response is due to the high dynamic strain on the NW arrays that induced an alternating piezoelectric charge accumulation at the two electrodes. High I_{sc} values from the NW arrays are noticed by exciting with a sine wave at resonant frequency (~160 Hz) with the peak to peak current (I_{MP}) of ~1.8 nA at 1 g RMS (Figure 7.6c). High I_{sc} is directly proportional to the piezoelectric charge generation from the poled ferroelectric $BaTiO_3$ NW arrays when increased strain is applied through the resonating indium beam structure. In short circuit electrical boundary condition, the voltage is theoretically zero, so again the AC power is zero. The AC power from the energy harvester is calculated by measuring the voltage, V_L, across several load resistors, R_L, ranging from 1 to 500 MΩ. The source impedance, Z_S, of $BaTiO_3$ NW arrays with a capacitance of ~8.21 pF at natural frequency ($\omega_n = 2\pi f_n$, where f_n ~160 Hz) was evaluated to be ~121 MΩ. The AC power from the $BaTiO_3$ NW NEMS energy harvester increased rapidly as R_L increases up to 50 MΩ reaching a uniform peak value of ~125.5 pW at an optimal R_L of 120 MΩ [10]. The peak power density across the optimal R_L is calculated to be ~6.27 μW cm⁻³ from 1 g RMS base acceleration (Figure 7.6d). In addition, it was observed that the peak open circuit voltage and peak short circuit current levels at resonant frequency measured from the $BaTiO_3$ NW-based NEMS energy harvester are demonstrated to be more than 5 times greater and power density ~16 times higher than the response recorded from a semiconductor based ZnO-based NEMS energy harvester. Author reported that the achieved power density of the BaTiO$_3$ NEMS energy harvester (~6.27 µWcm^{-3}) is comparable to several meso-scale and microelectromechanical system-scale resonant cantilever based energy harvesters driven by base vibration.

7.6.6 Research and Development Scope

In order to enhance the recent market potential, consideration of nanostructured piezoelectric ceramics has significant influence including their synthesis and fabrication of testing module. Apart from this class of ceramics, several materials are highly studied around the globe. Dramatically, new families of NEMS might also overcome most of the practical barriers to utilize the entire range of vibrations through 1D nanomaterials and their classic alignment during fabrication of NEMS. The application segmentation for the NEMS market is broadly divided into three groups; tools and equipment, sensing and control, and solid-state electronics. Each of the applications can be further concentrated into different applications such as microscopy, automotive, medical, sensors, and memories. In a current report, an interesting analysis predicts that the global NEMS market is expected to reach $108.88 million by 2022 at an estimated compound annual growth rate of 29.69%. Recently, North America leads the NEMS market. In any case, it seems clear that a creative breakthrough, probably focused within the boundaries defined by intriguing basic research and device fabrication conducted over the last decade, is needed to reach this target.

7.7 Concluding Remarks

The main perspective is to demonstrate how beginners can involve in nanomaterial research in specific nanostructured ceramics in order to obtain renewable or alternate energy. However, one can develop more advanced materials for renewable energy research with consideration of few essential factors:

1. Selection of materials, their crystal structure, defect chemistry
2. Selection of suitable synthesis conditions
3. Apposite shape and size, and preferential crystal plane growth
4. Proper characterization of nanomaterials in each aspect prior to fabricate any device
5. Fabrication of test module and obtained data interpretation in the target of different testing parameters in order to validate the performance

Despite these selective ceramics TiO$_2$, WO$_3$, ZrO$_2$, and BaTiO$_3$, there are extensive scope to develop different class of materials. A close look at Figure 7.1 illustrates a loop is missing, thus, an opportunity may open up to make hydrogen gas by waste vibrational energy through such an effective nanostructured ceramics, still it is a question mark, which may reduce the consumption of fossil fuel and supply of energy without solar light (at night) as well.

References

1. C. T. Yip, H. Huang, L. Zhou, K. Xie, Y. Wang, T. Feng, J. Li, and W. Y. Tam, Direct and seamless coupling of TiO_2 nanotube photonic crystal to dye-sensitized solar cell: A single-step approach, *Adv Mater* 23, 5624–5628, 2011.
2. S. Adhikari and D. Sarkar, High efficient electrochromic WO_3 nanofibers, *Electrochim Acta* 138, 115–123, 2014.
3. S. Adhikari and D. Sarkar, Synthesis and electrochemical properties of nanocuboid and nanofiber WO_3, *J Electrochem Soc* 162(1), H58–H64, 2015.
4. B. C. H. Steele and A. Heinzel, Materials for fuel cell technologies, *Nature (London)* 414, 345–352, 2001.
5. D. Sarkar, Synthesis and properties of $BaTiO_3$ nanopowders, *J Am Ceram Soc* 94(1), 106–110, 2011.
6. B. O'Regan and M. Grätzel, A low-cost, high efficiency solar cell based on dye-sensitized colloidal TiO_2 films, *Nature* 353, 737–740, 1991.
7. J. Gan, X. Lu, and Y. Tong, Towards highly efficient photoanodes: Boosting sunlight-driven semiconductor nanomaterials for water oxidation, *Nanoscale* 6, 7142–7164, 2014.
8. A. Hauch, A. Georg, U. Opara Krašovec, and B. Orel, Photovoltaically self-charging battery, *J Electrochem Soc* 149(9), A1208–A1211, 2002.
9. H. Huang, M. Nakamura, P. Su, R. Fasching, Y. Saito, and F. B. Prinz, High-performance ultrathin solid oxide fuel cells for low-temperature operation, *J Electrochem Soc* 154 (1), B20–B24, 2007.
10. A. Koka, Z. Zhoub, and H. A. Sodano, Vertically aligned $BaTiO_3$ nanowire arrays for energy harvesting, *Energy Environ Sci* 7, 288–296, 2014.
11. N. Alonso, V. M. Beley, P. Chartier, and V. Ern, Dye sensitization of ceramic semiconducting electrodes for photoelectrochemical conversion, *Rev Phys Appl* 16, 5, 1981.
12. U. Bach, D. Lupo, P. Comte, J. E. Moser, F. Weissörtel, J. Salbeck, H. Spreitzer, and M. Grätzel, Solid-state dye-sensitized mesoporous TiO_2 solar cells with high photon-to-electron conversion efficiencies, *Nature* 395(8), 583–585, 1998.
13. S. Licht, Multiple band gap semiconductor/electrolyte solar energy conversion. *J Phys Chem B* 105, 6281–6294, 2001.
14. A. Fujishima and K. Honda, Electrochemical photolysis of water at a semiconductor electrode, *Nature* 238, 5358, 37–38, 1972.
15. M. Grätzel, The artificial leaf, bio-mimic photocatalysis, Baltzer Science Publications, 3(1), 4–17, 1999.
16. M. Gratzel, Photoelectrochemical cells—Review, *Nature* 414(15), 338–344, 2001.
17. R. Abe, Recent progress on photocatalytic and photoelectrochemical water splitting under visible light irradiation, *J Photochem Photobiol C* 11, 179–209, 2010.
18. Y. Sasaki, H. Nemoto, K. Saito, and A. Kudo, Solar water splitting using powdered photocatalysts driven by z-schematic interparticle electron transfer without an electron mediator, *J Phys Chem C* 113, 17536–17542, 2009.
19. G. Hodes, J. Manassen, and D. Cahen, Photoelectrochemical energy conversion and storage using polycrystalline chalcogenide electrodes, *Nature (London)* 261, 403–404, 1976.
20. M. Yu, W. D. McCulloch, Z. Huang, B. B. Trang, J. Lu, K. Amine, and Y. Wu, Solar-powered electrochemical energy storage: an alternative to solar fuels, *J Mater Chem A* 4, 2766–2782, 2016.
21. S. Souza, S. J. Visco, and L. C. De Jonghe, Thin-film solid oxide fuel cell with high performance at low-temperature, *Solid State Ion* 98, 57, 1997.
22. SECA-Coal and Power Systems. Netl.doe.gov. Retrieved on 27 November 2011.

8

Nanostructured Ceramics for Environment

8.1 Introduction

Global warming and modern civilization utilizes day to day advance technologies and luxurious lifestyle that may responsible increasing pollution exponentially either in air or water, results Global warming. Recently, technocrats and professionals around the world are involved to develop technology for comfort without enough concern about nature. However, several advance technology as already being developed to solve this issue through upcoming or ongoing environmental research by different classic materials or device fabrication. Despite wide horizon of materials, herein, specific groups of ceramics are well-thought-out to discuss how their nanoscale could be a smart choice either look after or serve a better environment through:

1. Protective electrochromic glass to control the sunlight transmittance and solar heat into building that reduce the energy consumption for cooling system and increase the occupant comfort

2. Photoassisted degradation and decolorizing of industrial effluent through advanced oxidation process

3. Piezoelectrochemical assisted pollutant removal and eutrophication in waste water

4. An effective and inexpensive nanofiltration of high salinity waste water to drinking water

5. Photofunctional self-cleaning coating or cloth in order to minimization of water consumption and avoid water scarcity

Thus, discussion on electrochromic smart glass made of hexagonal nanofiber WO_3 coating, a mixed photocatalyst WO_3 (monoclinic, nanocuboid)-ZnO (hexagonal, spherical nanoparticles) for effective textile dye degradation, piezoelectro chemical efficient tetragonal $BaTiO_3$ dendritic fiber for dye degradation, nanopore consist tetragonal ZrO_2 membrane for desalinization, and photoinduced superhydrophobic TiO_2 (anatase, nanofiber) self-cleaning coating are the major apprehension in this chapter [1–5].

Brief illustration of aforesaid utility is summarized in Figure 8.1. However, one can opt out for multivariant applications from these oxides as well, and in fact, it is subject to research. Apart from these token materials, there is also extensive scope to promote different classes of novel nanostructured ceramics and management for environmental issues.

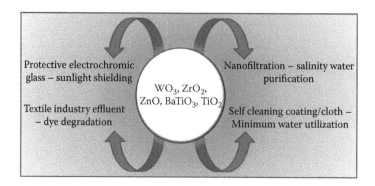

FIGURE 8.1

A schematic representation on the probable uses of semiconductor, piezoelectric, and ion conductive ceramics for environmental research. WO_3—electrochromic sunlight protective glass, WO_3/ZnO—dye degradation, $BaTiO_3$—textile industry effluent degradation, ZrO_2—saline water purification, and TiO_2—self-cleaning coating.

8.2 Nanostructured WO_3 for Electrochromic Smart Glass

8.2.1 Background

Electrochromism is a reversible color and optical change phenomenon influenced by electron charge, and that is the key feature to focus on improved electrochromic glass panels to protect the entering sunlight and thus radiation of solar heat in building through glass part. However, an optimum charge transported electrochromic window provide dynamic control over the transmission of solar radiation using a suitable material with high optical contrast, fast switching time, long cycle life, and low manufacturing cost. Unfortunately, the "smart glass" made of electrochromic material is not popular in commercial scale as conventional materials suffer either performance durability or manufacturing cost. Despite choice of different class of materials, crystal structure, morphology, and dimension of particles and there uniform coating on conductive glass are critical issues to get control over the ion intercalation followed by coloring and bleaching conversion during operational condition [6]. Assuming that a real-world electrochromic window undergoes one full cycle per day, a 30-year lifetime equates to a cycle life of about 11,000. Thus, an emerging industry standard includes testing to at least 50,000 cycles to serve consistent performance [7].

Usually, tungsten tri-oxide (WO_3) is the most widely studied electrochromic inorganic compounds having enhanced properties, including faster ON/OFF switching times and amenability to potentially low-cost solution processing. Upon cathodic charge injection, WO_3 changes color from a clear, transparent state to a dark blue and subsequently translucent state as tungsten ions are being reduced. It is opted as an efficient intercalation host to produce tungsten oxide bronzes by insertion of electrons and protons or metal ions like Li^+, Na^+, K^+, etc., into the WO_3 crystal structure. This class of material consisting of hexagonal crystal structures has considerable attention due to their dimensional confined transport phenomena and effective intercalation chemistry because of open tunnel like structure [8]. In the previous perspective, a brief illustration is highlighted on the probable utility of a representative one-dimensional hexagonal WO_3 nanofiber for the electrochromic application.

8.2.2 Materials and Method

In the present methodology, base precursor sodium tungstate ($Na_2WO_4 \cdot 2H_2O$), structure directing agent sodium chloride (NaCl), and catalyst HCl were used to synthesize WO_3 nanofibers. One-dimensional pure hexagonal WO_3 nanofiber has been successfully grown in the presence of optimum hydrothermal temperature of 180°C for 12 h and 4.5 M NaCl concentration. A classic explanation has been demonstrated to support this optimized condition in Adhikari and Sarkar [1]. Intermediate concentration develops a discrete nonconforming plate like particles, because of insufficient availability of Na^+ ions, whereas preferential nanofiber growth is evident when concentration enhancing to 4.5 M. A mixture of particles with spherical, plate, and rod-shaped morphology depicts at 160°C, whereas agglomerated rod-like particle is found at temperature of 170°C. This observation reveals that 180°C is the minimum required temperature for the growth of nanofibers. Hydrothermal duration of 4 and 6 h initially develops rod-like morphology with some nonuniform plates. Although, all the powders have hexagonal crystal phase, but initiation of nanofiber formation started from 8 h duration, and almost uniform nanorods are seen after 10 h duration, where beyond 12 h, fibers become agglomerated bundle. The average particle size is found to be length ~256 nm and thickness ~30 nm, respectively. A homogeneous dispersion of WO_3 nanofiber particles in ethanol was prepared and dip coated onto a transparent conducting oxide substrate having an electrode dimensions: 2×1 cm and area 2 cm² to evaluate the electrochromic effect obtained from the optimized WO_3 nanofibers. The dip coated samples were dried at 80°C for 30 min and measured their optical transparency and cross-sectional thickness. The three electrode cell configuration consist of Platinum (Pt) as a counter electrode, saturated Ag/AgCl as a reference electrode, and as prepared WO_3 films were used as the working electrode in 1 M H_2SO_4 electrolyte solution. Bare indium tin oxide (ITO) glass substrate was taken as a reference electrode for transmittance measurement followed by fabrication of cell and electrochromic study [1].

8.2.3 Working Principle

An emphasize has been given as to how nanostructured WO_3 become a potential candidate for electrochromic device, and consequently, a brief schematic and working principle of electrochromic set-up analogous to smart window is depicted in Figure 8.2a. Herein, a multi-layer device consisting of an active electrochromic electrode layer, a counter electrode layer, an electrolyte layer separating the two electrodes, two transparent conducting layers (ITO) serving as electrical leads, and the supporting substrates. This device structure is referred to as "battery type" and is the most common geometry for electrochromic devices. When the device is switched on, an applied voltage between the opposing conducting layers drive cations (e.g., Li^+ or H^+) to migrate from the counter electrode, through electrolyte, and into the electrochromic electrode through Faradaic process. This eventually changes its oxidation state and its optical properties. An effective electrochromic device will have fast switching between its "on" and "off" states, good durability characterized by a long cycle life, and a high optical contrast ratio. Finally, the durability of any electrochromic device is established by measuring the changes in charge capacity or coloration efficiency over many thousands of electrochemical cycles.

To compare performance among different electrochromic materials and devices, researchers used the degree of coloration efficiency as a key figure of merit. Coloration efficiency is governed by optical charge density (ΔOD), transmittance percentage during

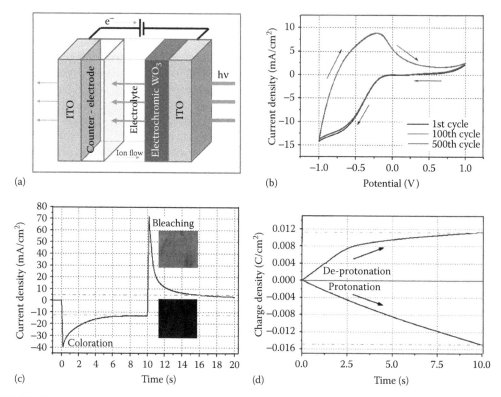

FIGURE 8.2
(a) An electrochromic device stack, in which electrons flow through an external circuit in the electrochromic WO₃ and ions flow through electrolyte to compensate the charge, (b) CV of 1st, 100th, and 500th cycles for nanofiber WO₃/ITO electrode performed at 100 mV/s, (c) chronoamperometry, and (d) chronocoulometry measurement of WO₃/ITO electrode in 1 M H₂SO₄ under potential sweep of ±1 V for 10 s each. Inset of Figure 8.2c represents the colored and bleached WO₃.

colored (T_c), and bleached (T_b) state, and amount of charge intercalated, (Q_i), and effective working area, (A), so one can estimate the efficiency by following equations:

$$\Delta OD = \log\left(\frac{T_b}{T_c}\right) \tag{8.1}$$

$$CE \text{ or } \eta = \frac{\Delta OD}{Q_i/A} \tag{8.2}$$

In general, materials with higher coloration efficiencies will have better durability and faster switching times, since less charge is required to produce a given optical response.

8.2.4 Characterization and Analysis

Cyclic voltammetry (CV), chronoamperometry, and chronocoulometry (CC) techniques were used to investigate the electrochemical properties for WO₃/ITO electrodes in 1 M H₂SO₄ under potential sweep of −1 V and +1 V at 100 mV/s. In addition, the electrochemical protonation and deprotonation study was carried out to understand the intercalation phenomena. Dip coated WO_3 thickness was optimized in the perspective of H^+ intercalation, charge transfer behavior, and optical transparency. High coating thickness comprises more charge intercalation, whereas both cathodic and anodic current peak move towards positive potential region. It suggests that the reduced charge transfer resistance in the interface during cathodic reaction and deintercalation during anodic reaction in thinner layer occurs faster at lower voltage. Interestingly, the current density, (8.9 mA/cm²), not varying beyond the optimum thickness ~11 μm and considered to determine next step electrochromic properties. The current response for such $nFWO_3$/FTO electrode remains near identical in shape up to 500 cycles and supports their excellent cyclic stability (Figure 8.2b). The hexagonal tunneling zone facilitate ion intercalation of electrons and H^+ ions into the electrode causes reduction from W^{6+} to W^{5+}, and, hence, the electrode gets colored due to formation of disordered tungsten bronze structure. The current is found to increase steadily from negative to a maximum value and then decreases to zero corresponding to the coloration and bleaching of the electrodes at varying voltages. During each of this scan, the electrode undergoes typical reversible color change mechanism from blue to colorless, whereas reduced state appears bluish in color. The obtained current is recorded is during H^+ intercalation/deintercalation and the electron transfer process according to the following electrochemical reaction:

$$WO_3 \text{ (colorless)} + xe^- + xH^+ \rightarrow H_xWO_3 \text{ (blue tungsten bronze)} \tag{8.3}$$

The bleached peak current density is found much higher than that of coloration effect, however, the current density during bleaching decays faster than coloration. The ease mode of ion penetration during repeated cycles resulting in high anodic peak current. Apart from the current density, coloration/bleaching duration and charge storage capacity for 10 s data obtained by chronoamperometry and CC are represented in Figure 8.2c and d, respectively. In the beginning, the negative biased voltage (–1.0 V) allows H^+ intercalation from the electrolyte and immediate reversed back to equivalent positive voltage for the deintercalation occurrence. Current density increment under negative voltage and subsequent relative faster decay phenomenon are recorded as 6.50 and 4.02 s after saturation, respectively. The coloration kinetics under negative voltage is slower than the bleaching kinetics, which is governed by the interfacial potential barrier generation between electrolyte and WO_3 and the space charge transfer through the electrode that controls the two processes separately. The coloration is faster for thick WO_3 coated electrode, but bleaching is faster for thin coated electrode. Negative potential scan facilitates the charge intercalate ion by reduction of W^{6+} to W^{5+} states and vice versa. The quantified data of charge intercalation and deintercalation from the plot have been used to calculate the reversibility of the intercalation/deintercalation methods (Figure 8.2d). The reversibility of the film can be calculated from the following relation mentioned in the following:

$$\text{Reversibility} = \frac{Q_{di}}{Q_i} \tag{8.4}$$

where Q_{di} (0.0115 C/cm²) and Q_i (0.0149 C/cm²) refer to the amount of charge deintercalated and intercalated in the $nFWO_3$/FTO electrodes, respectively. The percentage electrochromic reversibility is estimated near to 77.48% and exhibits consistent with respect to time and recycling. Therefore, it is obvious to believe that reversibility of the electrodes also remains stable till 500th cycle. Optical transparency of WO_3/ITO colored and WO_3/ITO bleached electrodes were measured under wavelength range between 200 and 800 nm. The optical transmittance has been found to decrease with the charge intercalation of the electrodes and subsequent increase upon the bleaching of electrodes. Herein, the obtained transmittance values for colored and bleached nanofiber WO_3/ITO electrode at visible wavelength of 550 nm are found as T_c (colored) = 22% and T_b (bleached) = 65%. The coloration efficiency for effective working area $A \approx 2$ cm^2 was calculated as 63.15 cm^2/C (Equation 8.1) because of WO_3 nanofiber has high efficiency to hold ions in its hexagonal crystal structure and optimum ($\sim$11 μm) thickness to control over the coloration/bleaching response and coloration efficiency.

8.2.5 Research and Development Scope

Despite conventional material synthesis and characterization, several possibilities opened up to explore the development of pilot scale module in target of the commercialization of electrochromic smart glass for building and construction purpose. However, some factors are essential to consider for the making effective color conversion and optical transparency as well. Many factors can influence switching speed, including the electronic conductivity of the electrode materials and the underlying conducting layers, the ionic conductivity of the electrolyte, the morphology of the electrochromic layer and the associated changes in ionic diffusion within that morphology, and ion insertion kinetics. Thus, WO_3 has promising future to make effective electrochromic device including large piece as smart glass. Recently, smart glass windows are manufacturing of any size glass panel up to 6 × 10 feet and ability to switch-over to the environment automatically.

8.3 Nanostructured WO_3-ZnO for Pollutant Dye Degradation

8.3.1 Background

Environmental pollution is the emerging and serious problem in most of the developing and developed nations. Anthropogenic activities contribute to the major imbalance of ecosystem via pollution made by air, water, and solid wastes. Most of the common pollutants include highly toxic organic pollutants consist of aliphatic and aromatic compounds with chlorine, agro-wastes like pesticides and insecticides, disinfected by-products, and more. Inorganic compounds like heavy metals, obnoxious gases like SO_x, NO_x, CO, and pathogen also contributes in pollution of the environment. Apart from these pollutants, untreated dyes, surfactants, and detergents are directly disposed into the water from different industries. Organic textile dyes came up as one of the many new chemicals which could be used in many industrial activities. Extensive use of these dyes in industries has become an integral part of industrial effluent. With escalating revolution in science and technology, demand for newer chemicals is rising, which could be used in various industrial processes as well. People living in undeveloped region are supposed to use effluent mixed polluted water for their daily household needs. Sometimes, due to scarcity of water, people are also using polluted water for drinking and irrigation purposes. There are legislations relating to the safe disposal with clean and green processes, but lack in implementation may cause disaster later in the environment. Thus, scientists are coming together and searching for methods that can treat pollutant like dyes mixed in industrial waste before releasing to the environment. In this perspective, photodegradation of textile dyes using nanoparticles is one of the best options and attracts much attention because of ease mode of application and economic as well. Thus, the metal oxide semiconductors are found as a potential candidate and most researched for photooxidation of dye under influence of natural or artificial source of light. In this section, brief insight of metal oxide semiconductors assisted photodegradation of typical cationic and anionic dye has been discussed in view point of environmental research.

8.3.2 Materials and Method

WO$_3$ nanocuboid was hydrothermally synthesized using base precursor Na$_2$WO$_4$.2H$_2$O and structure directing agent 4M fluoroboric acid (HBF$_4$) at 180°C for 6 h; different from precursors used to synthesis of WO$_3$ nanofiber. The basis of process optimization for WO$_3$ nanofiber has been discussed in previous section, and one can follow the identical optimization protocol for WO$_3$ nanocuboid in an extensive literature [9]. In another end, zinc oxide (ZnO) nanoparticles were synthesized by wet chemical combustion method. In a typical auto-combustion synthesis, stoichiometric quantities of zinc nitrate hexahydrate and oxalic acid were taken in crystallization dish (50 × 100 mm) and dissolved in minimum amount of water at room temperature to form clear solution. The prepared aqueous solution was kept in a preheated muffle furnace at 450°C ± 10°C until complete combustion. Initially, the furnace was opened to allow passage of sufficient air for the completion of chemical reaction and formation of white porous mass. Herein, the obtained ZnO quasi fiber process optimization is described in detail somewhere else [10]. The obtained dry, porous material was ground using mortar and pestle and used different weight percentage for the preparation of WO$_3$-ZnO mixed oxides for photocatalytic reaction. The mixture was then suspended in 20 mL water followed by ultrasonication for 1 h to form a colloidal suspension. The solution was oven dried at 80°C and subsequently heat treated at 450°C for 2 h in air to ensure the effective mixing of photoactive mixed oxide catalyst. The mixed oxide catalysts were characterized through XRD, TEM, UV-Diffuse Reflection Spectroscopy (DRS), and photoluminescence at 350 nm. Nevertheless, other characterizations and analysis including morphology and surface area are also discussed.

8.3.3 Working Principle

In practice, the combination of two different semiconductors has high degree of dye degradation efficiency compared to single photocatalyst, and thus a brief working principle of such system provide an awareness of the subject. Single semiconductor has either wider band gap restricts the electron-hole pair generation under visible region or narrow gap facilitate the rapid recombination of electron-hole pair and thus low degree of dye degradation performance.

In this context, herein, two nanostructured metal oxide semiconductors ZnO (~3.2 ev) and WO$_3$ (~2.6 ev) are synthesized and mixed to use as a combined catalyst

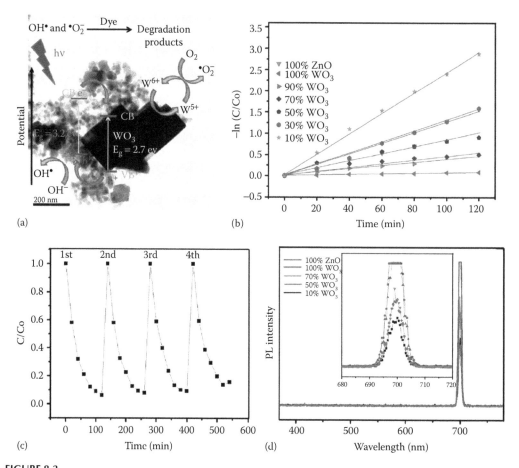

FIGURE 8.3

(a) Radical formation and dye degradation mechanism by 10 wt% WO₃/ZnO in presence of simulated solar light, (b) kinetic studies of MB with different mixed oxide composites, (c) MB degradation profile on reusability of the 10 wt% WO₃-ZnO mixed oxides, and (d) composite photoluminescence spectra of WO₃, ZnO, and WO₃-ZnO mixed oxide composites with different WO₃ loading.

for effective dye degradation. Figure 8.3a shows the TEM image of optimum content 10 wt% WO₃ cuboid mixed spherical ZnO and their plausible electron-hole formation and transformation followed by dye degradation under absorption of visible light. In this process, an optimum content of mixed oxide is beneficial for such concept and discussed in Section 8.3.5. The photogenerated electron (e^-) in conduction band of ZnO transfer to the lower energy conduction band of WO₃, subsequently the photogenerated hole (h^+) in valance band of WO₃ shift to the valance band of ZnO. The transfer of photo-generated carriers was accompanied by consecutive W^{6+} reduction into W^{5+} and captur-ing photogenerated electrons at trapping sites in WO₃. At the same time, W^{5+} ions on the surface of WO₃ nanocuboids were reoxidized into W^{6+} by oxygen that was subsequently reduced into $\bullet O_2^-$.

The recycling of the electron leads to an increase in the lifetime of the photogenerated pairs and as a consequence to promotion of the photonic efficiency for the degradation of organic pollutants. Holes were then captured by surface hydroxyl groups, (OH^-), on the ZnO photocatalyst surface and yielded hydroxyl radicals $\bullet OH$. The formed superoxide anions, (•O_2^-), may either attack the organic molecules of dyes directly or generate hydroxyl radicals, (•OH), by reacting with photogenerated electrons and hydrogen ions, (H^+). Finally, the strong oxidizing agents hydroxyl radicals, (•OH), degrade the both cationic and anionic organic dyes. The charge separation and probable photocatalytic reaction follows four steps reaction and can be expressed through Equations 8.5 through 8.11, as given in the following:

Step 1: Electron-hole pair generation:

$$WO_3 + h\nu \rightarrow WO_3 \, (e^- - h^+) \tag{8.5}$$

$$ZnO + h\nu \rightarrow ZnO \, (e^- - h^+) \tag{8.6}$$

Step 2: Charge transfer/recycling reaction:

$$\frac{WO_3 \, (e^- - h^+)}{ZnO \, (e^- - h^+)} \rightarrow \frac{WO_3 \, (e^-)}{ZnO \, (h^+)} \tag{8.7}$$

Step 3: Radical formation:

$$WO_3 \, (e^-) + O_2 \rightarrow WO_3 + \bullet O_2^- \tag{8.8}$$

$$ZnO \, (h^+) + OH^- \rightarrow ZnO + \bullet OH \tag{8.9}$$

Step 4: Dye degradation:

$$Dye + \bullet O_2^- \rightarrow Mineral \; Acids + CO_2\uparrow + H_2O \tag{8.10}$$

$$Dye + \bullet OH \rightarrow Mineral \; Acids + CO_2\uparrow + H_2O \tag{8.11}$$

An additional tertiary-butyl alcohol (•OH scavenger) and benzoquinone (•O_2^- scavenger) can change the rate of photodegradation compared to mixed oxide under visible light radiation. Details about scavenger are discussed in Section 4.2.3.

8.3.4 Characterization and Analysis

The photocatalytic experiment was carried in a set-up consisting of jacketed quartz tube with jacketed pyrex slurry reactor, and the entire assembly was enclosed in a rectangular hard wood casing. Water supply was provided through jacket of quartz tube and slurry reactor to avoid unwanted thermal degradation and maintained temperature at 29°C ± 2°C. The dye solution was kept in a slurry reactor and 400 W metal halide lamp was positioned 2 cm above from the just top level of the suspension of mixed semiconductor, (WO_3/ZnO), nanoparticles. A distinct cuboid morphology of soft agglomerated WO_3 nanoparticles consist of an average particle dimension ~152 nm length and ~120 nm width mixed with ~60 nm ZnO average particle size of the primary sphere was used in this research work (Figure 8.3a). In the present system, photocatalytic performance of representative cationic methylene blue (MB) and anionic orange G (OG) dye was studied in the presence of WO_3 (0, 10, 20, 30, 40, 50, and 100 wt%) mixed with ZnO photocatalyst. Further detail characterization and their analysis modules are described in the literature [2]. In the earlier combination, 10 wt% WO_3 loaded ZnO exhibits faster and highest decolorization activity compared to either pure ZnO or pure WO_3. High decolorization efficiency of 93% and 89% was observed for MB and OG, respectively. Among these two ionic dyes, a typical photocatalytic degradation kinetics behavior of cationic MB is shown in Figure 8.3b to elucidate the effect of mixed semiconductors. The photocatalytic degradation is governed by the kinetics of:

$$-\ln(C/C_o) = kt \tag{8.12}$$

where C is the concentration at any time (t), C_o is the initial concentration, k is the apparent reaction rate constant and t is the reaction time. The reaction followed pseudo first-order and kinetic rate constant (k, min^{-1}) was obtained from the slope of linear plot of $-\ln(C/C_o)$ with t. With decrease in WO_3 content in ZnO, the rate constant value increases, thereby increasing the photocatalytic efficiency. Maximum rate constant of 0.0231 min^{-1} and 0.0198 min^{-1} is found in presence of 10 wt% WO_3-ZnO mixed oxide composite for MB and OG, respectively. The probable reason for increased activity is attributed to the charge separation mechanism in the WO_3-ZnO mixed composites. The probability of electron-hole recombination reduces with an optimum WO_3 content. The reduction in recombination is further supported by the photoluminescence spectra of the mixed composites, as discussed in later. Although WO_3 has visible light absorption, but lower activity is observed for its narrow band gap resulting in high recombination rate. Therefore, coupling of WO_3 with other semiconductor like ZnO facilitate the photochemical activity because of optimum charge separation and thereby dye degradation. Multiple use assessment predicts the long term performance and economic viability of photocatalyst. The reusability of optimum 10 wt% WO_3 mixed ZnO describes the consecutive photocatalytic degradation efficiency for MB as shown in Figure 8.3c. Each time centrifugation of the catalyst was carried to remove the solid catalyst and the catalyst was dried at 100°C for further use. Initially, 93% MB degradation is found in 1st run, which gradually decreases as 92%, 90%, and 84% in 2nd, 3rd, and 4th run, respectively. The probable reason for the slight performance decrement can be summarized as:

1. Surface leaching during photocatalytic reaction attributing to the loss of active support sites without change of any crystal structures for both semiconductors

2. Consecutive heat treatment after each cycle enhances the agglomeration results in lowering the active surface area of the catalyst

3. Organic intermediates that are formed during the catalytic process can also adsorb on the surface, thereby reducing the overall efficiency of the photocatalyst

4. Mass loss of catalyst occurs during repetitive runs resulting in reduced photoreactivity

The photoluminescence (PL) spectrum in time predicts the rate of electron-hole pair recombination phenomenon during photocatalysis process that actually suppresses the dye degradation rate of reaction. Figure 8.3d represents the composite PL spectrum of ZnO, WO_3, 70 wt% WO_3, 50 wt% WO_3, and 10 wt% WO_3, respectively. A combination of peak shifting and intensity variation is observed with respect to addition of WO_3 photocatalyst. The shift of absorption intensity toward higher wavelength is observed well in the absorption spectra as discussed in the literature [2]. When the samples are excited at this particular wavelength, an intense emission peak appears at ~700 nm wavelength for all the samples. Unlike absorption spectra, the PL emission spectra do not shift to higher wavelength, but shows change in intensity with respect to WO_3 loading. The change in intensity follows the sequence of 10% WO_3 < ZnO < 50% WO_3 < 70% WO_3 < WO_3, respectively. The PL intensity is suppressed in the presence of 10% WO_3 with ZnO. Pure WO_3 shows the maximum intensity that depicts the highest recombination rate. Intensity reduction directly relates to the suppression of the electron-hole recombination rate. The reduced recombination rate reveals the efficient charge transfer within the catalyst mixture. This means that more hydroxyl radicals •OH can be produced in the system containing 10% WO_3-ZnO compared to pure WO_3, which is advantageous to the visible light photocatalytic activity of WO_3. Increasing WO_3 content may lead to low electron accepting efficiency of ZnO due to less amount of ZnO in nanocomposites. This can prove as a disadvantage to the visible light photocatalytic activity of the composites. Consequently, the visible light photocatalytic activity of the investigated composites increases at first and then decreases as the WO_3 content increases. It is observed that the absorption band is in the wavelength between 380 and 520 nm for all WO_3, ZnO, and WO_3-ZnO mixed oxide composites. Broad tails are observed for WO_3-ZnO mixed oxide samples. The mixed oxide composite shows a slight red shift of band gap absorption in comparison to the pure ZnO. With addition of WO_3 to ZnO, there is the formation of the defect energy levels within the forbidden band that initially increases the band gap energy to 2.98 eV and then decreases with increased loading of the WO_3 in the mixed oxide composites. Thus, the decreased PL intensity of the hierarchically structured ZnO/WO_3 nanocuboid indicates a high quantum efficiency and photocatalytic activity.

8.3.5 Research and Development Scope

Textile industries consumes billion gallon of fresh water and discharges near to same amount of waste water consisting of highly structured dye stuffs that are difficult to be broken down biologically. Apart from different techniques, waste water treatment by photocatalyst is a popular practice in presence of dispersed semiconductor particles in liquid medium that form thin films on the supported materials. Layer-by-layer low cost coating technique commensurate the formation of thin films on supported materials like glass plates, perlite granules, steel fibers, and aluminum foils. Despite such coating techniques, making an effective porous membrane made of semiconductor particles can promote pollutant removal when industrial effluent passing with an optimum turbulence and flow, although performance depends on several factors, and it is subject to identify. In addition, modification by noble and transition metals and their content, band gap design by different combination of semiconductors, particular dose, high surface area promote photoassisted degradation of waste water color pollutants. This process is eventually advantageous because of possibility of complete disintegration of the pollutants, utility of abundant solar light, no need of additional chemicals, and low operational cost at near room temperature. Thus, we need to focus and work together from multidiscipline communities for the development of effective material and technology to provide pollutant free water for different purposes and societies.

8.4 Mechanically Induced Waste Water Treatment

8.4.1 Background

The increased environmental consciousness demands for the pollutants (organic like dyes and inorganic constituents) in the water bodies to be removed/degraded as they may undergo chemical reactions to form harmful by-products. The traditional nondestructive techniques like activated carbon adsorption, ultrafiltration, reverse osmosis, etc. can only transfer the pollutants present in one medium to another, but cannot completely destroy those long chained molecules. Various other processes like ozonation, chlorination, and biological treatments are proposed for dye destructions, but frequently operational complicacy, relatively high operating costs, and ineffectiveness over large amounts of dye molecules limit their applicability. Recently, advanced oxidation processes and semiconductor photocatalysts, (TiO_2, ZnO, WO_3), are being utilized for dye (organic chemicals) degradation and are found to be effective, nontoxic, and inexpensive, but they too have some limitations like solar efficiency and corrosion under low pH conditions, which further reduce the degradation efficiency [11,12].

While discussing the dye removal through photocatalyst, another cheap and economic approach, namely, mechanically induced waste water treatment is now drawing attention in environmental research and development. In this context, a phenomenon called piezoelectrochemical (PZEC) effect that directly converts the mechanical energy to chemical energy is recently studied to know its potential use for dye decomposition. This mechanism depends on the piezoelectric properties of the catalysts for the dye degradation and can be utilized scavenging waste energy (noise, vibration, etc.) as well [13]. The process at a glance is shown in Figure 8.4a and described in detail in the following sections with help of a token material, ($BaTiO_3$), consist of PZEC property. $BaTiO_3$ has many advantages like tailor-made synthesis and control over appreciable amount of piezoelectric properties (as required by PZEC). Herein, the microdendrite $BaTiO_3$ structure has been attempted in view of more strain ability as well as high surface area adsorb the dye/organic impurities for the probable degradation.

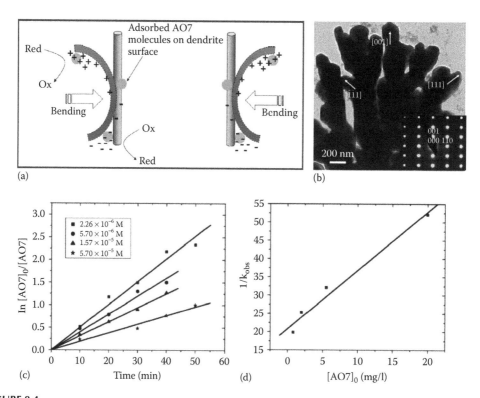

FIGURE 8.4
(a) Schematic representation on the charge generation by dendritic and piezoelectric $BaTiO_3$ and corresponding decomposition of AO7, (b) TEM image showing the $BaTiO_3$ dendrite structure, (c) plot for the pseudo first order kinetic rate constant, and (d) plot for the determination of adsorption equilibrium constant for the Langmuir-Hinshelwood kinetic model.

It is a stable compound over a wide range of pH and accordingly can be operated. The advantage of utilizing waste energy (vibrations/noise) to operate because of its piezoelectric properties makes it more economical for industrial/commercial purpose [3]. Advantage of $BaTiO_3$ as a piezoelectric material is discussed in Section 7.6.

8.4.2 Materials and Method

The synthesis of $BaTiO_3$ dendrite samples, (PZEC catalyst), was done by hydrothermal reaction method through 99.99% pure chemicals. Initially, $Ti(OH)_4$ precursor was prepared from $Ti(OC_2H_5)_4$ and acetic acid, where 25 mL of $Ti(OC_2H_5)_4$ was added drop wise to 1.0 M acetic acid and kept aside for 72 h to settle down the precipitate. After that, the product was rinsed with deionized (DI) water and dried at 60°C. The above synthesized $Ti(OH)_4$ and $Ba(OH)_2 \cdot 8H_2O$ (commercial grade) were added into 0.25 M NaOH in 1:1 molar ratio, (Ti/Ba). This mixture was then placed in a Teflon cup (60% capacity) and stirred. The cup was sealed tightly in a stainless steel autoclave called as digestion bomb. Hydrothermal reaction takes place as the sealed bomb (Parr-type) was kept at 200°C for 6 h. After that, it was air quenched at room temperature and a white precipitate formed. This white precipitate was then cleaned to remove adsorbed impurities by DI water and then cooled in air at room temperature. A known concentration (ranging from 10^{-6} to 10^{-4} mol/L) of commercial grade 4-(2-hydroxy-1-naphthylazo) benzene sulfonic acid sodium salt, (AO7), also known as C. I. Orange and predetermined quantity of synthesized $BaTiO_3$ microdendrite powder were mixed to estimate the degradation behavior of AO7. All of the reactions/experiments were carried out at 25°C temperature and in dark conditions to eliminate the chances of additional photocatalytic reaction [3].

8.4.3 Working Principle

The working of the $BaTiO_3$ on water purification is described based on the PZEC phenomenon. According to this hypothesis, the degradation of organic compounds in solution (depends on vibration used) takes place as the deformation (i.e., strain) of the piezoelectric material is started, followed by the accumulation of charge generation on the material surface. Therefore, the strained piezoelectric material triggers the redox reactions by which simultaneous decomposition of water and disintegration of AO7 by PZEC effect takes place in the system. For the purpose of understanding, this disintegration of AO7 by PZEC mechanism can be viewed as analogous to that of photo catalytic decomposition, but reality is different. The only difference being the origins of the electron-hole pairs, in former strain assisted charge separation in PZEC $BaTiO_3$ system, whereas later follows photocatalytic activation of photoelectrons in the conduction band. Strain assisted degradation process mechanism can be summarized as:

$$BaTiO_3 + vibration \rightarrow BaTiO_3 \ (e^- + h^+) \tag{8.13}$$

Reaction at anode:

$$4e^- + 4H_2O \rightarrow 4OH^- + 4H^{\bullet}$$

$$4H^{\bullet} \rightarrow 2H_2 \tag{8.14}$$

$$\overline{4e^- + 4H_2O \rightarrow 4OH^- + 2H_2}$$

Reaction at cathode:

$$4OH^- \rightarrow 4e^- + 4OH^\bullet$$

$$2(OH^\bullet + OH^\bullet) \rightarrow 2H_2O + 2O^\bullet$$

$$\underline{2O^\bullet \rightarrow O_2}$$ \hfill (8.15)

$$2OH^- \rightarrow 4e^- + 2H_2O + O_2$$

Overall water decomposition reaction:

$$2H_2O \rightarrow 2H_2 + O_2 \hspace{3cm} (8.16)$$

Dye disintegration reaction:

$$OH^\bullet + AO7/dye \rightarrow disintegration\ product\ of\ dye$$

$$e^- + AO7/dye \rightarrow disintegration\ product\ of\ dye \hspace{1cm} (8.17)$$

$$h^+ + AO7/dye \rightarrow disintegration\ product\ of\ dye$$

In presence of vibration, $BaTiO_3$ dendrite changes according to Equation 8.13 and can be presumed as two electrodes because of charge separation. The reaction at anode occurs according to Equation 8.14, as the electrons induced on the surface of negatively charged sides of $BaTiO_3$ microdendrite (due to PZEC effect) attracts the hydrogen from the water molecules and produce hydrogen radical ($H^\bullet$) and forms H_2 gas through combine with another $H^\bullet$. In the reaction at cathode, the $OH^\bullet$ are released by the holes that were accumulated on the positively charged side of the $BaTiO_3$ that attracts the electrons from hydroxyl group as shown in eqn. 8.15. These $OH^\bullet$ which could also be formed by the oxidation of organic molecules/OH^-/H_2O by holes induced from PZEC effect in the system are the main cause for the degradation of AO7/dye in the system. The partial/complete oxidation of AO7 according to Equation 8.17 leads to formation of several products along with benzene sulfonate and naphthoquinone as primary products [14].

8.4.4 Characterization and Analysis

This section describes different characterization techniques that are used to understand the working behavior of the prepared $BaTiO_3$ microdentrides. The high resolution TEM and Selected area electron diffraction (SAED) pattern of $BaTiO_3$ dendrite are shown in Figure 8.4b. A characteristic time dependent analysis depicts PZEC ($BaTiO_3$ with vibration) is near to 100 times more effective in 50 min compared to independent $BaTiO_3$ and vibration separately. However, the amount of AO7 degradation increased with the increase in amount of $BaTiO_3$ up to some extent (0.025 g) and reaches a flat region and then decreases slightly. This phenomenology can be explained as the total surface area increases with the increase in the amount of BaTiO$_3$ microdendrites and after certain amount there is a more possibility of opposite charged surfaces of microdendrites being in contact, which decreases the effective surface area left for sorption of AO7 and their degradation. Thus, there is a slight decrease in the degradation of AO7 after certain amount. The kinetics of the degradation reaction of AO7 by PZEC effect can be expressed by Langmuir-Hinshelwood model. According to this model, in this PZEC effect, the rate determining step involves the oxidants and reductants in a monolayer at the dendrite surface-liquid interface and all other oxidation and reduction reactions are considered as rapid equilibrium process. The decomposition/degradation of AO7 by PZEC effect exhibits pseudo first order kinetic with respect to AO7 concentration:

$$-\frac{d[AO7]}{dt} = k_{obs}[AO7] \tag{8.18}$$

$$\text{At } t = 0, [AO7] = [AO7]_0$$

$$\ln\frac{[AO7]_o}{[AO7]} = k_{obs} \times t \tag{8.19}$$

where, k_{obs} is pseudo first order rate constant.

From Equation 8.19, the rate constant can be obtained for different initial dye concentrations as shown in Figure 8.4c [3]. The initial degradation rate, (r), and initial dye concentration for PZEC degradation are related as given in Equation 8.20.

$$r = k_c \frac{K_{AO7}[AO7]}{1 + K_{AO7}[AO7]_o} = k_{obs}[AO7] \tag{8.20}$$

where K_{AO7} is the Langmuir-Hinshelwood adsorption equilibrium constant, k_c is the second order rate constant. These constants can be determined from Figure 8.4d and the value of K_{AO7} was 0.149 (mgL^{-1})$^{-1}$, while k_c was found to be 0.50 mgL^{-1}min^{-1}. However, details on the effect of initial AO7 dye concentration and pH on the decomposition efficiency can be found in a pioneer work done by Kuang-Sheng Hong et al. [3].

8.4.5 Research and Development Scope

While discussing the utilization of piezoelectrochemical effect of perovskite material for waste water treatment, semiconductor ZnO could be another choice of material if piezoelectric properties and photocatalytic properties coupled in a single internal physical-chemical process [15]. Apart from these class of materials, it is essential to consider different materials in order to synchronize the efficiency in specific crystal structure, morphology, strain effect, and mechanical to chemical energy conversion factors, etc. A persistent coating of suitable piezoelectric material on conductive glass and their placement in the bottom of waste water reservoir may produce consistent wave and vibration that induce the PZEC effect and dye degradation without any additional effort. Tailoring the nanoscale size and morphology of piezoelectric materials has economic potential in this aspect. Interesting to mention that a combined effect of photocatalytic and PZEC effect may further boost up the pollutant degradation process.

8.5 Purification of High Salinity Waste Water

8.5.1 Background

Nanofiltration (NF) is in forefront to reduce and provide more safe water and protect human as well as aquatic life when micro filters (MF) and ultra-filters (UF) are unable to purify the industrial waste like pharma, pesticide, agrochemicals, and more in completely. These NF's have a nominal pore size in range of 0.5–2 nm and a molecular weight cut-off (MWCO) of 400–500 Da (Daltons) [16]. High operation pressure in the range of 7–30 atmosphere (higher than MF and UF) is desirable to pass the water through these pores. NF's are capable of removing all cysts, bacteria, viruses, and humus materials and also hardness (due presence of ions such as Ca, Mg, etc.) from water as shown in Figure 8.5a, that enables to provide another name called "softening membranes." Thus, they are mainly used to produce soft water for industrial usage or potable water. Polymer NF membranes are commercially available in market, but may damage in elevated temperature, undergo biodegradation, and requires a narrow pH range (4–8) for their operation. To overcome this situation, recently, crack-free, pure TiO_2 NF membrane (MWCO-800 Da) was prepared via a colloidal sol-gel route [17]. Different types of ceramic materials like γ-Al_2O_3, TiO_2, SiO_2, TiO_2-ZrO_2, and SiO_2-ZrO_2 have also been used

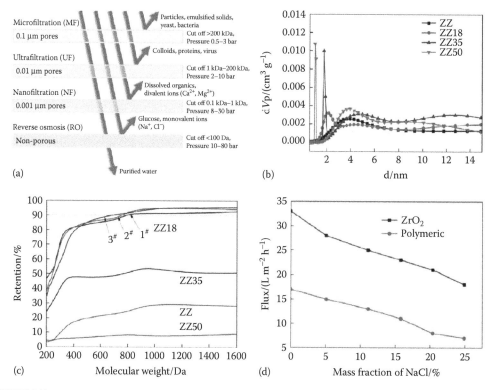

FIGURE 8.5

(a) Schematic representation on the step filtration processes along with their working pressure range and MWCO; (b) N_2 desorption curve of the ZrO_2 material calcined at 400°C; the increased desorption shows the presence of effective micro pores in the sub-nanometre range; (c) PEG retention curves for ZrO_2 membranes fabricated and calcined at 400°C; and (d) effect of mass fraction of NaCl on permeate flux.

to prepare NF membranes preferably to work in wide range of pH, good thermal, and chemical stability along with mechanical strength and can withstand high concentration of chlorine, as they are chemically more stable.

Among others, ZrO_2 membranes are considered one of the most promising candidates for harsh conditions, as they have good chemical stability and mechanical strength. A method of fabrication of ZrO_2 NF membranes and their working principle along with different characterizations are briefly discussed.

8.5.2 Materials and Methods

In order to prepare the ZrO_2 NF membrane, the colloidal sol-gel process was adopted, as it is a preferential protocol because of ease operation and low volatilization process facilitate to easy make of industrial scale component. Prior to fabrication of NF membrane, a stable sol of 0.05 M zirconyloxalate, $(ZrOC_2O_4)$, was prepared from mixture of 150 mL aqueous (deionized water) solution of zirconyl chloride octahydrate, $(ZrOCl_2.8H_2O)$, 50 mL aqueous solution of oxalic acid dihydrate, $(H_2C_2O_4.2H_2O)$, as complexing agent, and desired amount of glycerol. On the basis of glycerol content, the samples were designated as ZZ, ZZ18, ZZ35, and ZZ50 in which the glycerol weight percentage was varied 0, 0.18, 0.35, and 0.5, respectively. A typical tubular α-Al_2O_3 ultrafiltration layer (MWCO-30 kDa) was formed by colloidal sol-gel route and used as a support during fabrication of zirconia NF. Polymer additive was added into the sol as a binding agent and maintained the desired viscosity. Subsequently, the α-Al_2O_3 UF-membranes were coated by stable sol for 10 s followed by membranes kept in oven and cumulatively heated at 60°C, 100°C, and 110°C for 12 h. Finally, the membranes were sintered at 400°C for 2 h in a muffle furnace to produce ZrO_2 nanomembranes [4].

8.5.3 Working Principle

Physical sieving is the leading separation mechanism for large molecules and colloids, but diffusion mechanism and charge effect are predominant for low molecular weight and smaller substances. NF has advantages of lower operating pressure compared to reverse osmosis (RO), and higher organic rejection compared to UF, and their systematic mechanism can be presumed as [18]:

1. In the beginning, surface is getting wetted, where water molecules form hydrogen bond during membrane contact and transported through the membrane results wetted surface.

2. Surface behaves as preferential sorption and/or capillary rejection, where membrane is considered as heterogeneous and microporous, and electrostatic repulsion occurs between solution and membrane due to difference in electrostatic constants.

3. Solvent transportation, where the membrane is considered as homogeneous and nonporous, and the solute and solvent molecules are dissolved in the active layer of the membrane and the solvent transportation occurs by diffusion through the layer.

4. Charged capillary, where an electric double layer in the pores determines the rejection of impurities, where the streaming potential makes the ions of same charge as that of membrane to be attracted to the membrane and counter-ions to be rejected.

5. Finely porous, where the membrane is considered as a dense material punctured by pores and the transport is determined by partitioning between bulk and pore fluid.

Thus, the uncharged molecules are rejected mainly by size exclusion (like in UF) and ionic species are rejected by both size exclusion and electrostatic interactions. The basic transport of ions/solutes flux through the membranes is given by the extended Nernst-Planck equation (Equation 8.21) [19].

$$J = D_p \frac{dc}{dx} - \frac{zcD_p}{RT} F \frac{d\psi}{dx} + K_c cV \tag{8.21}$$

where J = ion flux based on membrane area (mol m^{-2}s^{-1}), D_p = hindered diffusivity (m^{2}s^{-1}), c = ion concentration in the membrane (mol m^{-3}), x = distance from the membrane (m), z = valence of ion, R = gas constant (J mol^{-1}K^{-1}), T = absolute temperature (K), F = Faraday constant (C mol^{-1}), K_c = hindrance factor for conversion, ψ = potential difference, and V = solvent velocity (m s^{-1}).

8.5.4 Characterization and Analysis

This section comprehends the different experiments that analyzed to quantifying the performance of membrane. Glycerol content has significant influence on the particle size distribution of zirconia sol. High glycerol content makes narrow particle size distribution of sols as it prevent the attachment of individual particles by steric stabilization. N$_2$ adsorption-desorption isotherms are computed using adsorption porosimeter on the ZrO$_2$ materials after calcined at 400°C that exhibits the influence of glycerol (inhibits tetragonal to monoclinic transformation during calcination) addition on the structure of pore (surface area of pore, pore size distribution, and pore volume), as shown in Figure 8.5b. To understand the filtration efficacy of the prepared ZrO$_2$ NF membrane, filtration tests were carried out to know the retention properties and water flux of membrane and compared with commercially available polyamide spiral membrane. The retention tests were conducted by using a feed mixed with polyethylene glycols (PEG) of different molecular weights (Da), and the results are as shown in Figure 8.5c. From Figure 8.5c, the MWCO of the prepared ZrO$_2$ membrane is taken as the value corresponding to 90% retention and is 750 Da. The high PEG retention values (after 60 min of filtration) are shown by ZZ18 compared to the other three samples. Less than optimum content (0.18 wt%) of glycerol is responsible for resultant ~5% volume expansion during tetragonal to monoclinic phase transformation of zirconia, and thus cracking, whereas high glycerol content initiate crack formation during high temperature fabrication. Finally, the flux of the ZrO$_2$ membrane was quantified by passing NaCl (feed water) and compared with polymeric NF having a similar MWCO of 920 Da. In this test, transmembrane pressure of 0.7 MPa was applied and solution was mixed with PEG 1000 (0.25%) and NaCl concentration was varied in the range of 0%–24.92%. The flux results of the performed test for both ZrO$_2$ and polymeric NF are shown in Figure 8.5d, respectively. The decrease in flux with increased mass fractions of the NaCl in solution can be explained by:

1. Effect of salting-out, which makes the PEG to be absorbed by the NF membrane as NaCl concentration increases in the solution

2. Solution viscosity increment due to excessive salt resulting hydraulic resistance in the pore

Other different tests were also performed to know the effect of membrane pressure and organic concentrations, and their results can be found in work carried out by Xiaowei Da et al. [4].

8.5.5 Research and Development Scope

From the earlier analysis, the advantages of ceramic NF membranes are clearly visible, but there is still need to increase the flux through the membrane, reduce MWCO, and, more importantly, stabilization of the ZrO_2 membranes against cracks that deteriorates the properties like retention of organic impurities drastically. A more insight of the preparation methods is needed for cost effective production and commercial/domestic use rather than a mere laboratory work. The usage of search for materials other than ZrO_2, which are more abundantly available in nature and also serve the same purpose is encouraged. Any other chemically and thermally stable ceramic can also be considered for this purpose through judicious processing conditions.

8.6 Self-Cleaning Coating

8.6.1 Background

Self-cleaning coating on cloth, glass, and building construction can be divided into two categories: hydrophilic and hydrophobic. Both strategy clean themselves through action of water, former by contacting water that carries away dirt and stains, and latter by rolling droplets as maximum as possible. Many available commercial products like water purifiers, anti-microbial surfaces, anti-fogging, and more are using photocatalyst to clean unwanted environmental substances [20]. Self-cleaning surfaces consist of photocatalysts are becoming popular and widely used in commercial constructions. Here, "self-cleaning" should not be misunderstood, rather only implies the extended time gap between the cleaning cycles. This minimizing the surface cleaning by additional detergents and cleaning substances, and subsequently their further exposer to the environment to create pollution. With minimization of dirt on the glass surface enhance the visibility and aesthetics, and in day it results in good illumination by sunlight. These self-cleaning surfaces are present in nature, for example: lotus leaf (super hydrophobic surface i.e., water contact angle > 150°). In recent, a self-cleaning glass was developed by coating a transparent titanium dioxide, (TiO_2), film on glass [21]. Followed by the titanium dioxide was used as active surfaces on different products, such as construction bricks, fabrics/textiles, ceramics, etc., to achieve self-cleaning properties, when it is exposed to the light. Self-cleaning properties of cotton fabrics coated by TiO_2 is also studied by Rahal et al. [22]. Apart from individual TiO_2, combination of TiO_2-ZnO and Ag doped TiO_2 are also studied to develop self-cleaning surfaces [23,24]. The TiO_2 can be used as hydrophilic when it consist of only hydroxyl group, as well as hydrophobic contains alkyl silyl ethers along with the hydroxyl group [25]. However, TiO_2 undergoes photoinduced wettability changes and there is effective switching of surfaces from hydrophobic to hydrophilic. Ultraviolet irradiation creates surface oxygen vacancies at bridging sites, resulting in the conversion of relevant Ti^{4+} sites to Ti^{3+} sites that are favorable to form hydrophilic domain for dissociative water adsorption. In this regard, an interesting study demonstrates about developing surfaces that have together superhydrophobic properties along with photocatalytic self-cleaning properties [5].

8.6.2 Importance of Nanostructured TiO$_2$ for Self-Cleaning

A self-cleaning coating made of TiO$_2$ has ability to adhere on the substrate surface, high photocatalytic activity, and resistance to abrasion along with high transparency and low reflectivity. The following are some of the properties of TiO$_2$ which makes it a potential material having applications in medicine, environmental protection, and in construction (facade paints, wall paper, tiles, etc.).

- At nanoscale range, TiO$_2$ does not appear white anymore and transparent.
- Under UV irradiation, the TiO$_2$ surface can effectively convert from hydrophobic to hydrophilic surface and initiate photocatalytic degradation.
- In presence of UV light, TiO$_2$ possesses photocatalytic activity resulting in the oxidative destruction of organic compounds (pesticides, dyes, pharmaceutics, alcohols, etc.) and biological species (bacteria, viruses, and fungi).

8.6.3 Materials and Methods

There are many methods to prepare thin films like reactive evaporation, sputtering, chemical-vapor deposition, pulsed-laser deposition, spray pyrolysis, and sol-gel processing. In this work, radio frequency-magnetron sputtering deposition was used to prepare a thin film of polytetrafluoroethylene (PTFE) and TiO$_2$ in an inert (argon) atmosphere. The substrate either quartz (Q) or titanium (Ti) was placed parallel at a distance of 100 mm from the sputtering target surface and coating process was performed at 298 K using a radio frequency power of 10 W. Before coating, the substrates were washed in acetone, ethanol, and distilled water and then dried in an oven for 1 h at 373 K. Different set of samples were prepared through permutation and combination of substrate and coating materials, and these are designated as: (a) PTFE/Ti, (b) TiO$_2$-PTFE/Ti, (c) TiO$_2$/Ti, (d) PTFE/Q, (e) TiO$_2$-PTFE/Q, and (f) TiO$_2$/Q, respectively.

8.6.4 Working Principle

The wetting of solid surface by a liquid takes place in such a way that total interfacial energy for all the phase boundaries becomes minimum. Accordingly, the solid surface with high solid–vapor interfacial energy makes the water to spread immediately on to surface so as to eliminate that interface (hydrophilic). The same reasoning can be applied to the solid–vapor surfaces with low surface energy (hydrophobic). By the Equation 8.22, the water contact angle, (θ), and shape of the water droplet on the surface determines the nature of surface (superhydrophobic ($\theta > 150°$) and superhydrophillic ($\theta < 5°$)); detail discussion available in Chapter 3:

$$\gamma_L \cos\theta = \gamma_S - \gamma_{SL} \qquad (8.22)$$

where γ_L is the liquid–vapor interfacial energy, γ_S is the solid–vapor interfacial energy, and γ_{SL} is the solid–liquid interfacial energy. In this paradigm, let's consider an interaction phenomenon of foreign particles with self-cleaning coated glass, as described in the

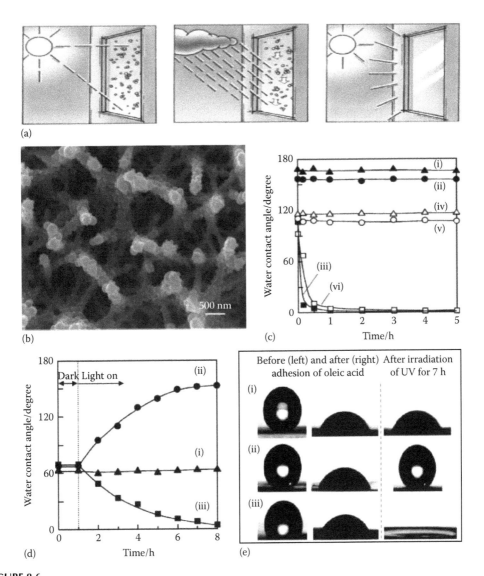

FIGURE 8.6
(a) Schematic representation on step-by-step photocatalytic induced self-cleaning of the surface under irradiation by UV light, (b) FESEM image showing the structure after uniform deposition of TiO$_2$-PTFE coating on Ti substrate, (c) effect of UV radiation over time on the water contact angle for different surfaces [(i) PTFE/Ti, (ii) TiO$_2$-PTFE/Ti, (iii) TiO$_2$/Ti, (iv) PTFE/Q, (v) TiO$_2$-PTFE/Q, and (vi) TiO$_2$/Q], (d) effect of surface dirt/contamination along with UV irradiation on water contact angle, and (e) photographic images showing the shape of water droplet before and after dirt adhesion to surface followed by UV irradiation for 7 h on to surface.

Figure 8.6a. In the beginning, the dirt/dust/organic materials are adhered to the surface that alters the original surface behavior (hydrophilic/hydrophobic) in the left most image. When this surface further exposed to sunlight/UV light, the photocatalyst breaks down those particles present on the surface by oxidation and same time photoinduced wettability changes the coating surface behavior. The moisture/rain water present on the self-cleaning surface easily forms a layer/spreads over the surface (as it was changed to hydrophilic by photoinduced wettability changes) and thereby takes away the dirt with it as the water runs off the surface. This results in a clear and clean surface, as shown in the rightmost image of Figure 8.6a. This phenomenology can be further improved by making a coat consist of additional superhydrophobic surface that capable to disintegrate the dirt by photoinduced process; TiO_2-PTFE composite is an example.

8.6.5 Characterization and Analysis

A representative uniform coating of TiO_2-PTFE on Ti substrate is represented in Figure 8.6b. In a similar fashion different set of materials were fabricated and used to correlate the wetting characteristics. Evaluation of the surface whether it is hydrophobic or hydrophilic is done by measuring the water contact angle with the surface using a pure water droplet. Among the various prepared samples, the water contact angle of PTFE/Ti and TiO_2-PTFE/Ti were quite high with 168° and 157°, respectively. The effect of the UV light irradiation on the water contact angle of the samples was investigated by exposing the samples for prolonged time and the results are shown in the Figure 8.6c. There is no change in the contact angle of the PTFE/Ti and TiO_2-PTFE/Ti samples even after UV irradiation for 5 h, as both the coating restricts the surface wettability change. But in the case of TiO_2/Ti surface, the surface gradually changed from hydrophobic to superhydrophilic due to the photo catalytically induced wettability changes of TiO_2 after 30 min exposed to UV irradiation (Figure 8.6c). The mixing of nanoscale TiO_2 with PTFE effectively prevented the hydrophilicity of surface and maintained the hydrophobic nature. In the same time, the effect of surface dirt/contamination on the water contact angle was estimated by using a model contaminant (oleic acid) on the surface and the results are illustrated in Figure 8.6d. In dark, the identical contact angle 65° maintains for all surfaces (PTFE/Ti, TiO_2-PTFE/Ti, TiO_2/Ti), however, dramatically changes when it is exposed to UV light. PTFE surface maintains same contact angle after even further exposing up to 8 h, but TiO_2-PTFE/Ti and TiO_2/Ti changes toward superhydrophobic and superhydrophilic, respectively. Former (TiO_2-PTFE/Ti) maintained the superhydrophobicity along the effective photocatalytic activity to disintegrate the foreign molecules, whereas later (TiO_2/Ti) exhibits complete spreading over the surface after UV irradiation, this approach is represented in Figure 8.6e. In order to establish this incidence further, a model contaminant, say MB was added to the water and its decolorization provides an indication of the photocatalytic performance of the TiO_2-PTFE/Ti surface under UV light. The disappearance of the blue color indicates the photocatalytic effect of the surface coating (TiO_2-PTFE). The change in concentration of MB over the time was analyzed by applying pseudo first order kinetics, as given in Equation 8.12. Even after five cycles of adhesions of oleic acid and UV light irradiation, TiO_2-PTFE/Ti restored its superhydrophobicity along with photocatalytic activity.

8.6.6 Research and Future Scope

Different additives that can control/hinder the wettability changes of nanosized TiO_2 particles without losing its photocatalyst behavior is a topic to look into. In the same time, the choice of different materials and increment of photocatalytic induced dirt/stain degradation efficiency is another concern. However, the ability of the coating to undergo repeated cycles of dirt adhesion and cleaning by UV irradiation without losing its surface wetting characteristics need to be increased. A rapid cleaning process is necessary and thus time required by the surface to clean itself under UV irradiation and retention of original surface characteristics can also be investigated further. Implementing this coating materials and technology, glass, textile, automobile industries, and civil engineers have created a chemical coating to clean themselves of stains and remove odors when exposed to sunlight. This technology is cheap, nontoxic, and ecologically friendly. Retailer thanks to new technology and exponential growing demand of such "functional clothing." Around the globe, gigantic garment market can initiate to take the advantage of self-cleaning concept that eventually minimize the environment pollution and consumption of water as well.

8.7 Concluding Remarks

Despite photocatalytic phenomena towards environmental issues, mechanical induced chemical conversion based portable depollution system, nanofiltration for desalination of water, and self-cleaning coatings are already commercialized to some extent, thus it is clear that the development of multivariant materials with improved outcome become extensive research interest in the near future. The synthesis of nanoscale WO_3, ZnO, $BaTiO_3$, ZrO_2, and TiO_2 are inexpensive and have relatively significant impact to protect environment in different aspects. In the perspective of environment protection, several directions are necessary to encounter and thus further development of materials. This may be the exact time to take advantage of all the mutual knowledge on such materials in order to keep clean our environment. However, the consideration of crystal structure, morphology, phase, surface modification, stability during synthesis of considered materials most likely is supposed to lead substantial breakthrough. Finally, the application of electrochromic, photocatalytic, and piezoelectrochemical conversion are most promising approaches, although much more research module is required, not only with regard to material development, but also in connection with device engineering. In this context, several organizations and government should support to pursue extensive materials research and development of devices for our societies.

References

1. S. Adhikari and D. Sarkar, High efficient electrochromic WO_3 nanofibers, *Electrochim Acta* 138, 115–123, 2014.
2. S. Adhikari, D. Sarkar, and G. Madras, Highly efficient WO_3-ZnO mixed oxides for photocatalysis, *RSC Adv* 5, 11895–11904, 2015.
3. K.-S. Hong, H. Xu, H. Konishi, and X. Li, Piezoelectrochemical effect: A new mechanism for azo dye decolorization in aqueous solution through vibrating piezoelectric microfibers, *J Phys Chem C* 116, 13045–13051, 2012.
4. X. Da, X. Chen, B. Sun, J. Wen, M. Qiu, and Y. Fan, Preparation of zirconia nanofiltration membranes through an aqueous sol–gel process modified by glycerol for the treatment of wastewater with high salinity, *J Membr Sci* 504, 29–39, 2016.
5. T. Kamegawa, Y. Shimizu, and H. Yamashita, Superhydrophobic surfaces with photocatalytic self-cleaning properties by nanocomposite coating of TiO_2 and polytetrafluoroethylene, *Adv Mater* 24, 3697–3700, 2012.
6. K. Lee, W. S. Seo, and J. T. Park, Synthesis and optical properties of colloidal tungstenoxide nanorods, *J Am Chem Soc* 125, 3408–3409, 2003.

7. E. L. Runnerstrom, A. Llordés, S. D. Lounis, and D. J. Milliron, Nanostructured electrochromic smart windows: Traditional materials and NIR-selective plasmonic nanocrystals, *Chem Commun* 50, 10555, 2014.

8. S. Adhikari and D. Sarkar, Synthesis and electrochemical properties of nanocuboid and nanofiber WO_3, *J Electrochem Soc* 162, H1–H7, 2014.

9. S. Adhikari and D. Sarkar, Hydrotermal synthesis and electrochromism of WO_3 nanocuboids, *RSC Adv* 4, 20145–20153, 2014.

10. S. Adhikari, D. Sarkar, and G. Madras, Synthesis and photocatalytic performance of quasi-fibrous ZnO, *RSC Adv* 4, 55807–55814, 2014.

11. A. A. Khodja, T. Sehili, J. Pilichowski, and P. J. Boule, Investigation of photocatalytic degradation of methyl orange by using nano-sized ZnO catalysts, *J Photochem Photobiol A* 141, 231–239, 2001.

12. C. Lizaman, J. Freer, J. Baeza, and H. Mansilla, Optimized photo degradation of reactive Blue 19 TiO_2 and ZNO suspensions, *Catal Today* 76, 235–246, 2002.

13. K. S. Hong, H. Xu, H. Konishi, and X. Li, Direct water splitting through vibrating piezoelectric microfibers in water. *J Phys Chem Lett* 1, 997–1002, 2010.

14. C. Galindo, P. Jacques, and A. Kalt, Photooxidation of the phenylazonaphthol AO20 on TiO_2 kinetic and mechanistic investigations, *Chemosphere* 45, 997–1005, 2001.

15. X. Xue, W. Zang, P. Deng, Q. Wang, L. Xing, Y. Zhang, and Z. L. Wang, Piezo-potential enhanced photocatalytic degradation of organic dye using ZnO nanowires, *Nano Energy* 13, 414–422, 2015.

16. L. Giorno, E. Drioli, and H. Strathmann, The principle of nanofiltration (NF), In: E. Drioli, L. Giorno (Eds.) *Encyclopedia of Membranes*, Springer, Berlin, Germany, 2015.

17. Y. Cai, Y. Wang, X. Chen, M. Qiu, and Y. Fan, Modified colloidal sol–gel process for fabrication of titania nano filtration membranes with organic additives, *J Membr Sci* 476, 432–441, 2015.

18. R. G. Macoun, The mechanisms of ionic rejection in nanofiltration, Chemical Engineering, PhD thesis, University of New South Wales, Sydney, Australia, 1998.

19. H. K. Shon, S. Phuntsho, D. S. Chaudhary, S. Vigneswaran, and J. Cho, Nanofiltration for water and wastewater treatment—A mini review, *Drink Water Eng Sci* 2013.

20. L. Brand, C.-S. Ciesla, and M. Werner, Chemistry & Materials: Application of Photocatalysis, 7th Framework Programme, Observatory Nano Briefing No. 10, 2011.

21. Y. Paz, Z. Luo, L. Rabenberg, and A. Heller, Photooxidative self-cleaning transparent titanium dioxide films on glass, *J Mater Res* 10(11), 2842–2848, 1995.

22. R. Rahal, T. Pigot, D. Foix, and S. Lacombe, Photo catalytic efficiency and self-cleaning properties under visible light of cotton fabrics coated with sensitized TiO_2, *Appl Catal B Environ*, 104(3–4), 361–372, 2011.

23. R. Fateh, R. Dillert, and D. Bahnemann, Self-cleaning properties, mechanical stability, and adhesion strength of transparent photo catalytic TiO_2–ZnO coatings on Polycarbonate, *ACS Appl Mater Interfaces* 6(4), 2270–2278, 2014.

24. B. Tryba, M. Piszcz, and A. W. Morawski, Photo catalytic and self-cleaning properties of Ag-Doped TiO_2, *Open Mater Sci J* 4, 5–8, 2010.

25. T.-Y. Yang, S.-J. Chang, C.-C. Li, and P.-H. Huang, Selectivity of hydrophilic and hydrophobic TiO_2 for organic-based dispersants, *J Am Ceram Soc* 100, 56–64, 2017.

9

Nanostructured Ceramics for Health

9.1 Introduction

Ceramics used to improve the quality of human life through development of specially designed and fabrication in the perspective of reconstruction and diagnosis of diseases is called bioceramics. Most clinical applications of bioceramics as either bioimplant related to the repair of the skeletal system, including bones, joints, and teeth to supplement both hard and soft tissues or extensive use of nanoparticles in medical treatment. In starting, a brief discussion has been emphasized on the fabrication and assessment of bioimplants in specific dental ceramics, femoral head for hip bone replacement, and hydroxyapatite-gelatin porous scaffold for soft tissue engineering [1–3]. However, several factors affect the implant-tissue interfaces, and thus artificial implant performance depends on composition, phase and phase boundaries, surface morphology and porosity, chemical reactions, blood and nutrient circulation, closeness of fit, and mechanical load. Furthermore, a critical particle and pore size can be varied from a few nanometers up to several micrometers to control the ease of drug adsorption, delivery, and dispersion of the materials to the targeted area. Despite ceramic based bioimplants, in recent focus on the synthesis of new nanoceramics is relevant to broad range of applications such as drug and gene delivery, treatment of bacterial and viral infections, cancer treatment, imaging, delivery of oxygen to damaged tissues, and materials for minimally invasive surgery to improve the diagnosis and patient life expectancy.

For example, there is a noninvasive photodynamic treatment technique in which only the cancer cells are destroyed by suitable drug loaded biocompatible ceramic nanoparticles [4]. In vivo localized temperature enhancement is also an effective management to destroy the cancerous tumor cell, as it sustains relatively low temperature (43°C) because of poor oxygen supply in the interior of tumor cell, whereas the normal healthy cells can alive up to 48°C [5]. Thus, introduction of ferro or ferromagnetic materials in patient and placing them under an alternating magnetic field facilitate local heat generation and destroying the cell [6]. In the earlier context and demand of ceramics for health research, a group of important materials including their synthesis, fabrication, and characterizations are emphasized systematically, these are:

1. Monolithic nanostructured zirconia for dental blank
2. Nanostructured zirconia toughened alumina based femoral head for hip replacement
3. Nano hydroxyapatite-gelatin porous scaffold for soft tissue and protein delivery
4. Nanoscale ZnO drug loaded for chemo-photodynamic cancer treatment
5. Nanoscale core-shell ferrimagnetic ceramics for hyperthermia

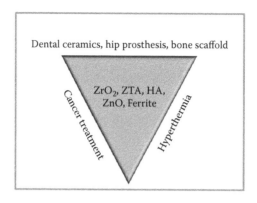

FIGURE 9.1

Schematic representation on the utility in different health sectors including dental ceramics for artificial teeth, femoral head for total hip prosthesis replacement, porous scaffold for tissue engineering, drug delivery for cancer treatment and hyperthermia resolution through nanostructured zirconia (ZrO_2), zirconia toughened alumina (ZTA), hydroxyapatite (HA), zinc oxide (ZnO), and core-shell ferrites ($CoFe_2O_4@MnFe_2O_4$), respectively.

Figure 9.1 illustrates a typical triangle representation for the use of different ceramics for health issues, and more details can be found elsewhere [1–4,6].

9.2 Nanostructured ZrO_2 for Dental Ceramic Blank

9.2.1 Background

Ceramics are mostly used biomaterials in prosthetic dentistry as it has attractive clinical clarity including good aesthetical properties like color, shade, luster, and chemical stability. In early days, the dental ceramic is consisting of Si based inorganic materials such as feldspar, silica, and quartz that comprises of multiphase systems. This ceramic exhibits low degree of mechanical response as the limited dispersed crystalline phase is surrounded by amorphous phase. Thus, a class of ceramics which consist of high crystalline phase is expected to achieve better performance, as well as biomechanical properties. In this context, the newly developed zirconia based ceramics used as restorative dental materials have much interest in the dental community, as they exhibit competitive mechanical properties and chemical stability. Herein, the basic paradigm of zirconia is athermal tetragonal to monoclinic phase transformation induced by the crack tip energy that eventually enhances the mechanical responses. However, this synthesized tetragonal phase is attributed in the presence of stabilizers such as Y_2O_3, MgO, CaO, and CeO_2 that permit retention of tetragonal phase at room temperature and effectively arrest crack propagation that leads to high fracture toughness under stress [7]. In fact, these materials have certain advantageous during in vivo applications, however, biomedical grade 3 mol% yttria, (Y_2O_3), stabilizer in 3Y-TZP is found as suitable for dental ceramics and extensively used for teeth restoration. The restorations are processed either by CAD-CAM assisted soft machining of presintered blanks followed by sintering at high temperature or by hard machining of fully sintered blocks. In this context, a recent study explores a new synthesis protocol of zirconia nanoparticles followed by fabrication of their cuboid blank by centrifuge casting, sintering, and their essential properties for dental restorations and applications [1].

9.2.2 Advantages of Nanostructured Zirconia

In recent years, 3Y-TZP ceramics has gained importance as restorative materials due to its good mechanical and tribological properties. Mechanical properties include high strength and toughness with good biocompatibility and esthetic properties are major concern for the development of dental ceramics. 3Y-TZP is less stable to spontaneous t → m transformation in presence of micron grain matrix, where in fact the smaller grain size (<0.2 μm) restricts the spontaneous transformation rate and improves high toughness and mechanical properties [8,9]. Guazzato et al. demonstrate that equiaxed (0.2–0.5 μm) sintered 3Y-TZP has a better microstructure and enhanced mechanical properties for bioimplant applications [10]. In convention, the hardness is inversely proportional to grain size, but fracture toughness depends on transformation toughening phenomenon of grain size, which is mostly controlled by yttria content or sintering atmosphere. Thus, integration of these properties favors the development of zirconia bioceramics with having nanostructured matrix and tetragonal phase.

9.2.3 Materials and Method

A rapid direct calcination of inorganic precursors was employed to synthesize 3Y-TZP nanopowders. A requisite amount of $Y(NO_3)_3.6H_2O$ and $ZrOCl_2.8H_2O$ inorganic salts are dissolved thoroughly in 250 mL distilled water, followed by step calcination up to 600°C in order to obtain 100 g 3Y-TZP nanopowders. Direct precursor calcination technique encompasses three step calcination process as follows: (a) mixed precursor solution becomes high viscous yellowish gel at 70°C, (b) gel changes to fluffy amorphous zirconia at 200°C, and (c) the ground amorphous zirconia transformed to 10–30 nm tetragonal zirconia nanopowders at 600°C.

In this newly developed process, an excellent yield up to 95 wt% was reported. A homogenous slurry of synthesized zirconia nanopowder was prepared by 1 wt% polyvinyl alcohol solution and fabricated the dental blank through homemade centrifuge casting method. It is worth to note that the required amount of solid NaOH and 1 wt% polyacrylic acid as a dispersant were added to control pH ~ 9 and of the homogenous and stable slurry. The slurry was homogeneously mixing in ball mill for 48 h. The ceramic blanks were prepared by centrifugal casting and measured their dimensions with having length 50 mm, width 25 mm, and height 20 mm (Figure 9.2a). The blanks were dried in control humidity to avoid generation of flaws. The blanks were sintered at optimum sintering temperature of 1350°C for 2 h.

9.2.4 Characterization and Analysis

In this research work, nanopowder characterizations to ranging microstructure, mechanical, and biological properties of sintered blanks were elucidated systematically to establish the potential use of developed blank as dental ceramics. A brief powder characterization emphasized the formation of near spherical and tetragonal phase with having BET specific surface area of 43.6 m^2/gm. The average zirconia particle size in this condition was observed to vary in a narrow range of 10–30 nm consist of interplanar distance in the crystal of 0.295 nm; that is well matched with [101] lattice spacing of tetragonal zirconia, as compared with XRD results. The Weibull size distribution analysis complying the narrow size distribution as it predicts the shape parameter, (α), scale parameter, (β), and location parameter, (γ), are 3.2, 21, and 8, respectively. Mercury porosimeter technique was used to

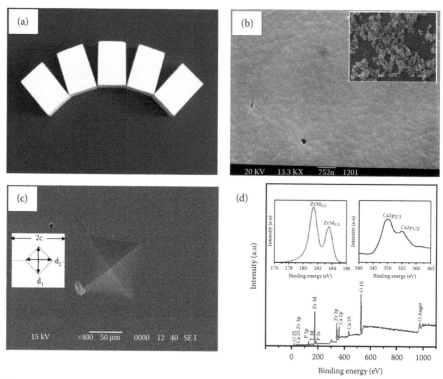

FIGURE 9.2
(a) Ceramic dental blanks made of centrifuge casting, (b) Vickers indented image under 10 kg on polished surface of sintered 3Y-TZP, (c) SEM image with average grain size ~250 nm, and (d) X-ray diffraction pattern after 28 days incubation of sintered zirconia in SBF solution. Inset represents the SEM image after 28 days incubation of sintered zirconia in SBF solution.

measure density, close porosity, and volume shrinkage of zirconia blank. The micrograph of the sintered and thermal-etched zirconia blank specimen reveals the equiaxed average grain size is near to 250 nm and few isolated pockets of closed pores (~2%), however, no abnormal grain growth or coarsening is detected all along the matrix (Figure 9.2b). Volume shrinkage and relative density of sintered blanks were measured as 24% and 98%, respectively. Hardness tests were carried under a load of 98.1 N (10 kg) with a holding of 10 s. The Vickers hardness, (Hv_{10}), and indented toughness were calculated by Equations 6.41 and 6.38, respectively. The high hardness of 13.5 GPa and the fracture toughness 3.5 MPa·m$^{1/2}$ were measured for zirconia nanopowder sintered at 1350°C for 2 h. Diamond indentation was perfectly rhombus in shape and uniform radial cracks propagated from the tip of the indentation zone without noticeable plastic zone (Figure 9.2c). In a recent article, 3 mol% yttria doped fully tetragonal zirconia has also been prepared through hot pressing at 1450°C for 2 h appears 0.3 μm average grain size and fracture toughness of 3.5 MPa·m$^{1/2}$, resembles with current set of data [11]. A systematic bioactivity assessment of centrifuge casted and sintered zirconia blanks was carried out in simulated body fluid (SBF) solution at constant pH up to 42 days to understand the bioactivity of this nano-structured microstructure. The SEM (inset of Figure 9.2b) and XPS analyses were used to understand the bioactivity phenomenon of zirconia surface. A close look at the morphology of foreign particles on zirconia surface reveals the presence of sponge-like deposition of spherical primary particles associated with agglomerated secondary particles.

Different contrast of exposed (dark grey) area from deposited (whitish bright) surface was attributed to the presence of two different materials. The spherical HAp was deposited and adhered on nanostructured zirconia surface after 28-days immersion in SBF that can be ascribed by the following reasons:

1. Adsorption of OH⁻ ions and formation of hydroxide bonds with zirconia
2. Creation of vacancy and increment of lattice spacing
3. Development of compressive stress zone
4. Formation of $Zr(OH)_6^{2-}$ at stress sites
5. Deposition of Ca^{2+} ions on exposed reactive sites

The interaction of 3Y-TZP with water leads to the slow hydration and formation of Zr-OH bonds, which have both bending and stretching mode of vibration. Trivalent yttrium ion creates oxygen vacancies, which assists to diffuse hydroxyl group within the lattice, increases the lattice spacing, and developed compressive stress zone within confined region. This selective area serves as an active nucleation site for the reaction and deposition of hydroxyapatite particles and identified after 4 weeks only. This information indicates that the prerequisite incubation time is required to develop stress-induced nucleation site for the deposition and subsequent growth of the HAp phase. The high negative charge is developed by $Zr[OH]_6^{2-}$, which assists to interact and leads to subsequent formation of HAp phase through deposition by Ca^{2+} ions. In order to detail understanding on the deposition of hydroxyapatite behavior and consist of surface elements like Zr, Y, Ca, P, and O, XPS was carried and represented in Figure 9.2d. The binding energy of $Zr-3d_{5/2}$ peak at 181.6 eV represents ZrO_2 and Y-3d peak represents Y_2O_3. The Zr-3d peaks have one doublet $3d_{5/2}$ and $3d_{3/2}$ spin-orbit splitting. Ca-2p spectrum represents doublet 347–357 eV for Ca^{2+} oxidation states in inorganic calcium-oxygen compound. The P-2p peak at 1.334 eV reveals stable apatite. The O1s peak at 531.8 eV belongs to oxide and phosphate. The XPS spectrum illustrates binding state of Ca, which coordinates with OH⁻ and PO_3^{4-} of HAP nanocrystals. Such bioactivity in SBF with having mechanical response of nanostructured zirconia blank demonstrates the plausible use as dental ceramics.

9.2.5 Research and Development Scope

A combined metal-ceramic dental fixture provides both extraordinary mechanical properties of metals and exceptional esthetic properties of ceramics. However, metals used as restorative material in dentistry may release metallic ions and initiate allergies and gum staining for some patients. These drawbacks are leading to search metal free alternative and extensively accepted the ceramics for dental applications. The use of zirconia ceramics for multiunit posterior fixed partial dentures has been facilitated by the advent of CAD/CAM systems. However, inaccurate scanning, software design, blank fabrication, machining, and shrinkage may lead to poor restoration fit and marginal misfit in the axial wall area and occlusal plateau can reduce the fracture resistance that seriously affect adjacent tissues. Thus, effective clinical trials is expected to improve the patient comfort. Behavioral science studies in clinical dental research is an important horizon to examine issues such as patient acceptability of dental treatment and/or preventive procedure with health promotion. Focus on major etiological factors like plaque control and tobacco deaddiction by appropriate behavioral methods are under foremost scope of clinical dental research.

9.3 Nanostructured Zirconia Toughened Alumina Femoral Ball Head

9.3.1 Background

Among all the articulating joints in a human subject, the hip joint is one of the most important flexible joints that allows a large range of motion and experience different degree of static and dynamic compressive stresses. For example, walking slowly, bending the knee, walking down stairs, climbing upstairs, standing up from a chair, and sitting down on a chair require respective compressive stresses of 3.3, 3.7, 3.8, 5.7, 8.8, and 9.4 MPa at the articulating joints [12]. Arthritis, injury, dislocation, or irregular activity bring wear and tear of the surrounding cartilage inside a hip joint and hence cause pain to the patients. This results in friction between the bones as they rub against each other and the hip joint becomes severely damaged in this process. The unexpected damage demands the replacement of the hip joint in a diseased patient by a synthetic biomedical device assembly made of individual components (stem, femoral head, and acetabular socket) [13]. From an engineering perspective, the ball-bearing mechanism comprises of a concave acetabular cup and a round convex femoral head with a conforming surface projecting the socket in a total hip joint replacement (THR) device. The THR assembly is further supported by the neck of the femur, while the spherical femoral head is placed upwards to maintain the convexity of the joint toward the pelvis, under normal health conditions [14]. With increasing life expectancy in many of the developing nations around the world, the demand for hip replacement surgery is on a rise [15]. A wide range of truncated spherical femoral ball head size and tapering angle is commercially available for different group of ethnics, and thus one can encounter different design to develop implant for THR. Thus, a number of new materials having competitive properties are being widely demanded and researched in the materials community. Apart from the material selection, nanostructured microstructure has several degrees of advantages to enhance the fracture strength under compressive load and reduce the wear rate that eventually controls the debris formation during articulating motion (details in Chapter 6).

9.3.2 Advantage of Nanostructured Zirconia Toughened Alumina

A conventional cobalt-chrome alloy femoral head-acetabular socket implants release elevated levels of metal ions in the urine and bloodstream and the wear-debris particles generated at local sites that may have adverse synergic effects after their bio-distribution. Thus, manufacture of ceramic prototypes with improved stability can be made possible with high fracture toughness, excellent wear characteristics, and low susceptibility to stress assisted degradation of zirconia toughened alumina (ZTA) over metal or conventional monolithic alumina. An optimum amount of zirconia (5 wt%) is encountered in order to enhance the fracture toughness compare to alumina matrix, whereas higher content avoided to minimize the in vivo degradation and reduction of mechanical properties. Despite composition, the fascination toward a definite nanoscale grain size window promotes most desirable high hardness and fracture strength for ceramic femoral head. An optimum amount of zirconia improves mechanical properties via two different mechanisms: stress induced transformation toughening and microcracking. Transformation toughening is induced from the stress-induced transformation experienced by t-ZrO_2 particles when interacting with an advancing crack. The effect of microcracks intrinsically related to such transformation, which then promote a toughening effect through the permanent dilatation and reduced modulus that they imply around the crack-tip. A generalized approach, based on the energy absorbed ahead of a crack movement, has been provided during loading from zero to the propagation stress and the following equation has been proposed [16]:

$$K_{IC} = K_o + \sqrt{2E\,\eta_T\,r_T} + \sqrt{2\,E\,\eta_m\,r_m} \tag{9.1}$$

where K_{IC} = fracture toughness, K_o = fracture toughness of the matrix, E = elastic modulus, η = energy density absorbed ahead of a crack, r = radius of the "process zone", subscript T = stress induced transformation mechanism, subscript m = microcracking nucleation mechanism.

9.3.3 Materials and Method

Prior to fabrication of nanostructured ceramic based ZTA femoral head, a typical mold was designed with consideration of shrinkage volume (18%–20%) associated with the transformation from the green compact to the final sintered compact as well as the machining (4%–5%) and polishing (1%–2%). The optimum composition (120 nm, 95 wt% Al_2O_3; 50 nm, 5 wt% ZrO_2; and 800 ppm magnesium oxide (MgO)) was mixed with 3 wt% polyvinyl alcohol (PVA) organic binder, and the powder mix was uniaxially pressed under a compaction force of 200 kN for a dwell time of 120 s. The demolded green femoral ball head was presintered at 1200°C for 2 h with a slow heating and cooling rate of 2°C/min. The truncated presintered ZTA spherical femoral head was computer numerical control (CNC) machined up to certain extent to maintain the dimension accuracy. This was then further sintered at 1600°C for 6 h to get a highly dense compact. Polishing was then done with a sequentially varying diamond paste up to 1 μm to obtain a smooth outer surface for the spherical dome-shaped object and measured the 2D microstructure by SEM analysis. 3D microstructural analysis of the fabricated femoral heads (26 mm OD) was done using microcomputed tomography (μCT) to investigate: (a) the structural integrity in case cracks/pores are present in the bulk, with a low resolution scan and (b) quantify the microstructure in order to assess the distribution of phases present in the region of interest, with small pixel size. The dimensional stability of the fabricated 26 mm OD femoral heads was measured using the coordinate measuring machine. Mechanical properties including hardness, indented toughness, and burst strength under uniaxial compressive mode of loading, (ISO 7206-10), was measured and analyzed systematically. Cytocompatibility was measured in presence of C2C12 mouse myoblast cells followed by cell viability was analyzed using colorimetric MTT assay. Cell morphological analysis was performed by standard process and images were observed under a fluorescence microscope and eventually performed the statistical analysis of cell viability [2].

9.3.4 Characterization and Analysis

The dimensional stability of the prototype, without any distortion or deformation, was ensured after polishing the near net-shaped zirconia toughened alumina femoral head and represented in Figure 9.3a. A representative SEM image of the polished ZTA femoral head surface is shown in inset. The alumina (3.98 g/cc) grains appear in medium grey, zirconia (6.07 g/cc) grains in bright contrast, and a few pores (1.22×10^{-3} g/cc) in dark contrast. Also, the zirconia particles are distributed uniformly either at the triple points or grain boundaries of alumina, thereby inhibiting the growth of alumina grains during sintering. The extensive quantitative image analysis reveals the average grain size of alumina to be ~3.2 μm. The detailed surface analysis of the polished ZTA femoral

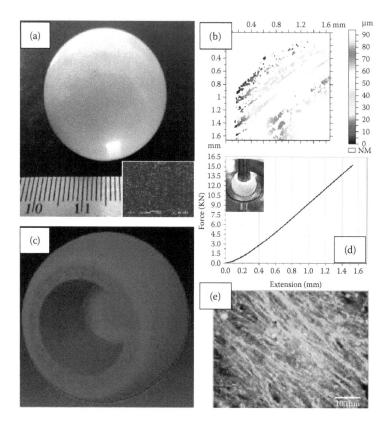

FIGURE 9.3
(a) The top view of sintered ZTA femoral head after machining followed by polishing, (b) 2D optical profilo-
metric image for the specific area of convex head of 26 (OD) after machining, sintering, and polishing, (c) 3D
μCT image on the perspective view of total femoral head (blind hole projection without any cracks and defects),
(d) force versus displacement plot during burst strength measurement of femoral head, and (e) representative
fluorescence microscopic images of C2C12 myoblast cells cultured on ZTA. Inset of 9.3a and 9.3d indicates the
SEM image of nanostructured ZTA of femoral head and front view of the assembly prior to conduct the burst
strength of fabricated femoral head, respectively.

head prototype using a noncontact optical profilometer confirmed that a convex mating
surface of the femoral head, with clinically relevant surface roughness ($R_a \sim 0.2$ μm), as
illustrated in Figure 9.3b.

A significant improvement in surface finish was achieved after polishing the femoral
head. The root mean square roughness of the polished head is 0.26 μm, which is close
to the value of R_a. The maximum height of the peaks and valleys are nearly equal and
therefore, the microasperities on the polished functional mating surface of the convex
head are similar to those of an isotropic surface. The μCT image of the full ZTA femoral
ball head was visualized in 3D space using an isosurface module with a lower resolution
scan (voxel size: 29.028 μm), as represented in Figure 9.3c. The perspective view shows
the convex surface including the circular blind hole and edges without any visible distortion of the circularity and truncated shape of the zirconia-toughened alumina femoral head. Microcomputed tomography and subsequent analysis confirmed the presence
of well-dispersed ZrO_2 particulates in an alumina matrix of the defect-free ZTA femoral ball head prototype. Specifically, high resolution tomograms of the 3D bulk microstructure confirmed that this particular manufacturing technique results in a unique

zirconia microparticle morphology and orientation in different parts of the femoral head. Apart from the surface and microstructural analysis, brief mechanical properties are also analyzed to justify the significance of nanostructured ceramics. The indent diagonal lengths were precisely determined from SEM images of the indented surfaces to reliably measure the Vickers hardness of ~19 GPa. The cracks originating from the indent edges were used to evaluate the fracture toughness, which is estimated to be 5.2 MPa·m$^{1/2}$. The use of nanosized powder particles as well as the retention of the finer scale microstructure are attributed to the attainment of higher hardness in the present case (see inset of Figure 9.3a). The achieved toughness can be primarily attributed to the ZrO$_2$ content or more specifically to the volume fraction of the transformable ZrO$_2$, as well as its transformability [11]. The measurement of a broad spectrum of mechanical properties reveals that the optimized coupon samples of ZTA composites have the best combination of compressive strength (1100 MPa) and tensile strength (200 MPa). The evaluation of wear resistance properties under unlubricated sliding conditions with commercial zirconia/alumina and steel counterbody reveal that a good combination of mechanical properties with a steady-state coefficient of friction of 0.4 and wear rate of 10^{-9} mm^3/Nm could be achieved with ZTA composites. The machined and sintered ZTA femoral head (without polishing) was placed on a copper ring cone during the measurement of burst strength (see inset of Figure 9.3d). As shown in Figure 9.3d, a largely linear force-extension response (equivalent of stress-strain response) was recorded with a typical fracture load of 15 kN. The initial stage of the nonlinear deviation is associated with the ductile behavior of copper compared to mild steel [17]. The obtained fracture or damage load of 15 ± 3 kN during the measurement of burst strength, (n = 5), is slightly lower when compared to the minimum acceptable fracture load limit of 20 kN, as recommended by the US Food and Drug Administration guidance draft [18].

The viability and phenotypical behaviors of C2C12 myoblast cells were analyzed after seeding them on ZTA ceramic substrates for a period of 72 h and are shown in Figure 9.3e. On observing them under a florescence microscope, the myoblasts showed enhanced cell proliferation, cell attachment, cell spreading, and cell-to-cell interaction with specific alignment and organization of cytoskeleton elements. The sequence of cellular response is characterized by initial cell attachment followed by cell spreading and morphological adaptation to the substrates. This is followed by long term expression of the cellular phenotype. The observations reveal that the fabricated ZTA prototypes could support favorable cell growth of the myoblasts. The MTT results confirm that the ZTA composition of the femoral head can support the growth of mitochondrially viable mouse myoblast cells in vitro, with the viability increasing in a statistically significant manner for a period of up to 72 h in culture. The comprehensive measurement of the properties of the test samples with identical composition as well as the cytocompatibility assessment suggest that the ZTA femoral head prototypes have the desired property spectrum. The compression strength measurement at the device level indicates a moderate burst strength property. Taken together, the efficacy of the manufacturing protocol in fabricating femoral head prototypes has been comprehensively validated and established. The adopted fabrication strategy could be implemented in the case of other bioceramics with material-specific process parameters.

9.3.5 Research and Development Scope

The demand of artificial implant is on rise, and an effective material is vital to fulfill the demand. Single matrix ceramics may not be enough good to serve cumulative mechanical response and thus composite within a particular grain size window attaining high density (>98%) is beneficial to sustain under static and dynamic load, and relative motion. Same time, material should possess dimensional stability and biocompetent to avoid revision surgery. In recent trend, material scientist seriously focus toward new class of prototypes made of ceramics or their composites in target of prolong life without formation of debris or ion leaching cause of early failure. Apposite design of experiment, material selection, patient specific component design, adequate mechanical response, and up to nanoscale roughness, can enhance the reliability and life expectancy. Despite development of bioimplant, factors that influence the longevity of the device include the patient's age, sex, weight, diagnosis, activity level, and conditions of the surgery. However, the top hip and knee implant manufacturers have cultivated the belief that these surgeries can restore close to normal joint function. The latest technological advancements in minimally invasive surgical procedures, novel materials leading to better performance and enhanced durability, and advanced designs of joint stability and functionality have made these implants attractive to more and younger adults, particularly those who want to maintain an active lifestyle [19].

9.4 Nano Hydroxyapatite-Gelatin Porous Bone Scaffold

9.4.1 Background

Bone tissue defects and disease are the major reason for trauma, infections, injury, and bone loss. Recently, near to multimillion cases of the bone injuries are reported around the globe and the number is increasing day by day, the demand for bone substitutes is also parallel increasing [20,21]. The clinical possibility for repairing bone disease and defects is bone grafting and these are autografts and allografts. The artificial use of scaffold are increased due to less access of autograft and consequential disadvantages of allograft, which induces the risk of transmission of diseases to the recipient [22]. To overcome these disadvantages, synthetic scaffold is used for repair bone defect, bone fracture repair, dental and ear implants that possess the bonding properties, and morphology of natural bone. The scaffold must have desired properties, such as moderate mechanical strength, interconnected pores with optimum porosity, and biocompatibility with the bone. Furthermore, the scaffold must promote bone formation while increasing cell proliferation, adhesion, and controlling osteogenic differentiation of host cells. The natural bone is made of organic, inorganic materials with proteins, mainly hydroxyapatite ($Ca_{10}(PO_4)_6(OH)_2$), collagen and apatite as biological mineral [23]. As collagen is costly and denatured while processing, gelatin is used as an alternative to collagen because it is easily available with low cost and having the same chemical composition as collagen. Gelatin has additional advantages like wound dressing and temporary defect filler. In vivo, the first constituent in living body interacts with the surface of the biomaterial is protein. Thus, after implantation, protein layer initially adsorb on foreign material surface and interacts with cell receptors around cell membrane surface, as the cells do not interact with the bare material surface [24]. Protein molecules facilitate the fixing and proliferation of cells onto the biomaterial surface. Hence, proteins form a bridge between biomaterial surface and cells, the adsorption of proteins depends on the size, surface porosity, surface area, morphology, and crystallinity of biomaterials. In this backdrop, continuous research is a challenging task to meet the requirements of the porous scaffold with new materials and desired properties.

9.4.2 Advantage of Nano Hydroxyapatite in Scaffold

Bone has a complex structure of hydroxyapatite crystals with an organic matrix collagen. The main advantages of the nanosize hydroxyapatite particles are compatible with inorganic bone crystal of human body and immensely considered in tissue engineering. The commercial available synthetic hydroxyapatite is effectively using for drug delivery media and bone growth factors. Hence, continuous research is in focused to synthesize and understand the structural properties of nano size hydroxyapatite (HA) particles that is also extensively used for biocoatings to reduce the failure of the implants [25,26]. Owing to high surface to volume ratio and resembles with natural bone constituent, the nanoscale HA can create easy cell accommodation and bonding with host bone and healing the defective spongy bone. However, their morphology, crystallinity, and surface characteristics influence the adsorption and desorption behavior of drug molecules. The active interface results in interaction between biological fluid and cells that form carbonated HA and analogous to bone material. While considering the porous scaffold made of hydroxyapatite, a slow and continuous and localized release of drug concede the better choice for ailment treatment. Thus, the drug delivery system is extensively developed using porous nano HA ceramic and biodegradable polymer matrix that eventually conjugated with different drugs include proteins, antibiotics, etc. Hence, porous nano HA scaffold has multivariant potential as an artificial bone substitute.

9.4.3 Materials and Method

Prior to fabrication of HA-gelatin porous scaffold, HA were synthesized using a chemical precipitation method, which requires precise control over solution pH ($9.5 \geq pH \geq 7.75$), temperature (303 K), and the Ca:P ratio (1.669) from calcium acetate (($CH_3COO)_2Ca$) and potassium dihydrogen phosphate (KH_2PO_4) precursors. The precipitate was washed and separated through a centrifuge at 14,000 rpm, and finally freeze dried at 220 K and 120 torr pressure. A large batch size of 60 g of nanosized HA was prepared for the fabrication of scaffolds as described elsewhere [27]. Gelatin-HA scaffolds were prepared from freeze dried HA powder (40 wt%) with having rod-shaped morphologies. The solid content was optimized by considering the combined properties of pore volume, pore diameter, and adequate strength of 30 wt%, 40 wt%, and 50 wt% solid loaded porous scaffolds. A slurry was made by adding PVA (10 wt%) and gelatin (15 wt%) to the HA powder. Highly dispersed HA nanorods were added to the mixed solution of gelatin and PVA through continuous magnetic stirring. The solid–liquid mixture turned into highly viscous slurry at 40°C. The slurry was then poured in a cylindrical glass mold and refrigerated at 268 K for 12 h to avoid sudden liquid solidification and evaporation. The semidried mass was freeze dried at 221 K and 77 torr pressure. Demolded freeze dried gelatin-HA cylindrical scaffolds were further cured in a liquid nitrogen cryochamber for different time intervals up to 5 h at atmospheric pressure. The scaffold was designated as cryo-treated nanohydroxyapatite-gelatin micro/macroporous scaffold (CHAMPS). The as-synthesized gelatin-HA scaffold and specimens cryo-treated for 0, 2, 5, 8, and 11 h were designated as HG00, HG02, HG05, HG08, and HG11, respectively.

9.4.4 Characterization and Analysis

To estimate the pore architecture, the microcomputed tomography (micro-CT) technique is used largely for protein loaded gelatin HA based porous nano biocomposite scaffolds. Details of protein adsorption and release technique are discussed in details somewhere else [3]. The micro-CT analysis is carried out in terms of strut thickness, pore volume distribution, and pore size distribution across the entire three-dimensional space of the nano HA-gelatin scaffold before and after protein adsorption experiments. The total open porosity is increased in volume percent from 63.5% to 71.8% for before and after protein adsorption makes scaffold more porous and enhances cell adhesion and proliferation. The distribution of pore sizes obtained from extensive analysis of entire 3D scaffold for protein loaded HA-gelatin scaffold is presented in Figure 9.4a.

The protein adsorption and release study identifies that the scaffold can be used as bone growth for eukaryotic cells as well as delivery of protein. From the results of in vitro bioactivity, biodegradability it is confirm that the protein interaction favorable of CHAMPS scaffold. The compressive strength of the cylindrical (14 mm diameter and 15 mm thickness) porous scaffolds was measured with a universal testing machine using a load cell of 10 kN with a cross head speed of 2 mm min^{-1}. Nanorod scaffold are considered for optimum cryogenic treatment because nanorod HA has maximum adsorption efficiency of Bovine Serum Albumin (BSA) protein and their HA-gelatin scaffold has high compression strength. Figure 9.4b shows the representation of compressive stress-strain curves of original and modified HG00 porous scaffolds with 5%–10% strain. The original HG00 represents plastic yield in absence of linear elastic region. HG00 has peak stress of 2 MPa causes the plastic collapse of hydrocarbon matrix, which results in cell walls

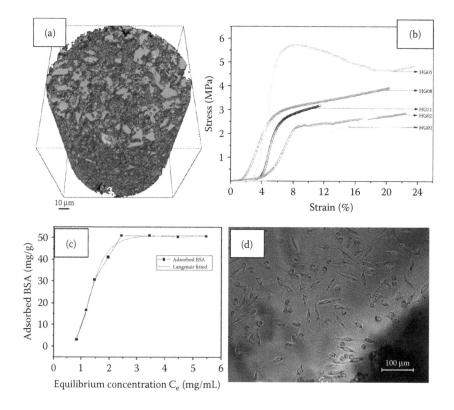

FIGURE 9.4
(a) 3D micro-CT scan image of BSA protein loaded CHAMPS, (b) stress-strain response of cryotreated scaffolds under compression, (c) adsorption isotherm of BSA on CHAMPS, and (d) fluorescence microscopic image revealing the adhesion of cultured L929 cells after 1 day of incubation of CHAMPS.

touching each other. The stress value slowly improved with 2 h liquid nitrogen treatment and compression strength is improved up to 6 MPa with 5 h liquid nitrogen treatment at 77 K. The mechanical properties of the material depends on the bond and strength of hydrogen matrix; HG05 scaffold exhibits maximum strength compared to others. However, the HG00 scaffold has high strain tolerance behavior compared with HG05. The strain tolerance is uniform after 5 h and gently decreases in the compression strength of scaffolds. This characteristic feature in liquid nitrogen is attribute to the change of bond behavior and internal stress related with thermal conductivity mismatch at low temperature. Hence, microcracking is caused due to the generation of internal stress and mismatch of thermal coefficient of HA and gelatin at 77 K. HA nanorod particles are tightly clamped with hydrocarbon matrix at 77 K that results in high compression strength compared to freeze casted scaffold. The hydrogen bond at 77 K is strong due to shorter bond length caused by thermal shrinkage. Due to thermal shrinkage, the free volume and free spaces are cured in scaffold, which results in increase intermolecular force and high strength at 77 K compared with the fabricated scaffold. Isotherm adsorption of BSA protein on HA-gelatin scaffold is shown in Figure 9.4c, that follows Langmuir isotherm and monolayer adsorption behavior. From Langmuir equation, the maximum adsorption capacity can be represented as:

$$q = \frac{q_{max} \, C_e}{(1/a_L + C_e)} \tag{9.2}$$

where q = the amount adsorbed of BSA on HA scaffold (mg/g), q_{max} = maximum adsorption capacity (51.3 mg/g), C_e = at equilibrium the concentration of BSA solution, a_L = Langmuir constant (0.213 mL/mg), and correlation factor (R^2) ~ 0.9902 in this experiment. Kim et al. revealed that adsorption BSA on collegen HA-gelatin nanocomposite occurs within 24 h, but in the case of the chitosan-gelatin scaffold, BSA adsorption reaches constant after 15 days [28]. The present study reveals that the adsorption of BSA protein in CHAMPS scaffold occurs within 48 h. In vitro cytotoxicity test of HA scaffolds are also carried with straight contact of L929 mouse fibroblast cells, and their observation is represented in Figure 9.4d. The scaffold cytotoxicity is evaluated by ISO-10993-5. The spindle-shaped L929 cells undergo apoptosis and necrosis, which found adhere and expanded on HA-gelatin scaffold. The globular shaped L929 cells result in degradation of HA-gelatin scaffold and release of polymeric small chains. The cytotoxicity test concedes the extract of HA scaffold is not affect the cell viability and proliferation. After 24 h of contact, confluent monolayer cell lines are observed with good morphology and proliferation of cells are maintained. The obtained results by extraction method indicate proliferation and viability of L929 fibroblast with the biocompatibility of HA scaffold.

9.4.5 Research and Development Scope

The scaffold can be effectively used for the bone tissue defects and disease, as it has competitive properties including mechanical strength, interconnected pores with optimum porosity, and biocompatibility with the bone. Effective BSA protein adsorption and release makes it perfect drug carrier/delivery media as well as cell adhesion, proliferation encompasses the bone growth capability of developed scaffolds. However, this analysis is not enough good to establish a viable bioimplant, as bone tissue engineering is a complex phenomenon. At this moment, this is probably one of the most challenging aspects to develop a bioartificial bone. An interesting approach known as extracellular matrix can be employed to make synthetic scaffold analogous to soft tissue. Surface modification of porous scaffold is a tactic to make it more osteogenic phenotype. Although a great advance in the knowledge of bone biology has been established until now, further steps need to be taken in order to better understand what is desired to develop a commercial tissue engineered bone [29].

9.5 Nanostructured ZnO for Cancer Therapy

9.5.1 Background

A drug carrier is one that is used to deliver the drug to the disease-affected cells in the body, while providing encapsulation so that drug does not affect the healthy cells of body. This state of carrier controls the release of drug by diffusion or by triggered release by some stimulus (changes in pH, application of heat, and activation by light). Drug carriers include liposomes, polymeric micelles, microspheres, and nanoparticles are getting importance for various treatment including cancer therapy [30]. The drug carriers are very important in case of cancer therapy because drugs used to cure cancer (chemotherapeutics) are highly toxic, lack proper distribution, and acquired drug resistance often fail them in eradicating cancer completely. Therefore, there is always a need for better drug carriers having micro/ nanostructures with superior drug loading ability and reduced system toxicity. Nanoscale polymersomes are evaluated as drug carriers, however, biocompatibility is an issue in vivo [31]. In the same time, extensive research work was done how to use carbon nano horns as drug carriers [32]. With these challenges, drug carriers with multifunctional nanostructure are attractive as an effective strategy and for multimodal therapy, defines the synergistic effect of two different therapeutic modalities. Photodynamic therapy (PDT) can be utilized for enhanced anti-cancer treatment, as it shows less systemic toxicity and side effects when compared to other cancer therapies (chemotherapy and radiation therapy). In this context, an effective research data shows that TiO_2 whiskers has less drug-encapsulation efficiency of 76.41% ± 5.29%, and thus further modification is required [33]. In order to enhance the efficacy, the prepared nanostructured hollow spherical and nanorods ZnO were loaded with a token drug (daunorubicin) and justified the photodynamic therapy.

9.5.2 Advantages of Nanostructured Hollow ZnO

There are many other materials like TiO_2, polymers are being tested and used as a drug delivery units. However, some of the advantages of the nanostructured ZnO hollow spheres (ZHS) that make them potential drug delivery units; are summarized below:

- Chemically, the ZnO surface is mostly terminated by OH groups that makes it an amenable material for enhanced drug-loading by functionalization with various drug molecules.
- Chemotherapeutics (drug, daunorubicin [DNR]) loaded ZnO nanoparticles experience photodynamic phenomenon, resulting in cancer cell destruction during ultraviolet (UV) radiation exposure.
- ZnO nanoparticles induce an excessive production of reactive oxygen species in different cell lines that results in cancer cell destruction through apoptosis; key target in cancer therapy.

- The drug (DNR) release from ZHS-DNR complex is pH dependent. This enhances the drug efficiency per dosage, and the destruction of cancer cells as therein continuous and prolonged drug release into the acidic medium of cell structure.
- The large specific surface area, hollow interior, and porous wall of ZHS results in high drug loading and good encapsulation while carrying the drug to cancer cells. This decreases the risk of drug exposure to the healthy cells.
- ZnO nanoparticles assisted PDT process has preferentially more ability to kill cancerous T cells when compared to normal cells.

9.5.3 Materials and Method

Two steps low temperature wet chemical method was employed to prepare the drug carrier units (ZHS). Initially, zinc nitrate hexa-hydrate (0.007 M) and hexa-methylene-tetramine (HMTA; 0.007 M) were dissolved in 100 mL distilled water. Followed by sodium citrate tribasic (0.002 M) was added dropwise to the above prepared solution to obtain a clear solution (sol-I). Initially, zinc oxide nanospheres were deposited on cleaned Si (100) substrate during reflux of sol-I at 80°C for 1 h. Followed by the substrate was carefully cleaned through deionized (DI) water to remove all other impurities. In next step, another set of solution (sol-II) was prepared with zinc nitrate hexa-hydrate (0.01 M), HMTA (0.01 M), and sodium citrate tribasic (0.001 M) in 100 mL DI water and maintained pH ~ 9 by adding NH_4OH to it. The sol-II along with the initially prepared zinc oxide nanospheres on Si substrate were heated in a three-necked refluxing reactor at 80°C for 1 h. This resulted in the formation of ZHS on the substrate. The substrate was then washed several times with DI water and dried at room temperature. Zinc oxide nanorods (ZNR) was also prepared in order to compare with ZHS performance. For preparation of ZNR, a solution of zinc nitrate hexa-hydrate (0.05 M) and HMTA (0.05 M) was prepared using DI water (50 mL) and then heated in an oven at 80°C for 3 h. Different concentration of drug (DNR, a model cancer drug) was loaded onto the synthesized ZHS and ZNR. The particle and drug mixed solution was stirred for 12 h in dark conditions to form ZHS-DNR and ZNR-DNR complexes, followed by centrifuged and washed with PBS buffer solution (till red color in the supernatant was disappeared) to remove unbound DNR molecules. To determine the intracellular localization, cells were stained with 50 nM LysoTracker Red DND-99 for 30 min. The formed complexes were again suspended in PBS at 4°C for subsequent in vitro DNR release and cytotoxicity tests.

9.5.4 Working Principle

The working mechanism of the drug carrier along with the model anti-cancer drug is schematically shown in Figure 9.5a and is self-explanatory.

The synthesized ZHS and ZNR, when loaded with the drug (DNR), forms a complex called ZHS-DNR and ZNR-DNR, respectively. This complex, when injected into the body, goes to cancer affected cell and gets absorbed by the cell by a process called endocytosis. Inside the cell, the endosomes/lysosomes (pH = 5.0) triggers the dissolution of ZHS, thereby releasing the DNR into the cell structure. The released drug gets transported to the nucleus by the proteasome and results in the damage of the cell. Simultaneously, the release of the drug takes place inside the cell as it gets transported to the nucleus and results in the prolonged exposure/retention of drug in the nucleus (chemotherapy). Now, as the drug was released completely from the complex, the

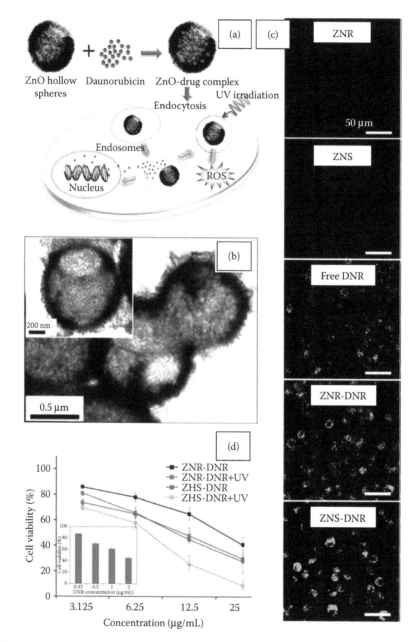

FIGURE 9.5
(a) Schematic representation of ZHS drug carriers with synergetic chemo-photodynamic cytotoxicity, (b) TEM image of the ZHS, (c) CLSM images of the intracellular distribution of ZNR, ZHS, free DNR, ZNR-DNR, and ZHS-DNR complexes in A549 cells, and (d) in vitro cell viability studied by MTT assay. Cells were treated for 48 h with ZHS-DNR, ZNR-DNR complexes and free DNR (inset) (presence/absence of UV irradiation).

remaining ZHS under UV irradiation induces the production of excess amounts reactive oxygen species in the cell thereby increasing the cellular toxicity (photodynamic action). Thus, this entire method of treating the cancerous cell with ZHS-DNR complex can be termed as a chemo-photodynamic action, as it consists of both chemotherapy and photodynamic therapy.

9.5.5 Characterization and Analysis

Particle size distribution and morphological study were conducted for both spherical (ZHS) and rod (ZNR) particles, however, optimum spherical particles have only been considered and discussed in the perspective of prime objective of the research, details are somewhere else [4]. Morphological analysis was conducted by TEM and represented in the Figure 9.5b. The classic image shows the existence of well-designed hollow spherical particles consisting of porous and nanostructured wall. Average particle diameter is found near to 800 nm and the hole opening diameter is ~350 nm. A competitive drug (DNR) encapsulation efficiency of the ZHS and ZNR are estimated as 87.7% ± 4.10% and 53% ± 5.76%, respectively and calculated by Equation 9.3.

$$\text{Drug Encapsulation Efficiency (DEE)} = \frac{\text{total DNR} - \text{DNR in supernatant}}{\text{total DNR}} \times 100 \qquad (9.3)$$

The higher specific area, hollow interior, and the permeable (porous) nanostructured wall of ZHS facilitates the DNR to get adsorbed/absorbed onto/into the structure resulting in high DEE in ZHS-DNR complex. It has observed that the DNR release from ZHS-DNR complex at 37°C (nearly 93% ± 4.8%) was higher at pH = 5.0–5.5 (lysosomal/endosomal pH range) than at physiological pH = 7.4 (17.9% ± 2.3%). This higher release rate of DNR from ZHS-DNR is due to the solubility of ZnO in acidic environment thus releasing 40% of DNR in nearly 6 h and the rest in the next 36 h. The intracellular drug release pattern of the ZHS-DNR and ZNR-DNR complexes is examined by CLSM, and their image is shown in Figure 9.5c. It is very clear that the ZHS-DNR complex shows a brighter fluorescence than the ZNR-DNR complex. This behavior (high intake of the drug by cell i.e., bright fluorescence) might be due to high drug encapsulation efficiency of ZHS and higher metabolism of cancer cells. Also, these complexes are mostly located in the lysosome/endosome and cytoplasm where the acidic environment (pH = 5.0–5.5), lysosome compartments) triggers the accelerated dissolution of ZnO, as described earlier. This increased amount of DNR is thus transported to the cell nucleus, and it ensures continuous drug release from complex and retention in the nucleus, which enhances the efficacy of cancer therapy. The chemo-photodynamic cytotoxicity of the complexes (ZHS-DNR and ZNR-DNR) were evaluated and shown in Figure 9.5d. The drug-encapsulated complexes exert prominent different degree of cytotoxicity in a dosage-dependent manner on cancer cells. An identical cell viability trend follows with respect to dose content. The cell viability is more reduced in the case of ZHS-DNR complex when compared with ZNR-DNR complex at a concentration of 3.125 µg/mL having 0.27 µg/mL DNR drug content. The cell viability decreases still when high concentrations of the ZHS-DNR complex were used. The higher cytotoxicity of ZHS-DNR complex is due to its ability to get into cancer cells and acid responsive drug release. After UV exposure, the ZHS-DNR complex shows significantly enhanced cancer cell destruction at all tested concentrations than the cells treated with chemotherapy only (ZHS-DNR and ZNR-DNR complexes without UV irradiation) and free DNR. For instance, the cell viability observed at the highest concentration (25 µg/mL with 2.19 µg/mL DNR) for the ZHS-DNR and ZNR-DNR complexes are ~9% and ~30%, respectively upon UV irradiation. This indicates that the high-performance of the ZHS-DNR complex could be achieved by a successful combination of chemotherapy and PDT to facilitate extensive cell death at various concentrations. Apoptosis (programmed cell death in multicellular organism) is a key process in cancer development and is important in developing a cancer therapy. Based on the previous analysis, a flow cytometric analysis of apoptotic cells over 24 h was performed by using various drugs along with double staining with Annexin V-FITC/PI. The apoptotic cell percentage is 17.62% for free DNR, 41.71% for ZHS-DNR complex, and 91.76% for ZHS-DNR complex with UV irradiation making ZHS-DNR complex (with UV irradiation) an efficient apoptotic inductor.

9.5.6 Research and Development Scope

Precise research planning and activity are required to increase the drug encapsulation efficiency, apoptotic cell percentage, and decreasing the toxicity on the body tissues (cell viability) while using the ceramic nanoparticles like ZnO drug carriers. In view of ZnO biomedical potential, zinc hollow sphere featured with hollow interior and inherent photodynamism were anticipated as to be beneficial for theranostics development. Furthermore, new strategies including ZnO structures doping/labeling with active materials could be studied in the future to achieve visible-to-near infrared irradiation wavelength and image-guided drug delivery. Various other materials that are effective in photoexcitation, but biocompatible may attempt as drug carriers for photodynamic therapy. Recently, researchers have discovered a new protein known as cadherin-22 that could reduce the spread of cancer by binding the cancer cells together and allowing them to invade breast and brain cancer cells by up to 90%. Thus, protein carrier like hydroxyapatite nanoparticles can conjugate with drug molecule and release effectively and can be attempted for such drug management. Not only materials, somewhat their morphology, crystallinity, hydrophilic/hydrophobic domains, and dispersion capability are also needed to encounter to develop effective target drug carrier.

9.6 Nanoscale Core-Shell Ferrite for Hyperthermia

9.6.1 Background

The effective thermal energy motivated conversion is gaining importance and focused more research on the materials. For example, when photons strike gold nanoparticles, thermal energy is released and used for photo thermal cancer therapy. However, the surrounding body tissues that interfere the photons and penetration ability into deep body tissues hinder the effectiveness. Thus, ability of the penetration of such energy into deeper targeted body tissues that can effectively convert into thermal energy by apposite foreign material is one of the important aspects to look into. In this regard, nanoparticles that have the capacity for the electromagnetic energy (radiofrequency waves have no tissue penetration issues) to thermal energy conversion are very useful in biological applications. Synchronization of this conversion is relatively easy to control and operate, and thus attract many applications like hyperthermia treatment, thermal imaging, etc. Generally, these materials are magnetic, but they have very poor conversion efficiencies. To overcome this problem, development of new nanoparticles which have high energy transfer capability (high specific loss power-SLP) is necessary. Nanoparticle size, composition, magnetic field, magnetic spin relaxation, and particle diameter are major controlling parameters to originate high specific loss power-SLP values [34–36]. The exchange coupling between magnetic nanoparticles consisting of hard core and soft shell helps in tuning the magnetic properties (magneto crystalline anisotropy, an inherent material property) and thereby increasing SLP (acts as gauge for conversion efficiency). These classes of nanoparticles are referred as core-shell nanoparticles and have properties superior to that of the individual counterparts. In this present case study, core ($CoFe_2O_4$)-shell ($MnFe_2O_4$) nanoparticles were synthesized and analyzed along with their ability to be utilized in hyperthermia treatment as mentioned in the below section.

9.6.2 Advantages of Core-Shell Nanoparticles

The following are some of the advantages that make the nanoparticles especially core-shell type a most sought in applications that require the conversion of electromagnetic energy into heat energy [37,38]:

- Core-shell nanoparticle allows the tuning of material properties like magneto crystalline anisotropy and magnetization, while maintaining combined properties that eventually promote high SLP values (prevents aggregation of nanoparticles by spin relaxation and demagnetization at room temperature).
- Core-shell nanoparticles have SLP values in the range of ferrimagnetic nanoparticles and superior to that of superparamagnetic nanoparticles.
- Core-shell nanoparticles provide classical magnetic properties that is required for biomedical applications.
- Proper selection of shell composition, it can be designed as less cytotoxic.
- It produce continuous thermal energy and near to 99% supplied external energy can penetrate even at depths of 15 cm inside the body.

9.6.3 Materials and Method

Core ($CoFe_2O_4$)-shell ($MnFe_2O_4$) nanoparticles of 15 nm size denoted as $CoFe_2O_4$@$MnFe_2O_4$ were synthesized by using 9 nm $CoFe_2O_4$ nanoparticles as seed and $MnFe_2O_4$ was grown over seed surface by thermal decomposition as described in literature [5]. Initially, to prepare $CoFe_2O_4$ nanoparticles, 3.25 mmol of $CoCl_2$ and 5 mmol of $Fe(acac)_3$ along with oleic acid, oleylamine, and octyl ether were placed in a 250 mL round-bottomed (3 neck) flask. This mixture in the flask was heated at 300°C for 1 h, and the obtained products were cooled to room temperature. In this mixture, ethanol was added and results in a black powder precipitate of $CoFe_2O_4$. The precipitated powder was centrifuged to separate nanoparticles of $CoFe_2O_4$ and the optimized 9 nm separated nanoparticles were dispersed in hexane. Optimized 9 nm $CoFe_2O_4$ nanoparticles was dispersed in hexane and used as seed. Followed by, a reaction mixture was prepared by adding 3.25 mmol of $MnCl_2$ and 5 mmol of $Fe(acac)_3$ in a 250 mL round-bottomed flask (3 neck) along with oleic acid, oleylamine, and trioctylamine to make $MnFe_2O_4$ shell over the core $CoFe_2O_4$ nanoparticles. The resultant reaction mixture was heated at 365°C for 1 h and cooled down to room temperature as mentioned above, followed by ethanol added and centrifuged to separate 15 nm core-shell ($CoFe_2O_4$@$MnFe_2O_4$) nanoparticles. A number of core-shell nanoparticles can be synthesized using the above-mentioned technique with different materials like $CoCl_2$, $FeCl_2$, and $MnCl_2$. The used organic surfactants (oliecacid and aleylamine) during preparation of nanoparticles were removed by using 2,3-dimercaptosuccinicacid and eventually used for hyperthermia experiments [39].

9.6.4 Working Principle

When an external magnetic field (using alternating current) is applied to the magnetic nanoparticles, a magnetization reversal process occurs resulting in the production of thermal energy as particles return to their relaxed states [31]. This principle is used to treat the diseases with hyperthermia (treatment in which cancer affected body tissues are exposed to temperatures as high as 45°C). In this method, mostly the heat is applied locally onto the small area by injecting core-shell nanoparticles into the tumor. Prior to magnetization of infected subject (in this case mouse), the synthesized core-shell nanoparticles were injected and placed inside the water cooled magnetic induction coil (diameter of 5 cm in this case) as shown discussed in details [5]. Through this coil an alternating current magnetic field (500 KHz) was applied for a certain time usually 10 min. This results in the generation of thermal energy by the nanoparticles, whereby the temperature of the surroundings increases resulting in the heating of tumor and there by killing/damaging the cancer cells/tumor.

9.6.5 Characterization and Analysis

The prepared core-shell nanoparticles are analyzed using transmission electron microscope and the images are shown in the Figure 9.6a. Synthesized core-shell particles are uniform in size without any soft or hard agglomeration that supports the reliability of the synthesis protocol and performance, as specific power loss depends on particle size. For example, the change in shape factor may reduce the SLP value up to 15% compared to monodisperse particles. The coupled magnetism of the core shell nanoparticles was investigated by measuring magnetization (M) versus coercivity (H) curve using magnetometer (superconducting quantum interference device) as shown in Figure 9.6b. The obtained M-H curve shows smooth hysteresis loop at low and ambient temperature. From the graph, the coercivity value (H_c) at 5 K for $CoFe_2O_4@MnFe_2O_4$ nanoparticles is 2530 Oe, whereas a distinct difference in between $CoFe_2O_4$ (11,600 Oe) and $MnFe_2O_4$ (0 Oe) is observed. This shows that the prepared core shell particles are magnetically exchange coupled. Inset in Figure 9.6b represents the M–H curve of $CoFe_2O_4@MnFe_2O_4$ at 300 K, showing its intermediate magnetic nature of both particles. Figure 9.6c shows the SLP values of the prepared core-shell nanoparticles and its original individual components that reveal SLP values of core-shell particles are much higher (2000–2500 w.g⁻¹) than the corresponding individual components.

The SLP was calculated by using the Equation 9.4.

$$SLP = \frac{CV_s}{m} \frac{dT}{dt} \tag{9.4}$$

where dT/dt is the initial slope of the graph of change in temperature versus time, C = specific heat capacity (const. volume) of the sample solution, V_s = sample volume, m = mass of the sample (magnetic material). For detailed method of measuring as well as calculating SLP value, see supplement provided in Jae-Hyun Lee et al. [5]. Owing to this classic magnetic behavior, the core-shell particles are used for anti-tumor hyperthermia therapy, as well as competitive analysis enables to demonstrate the efficacy of such research importance. This study was performed by human brain cancer cells (U87MG) and used as xenograft to the mouse abdomen. $CoFe_2O_4@MnFe_2O_4$ nanoparticles in weight of 75 μg were dispersed in 50 μl of saline water and injected into the tumor site. The method of treating is explained in Section 9.6.3, which shows the mouse

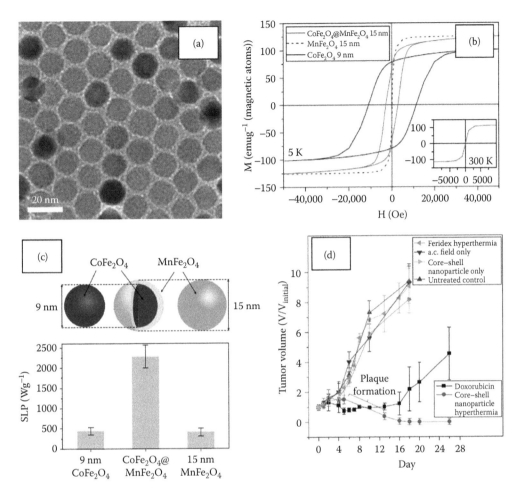

FIGURE 9.6
(a) HRTEM image of $CoFe_2O_4$-$MnFe_2O_4$ nanoparticle (15 nm) showing the uniform size distribution and single crystallinity. (b) M-H curve of $CoFe_2O_4$-$MnFe_2O_4$ (15 nm), $MnFe_2O_4$ (15 nm), and $CoFe_2O_4$ (9 nm) nanoparticles measured at 5 K using a superconducting quantum interference device magnetometer. The magnetization curve of the core-shell nanoparticle (light gray curve) shows the hard-soft exchange-coupled magnetism with a smooth hysteresis curve. (c) Schematic of 15 nm $CoFe_2O_4$-$MnFe_2O_4$ nanoparticle, and its SLP value in comparison with the values for its components (9 nm $CoFe_2O_4$ and 15 nm $MnFe_2O_4$), and (d) graph showing the change of tumor volume (V/V initial) as treatment proceeds with core-shell nanoparticle hyperthermia, doxorubicin, Feridex hyperthermia, only a.c. field, only core-shell nanoparticles, and untreated control.

being placed in a magnetic induction coil (diameter is 5 cm and water cooled). In order to understand the efficacy of such hypothesis, several groups of mice were treated with feridex (hyperthermia treatment), ac filed only, core-shell nanoparticles only, untreated control, doxorubicin (chemotherapeutic drug with equal dosage of core-shell nanoparticles), and their competitive analysis were plotted in a graph (tumor volume versus days) in Figure 9.2d. The obtained result clearly shows the tumor elimination in the mouse group treated with core-shell nanoparticle hyperthermia in 18 days, while its tumor growth increased near to 9 times when it is completely untreated. Interestingly, there is no effect of treating with feridex hyperthermia, only ac field, only core-shell nanoparticles on the tumor, as the tumor grew continuously over the period of treatment.

Although, the doxorubicin initially suppress the tumor growth up to 16 days, but cannot cure completely, rather grown has also been noticed. This analysis enables to justify the influence of nanostructured core-shell ceramic magnetic nanoparticles are an important domain in hyperthermia therapy, in specific cancer research.

9.6.6 Research and Development Scope

Core-shell magnetic particles assisted hyperthermia is both the ideal complementary treatment to a strong sensitizer of radiotherapy and many cytotoxic drugs. Herein, the highlighted methods used is very simple as it is transmitting electromagnetic energy as close as possible to tumor site. The combined paramagnetic and magnetic nanoparticles have the potential to reduce or eliminate the possible side effects by achieving accurate target doses. The concept of exchange coupling between magnetic particles can be extended to other materials and their conversion efficiency (high SLP value) is desired to enhance, accordingly. The ability of other biocompetitive materials showing heat generation property is in also forefront for various hyperthermia treatment. With the ultrasonic hyperthermia treatment of superficial and deep region tumors that include surface lesions, head and neck is in advance stage. Using this technique, heating is possible up to 5–10 cm depth with the single transducer and up to 20 cm depth with the multiple transducers. In order to treatment of the prostate cancer, hyperthermia with external radio frequency devices is a good alternative although the methods are nonspecific for tumor cells. Furthermore, the hyperthermic perfusion is used to treat cancer in arms, legs, or in some organs of body. To increase the treatment response and reducing side effects, the frequency enhancers and catheters are also used. Microwave hyperthermia is another promising technique for the treatment of superficial tumors in breast, limb, prostate, and brain. It is one of the advance technologies where large volume heating is possible with the help of a specialized antenna. However, at high penetration localized heating, temperature measurement is difficult. While discussing the possibility of new class of nanostructured materials and their use for hyperthermia treatment, additional side effects of patients should keep in mind, as it is one of the important criteria to select the materials and doses.

9.7 Concluding Remarks

In recent trend, "Ceramic Engineering" branch may not be a first choice among under graduate student, although we are comprehensively reliant on ceramics for living, curing, and comfort. In this regard, the prime motive is to emphasize on the importance of this class of materials and how their nanostructured particles and compacts (indirect use of nanoparticles) can be used as bioceramics for dentistry, orthopedic, and cancer treatment. Recent research trend is not only confined in the development of analogous teeth material, rather cumulative success depends on multidisciplinary activity of cell biologist, microbiologist, molecular biologist and biomechanics, and dentist. Despite aforesaid interest of either materials or medical issues, orthopedic extensively involves the musculoskeletal treatment that is made of bones, joints, ligaments, tendons, and muscles. On the basis of human body characteristics and activity, the bone problems may include deformities, infections, tumors, and fractures. Joints experience arthritis, bursitis, dislocation, pain, swelling, and ligament tears. So, common orthopedic-related diagnoses focus on issues with the ankle and foot, hand and wrist, shoulder, hip, knee, elbow, and spine. New generation computerized evolution helps to design patient specific orthopedic implants and dental restorations associated with nanostructured microstructures and mechanical properties of ceramics that caused in the clinical better workflow, as well as treatment options offered to patients. The photodynamic cancer therapy by photoexcited ceramic semiconductor based nanotechnology is advantageous over conventional chemotherapy. This mediated thermal therapy promises to change very foundation of cancer diagnosis, treatment, and prevention. An increasing research relates to targeting techniques, delivery strategies, and radiation dose enhancement are likely to energies this field in the coming years.

Material scientists, biotechnologist, clinicians, mechanical engineer, electronics engineer, and others came forward together to resolve and cure different diseases and problems. Thus, several materials and wide horizon of opportunities are opened up in biomaterial research and development that eventually help doctor to diagnosis and treat in better way, although society is expecting new generation materials as an answer and remedies of uncured diseases. To get more promising results, clinical trials still require and refinement of techniques to increase the efficiency. In fact, this horizon boosted up the socio-economic benefit and interact with social characteristics such as ethnic group and sex around the globe. Apart from lab scale research, intensive strategy is required to establish a successful product through high quality translational research for patient benefit.

References

1. D. Sarkar, S. K. Swain, S. Adhikari, B. S. Reddy, and H. S. Maiti, Synthesis, mechanical properties and bioactivity of nanostructured zirconia, *Mater Sci Eng C* 33(6), 3413–3417, 2013.
2. D. Sarkar, S. Mandal, B. S. Reddy, N. Bhaskar, D. Sundaresh, and B. Basu, ZrO$_2$-toughened Al$_2$O$_3$-based near-net shaped femoral head: Unique fabrication approach, 3D microstructure, burst strength, *Mater Sci Eng C* 77(1), 1216–1227, 2017.
3. B. Basu, S. K. Swain, and D. Sarkar, Cryogenically cured hydroxyapatite-gelatin nanobiocomposite for BSA protein adsorption/release, *RSC Adv* 3, 14622–14633, 2013.
4. N. Tripathy, R. Ahmad, H. A. Ko, G. Khanga, and Y. B. Hahn, Multi-synergetic ZnO platform for highperformance cancer therapy, *Chem Commun* 51, 2585–2588, 2015.
5. B. B. Nissan, Nanoceramics in biomedical applications, *MRS Bull* 29, 28–32, 2004.
6. J. H. Lee, J. Jang, J. Choi, S. H. Moon, S. Noh, J. Kim, J. G. Kim, I. S. Kim, K. I. Park, and J. Cheon, Exchange-coupled magnetic nanoparticles for efficient heat induction, *Nat Nanotechnol* 6, 418–422, 2011.
7. R. C. Garvie, R. H. Hannink, and R. T. Pascoe, Ceramic steel, *Nature* 258, 703–704, 1975.
8. A. H. Heuer, N. Claussen, W. M. Kriven, and M. J. Ruhle, Stability of tetragonal ZrO$_2$ particles in ceramic matrices, *Am Ceram Soc* 65, 642–650, 1982.
9. J. Chevalier, What future for zirconia as a biomaterial? *Biomaterials* 27, 535–543, 2006.
10. M. Guazzato, M. Albakry, S. P. Ringer, and M. V. Swain, Strength, fracture toughness and microstructure of a selection of all-ceramic materials. Part I. Pressable and alumina glass-infiltrated ceramics, *Dent Mater* 20, 449–456, 2004.
11. B. Basu, Toughening of yttria-stabilised tetragonal zirconia ceramics, *Int Mater Rev* 50, 239–256, 2005.

12. H. Yoshida, A. Faust, J. Wilckens, M. Kitagawa, J. Fetto, and E. Y. S. Chao, Three-dimensionaldynamic hip contact area and pressure distribution during activities of daily living, *J Biomech* 39, 1996–2004, 2006.

13. C. L. Abraham, S. A. Maas, J. A. Weiss, B. J. Ellis, C. L. Peters, and A. E. Anderson, A new discrete element analysis method for predicting hip joint contact stresses, *J Biomech* 46, 1121–1127, 2013.

14. V. Pakhaliuk, A. Polyakov, M. Kalinin, and V. Kramar, Improving the finite element simulation of wear of total hip prosthesis' spherical joint with the polymeric component, *Procedia Eng* 100, 539–548, 2015.

15. D. Sarkar, B. Sambi Reddy, S. Mandal, M. RaviSankar, and B. Basu, Uniaxial compaction-based manufacturing strategy and 3D microstructural evaluation of near-net-shaped ZrO_2-toughened Al_2O_3 acetabular socket. *Adv Eng Mater* 18, 1634–1644, 2016.

16. N. Claussen and M. Rühle, Design of transformation toughened ceramics, advances in ceramics, In: *Science and Technology of Zirconia*, Vol. 3, The American Ceramic Society, Columbus, OH, p. 137, 1981.

17. J. Gührs, A. Krull, F. Witt, and M. M. Morlock, The influence of stem taper re-use upon the failure load of ceramic heads, *Med Eng Phys* 37, 545–552, 2015.

18. U.F.A.D. Administration, *Guidance Document for the Preparation of Premarket Notification for Ceramic Ball Hip Systems*, US Food and Drug Administration, Medical Devices, pp. 5–13, 2009.

19. FDA Case Study, An innovative "Me-Too" idea: Premarket notification – 510(K) medical device submissions, *This Fictionalized Case Study is the Fourth in An Educational Series*, US Food and Drug Administration, 2014.

20. G. Willmann, Ceramic femoral head retrieval data, *Curr Orthop Pract* 379, 22–28, 2000.

21. The Market Reports WhaTech Agency. Ceramic femoral head market to 2021: Consumption volume, value, import, export and sale price analysis illuminated by new report, WhaTech Channel: Materials & Chemicals Research, News from The Market Reports - Industry & Market Reports at its BEST, October 20, 2016.

22. R. M. Deijkers, R. M. Bloem, P. C. Petit, R. Brand, S. W. Vehmeyer, and M. R. Veen, Contamination of bone allografts: Analysis of incidence and predisposing factors, *J Bone Joint Surg* 79, 161–166, 1997.

23. B. Basu et al., *Advanced Biomaterials: Fundamentals, Processing, and Applications*, John Wiley & Sons, Hoboken, NJ, 2010.

24. B. D. H. Ratner, A. S. Hoffman, F. J. Schoen, and J. E. Lemons, *Biomaterials Science: An Introduction to Materials in Medicine*, Academic Press, New York, 2004.

25. S. V. Dorozhkin, Bioceramics of calcium orthophosphates, *Biomaterials* 31, 1465–1485, 2010.

26. L. L. Hench, Biomaterials: A forecast for the future, *Biomaterials* 19, 1419–1423, 1998.

27. S. K. Swain, S. V. Dorozhkin, and D. Sarkar, Synthesis and dispersion of hydroxyapatite nanopowders, *Mater Sci Eng C* 32, 1237, 2012.

28. H. Kim, H. E. Kim and V. Salih, *Biomaterials* 26, 5221, 2005.

29. A. J. Salgado, O. P. Coutinho, and R. L. Reis, Bone tissue engineering: State of the art and future trends, *Macromol Biosci* 4(8), 743–765, 2004.

30. R. Chadha, V. K. Kapoor, D. Thakur, R. Kaur, P. Arora, and D. V. S. Jain, Drug carrier systems for anticancer agents: A review, *JSIR* 67(3), 185–197, 2008.

31. K. Ajima, M. Yudasaka, T. Murakami, A. Maigne, K. Shiba, and S. Iijima, Carbon nanohorns as anticancer drug carriers, *Mol Pharm* 2(6), 475–480, 2005.

32. R. G. Tuguntaeva, C. I. Okekea, J. Xu, C. Li, P. C. Wang, and X.-J. Liang, Nanoscale polymersomes as anti-cancer drug carriers applied for pharmaceutical delivery, *Curr Pharm Des* 22(19), 2857–2865, 2016.

33. Z.-Y. Zhang, Y.-D. Xu, Y.-Y. Ma, L.-L. Qiu, Y. Wang, J.-L. Kong, and H.-M. Xiong, Biodegradable ZnO@polymer core-shell nanocarriers: pH-triggered release of doxorubicin in vitro, *Angew Chem* 125, 4221, 2013.

34. J. P. Fortin et al., Size-sorted anionic iron oxide nanomagnets as colloidal mediators for magnetic hyperthermia, *J Am Chem Soc* 129, 2628–2635, 2007.

35. R. E. Rosensweig, Heating magnetic fluid with alternating magnetic field, *J Magn Magn Mater* 252, 370–374, 2002.
36. A. H. Habib, C. L. Ondeck, P. Chaudhary, M. R. Bockstaller, and M. E. McHenry, Evaluation of iron–cobalt/ferrite core shell nanoparticles for cancer thermotherapy, *J Appl Phys* 103, 07A307, 2008.
37. Q. A. Pankhurst, J. Connolly, S. K. Jones, and J. Dobson, Applications of magnetic nanoparticles in biomedicine, *J Phys D* 36, R167–R181, 2003.
38. A. M. Derfus et al., Remotely triggered release from magnetic nanoparticles., *Adv Mater* 19, 3932–3936, 2007.
39. J.-T. Jang et al., Critical enhancements of MRI contrast and hypethermic effects by dopant-controlled magnetic nanoparticles, *Angew Chem Int Ed* 48, 1234–1238, 2009.

10

Problems and Answers

PROBLEM 1:

a. How Fermi energy is dependent on available of number of free electrons per unit volume of gold nanoparticles, state and establish the relation? Is it possible to achieve nonmetallic behavior of metallic gold at 311 K through quantum confinement? If yes, what would be the expected diameter of the Au particle? Given, density of Au = 19.32 g/cc, atomic mass = 197 g/mol, Fermi energy = 8.78×10^{-19} J, Boltzmann's constant = 1.38×10^{-23} J/K.

b. Is it feasible to enhance the band gap of $BaTiO_3$ through particle size management? Explain and justify with suitable illustration.

c. How do surface area, surface area to volume ratio, and surface atom density vary with particle size? Explain with a suitable example and illustration.

HINTS 1:

Before taking into account, please look on the topic discussed in Section 1.2.1 through 1.2.4, and follow the steps as described in the following:

a. Consider Equation 1.10, the probable valence electrons, (N_v), that able to transfer if $\Delta E > k_B T$, we can obtain $N_v \approx 272$. So, if system has less than 272 number of valence electrons, it cannot mobilize effectively toward conduction band at 311 K. Now we have to find out the volume of particles that have less than 272 number of electrons to achieve nonmetallic properties. As we know, Au possesses $[Xe]4f^{14}5d^{10}6s^1$, so only one electron in 6s sublevel, so, one atom provides one valence electron only.

$$\text{The density of atoms in a given volume} = \frac{\left(19320 \text{ kg/m}^3\right)\left(6.023 \times \dfrac{10^{23} \text{ atoms}}{\text{mol}}\right)}{0.197 \text{ kg/mol}}$$

$= 5.90 \times 10^{28}$ atoms/m³. In another way, volume of the particle,

$$V_p = \frac{272 \text{ atoms}}{5.90 \times 10^{28} \text{ atoms/m}^3} = 4.61 \times 10^{-27} \text{ m}^3.$$

In consideration of spherical particle diameter D, less than 2 nm Au particle can behave as nonmetallic.

b. Yes, Figure 1.2 and Equation 1.11 can explain this phenomenon.

c. Consider a cubic particle with having 1 cm edge and 1g weight, and explain these features by Figure 1.3 and Table 1.1.

PROBLEM 2:

a. What is dangling bond? How do closest packing and broken bond influence the surface energy of crystal plane? Describe with a relevant schematic and example.

b. What are the basic strategies that are essential to encounter for the controlling of the critical radius of nanoparticles? How does heterogeneous nucleation depends on impurity and contact angle during synthesis of nanoparticles?

c. How cohesive energy depends on surface atom density, and what is the influence on Debye and melting temperature of material? Calculate the cohesive energy, Debye temperature, and melting temperature of Ag, where particle diameter "D" value varies from 1 to 50 nm, and plot all results. Given, $E_{bs} = -295.9$ kJ/mol, $T_{bs} = 961.8°C$, $\theta_{bs} = 235$ K, atomic radii = 165 pm.

HINTS 2:

Before taking into account, please look on the topic discussed in Sections 1.2.5 through 1.2.7, and follow the steps as described in the following:

a. Calculate the planar density of fcc and bcc crystal plane of interest, make them order, the surface energy become just reverse, as highest atomic packing has lowest number of unsaturated bond. Hence, result is fcc $\gamma(110) > \gamma(100) > \gamma(111)$, but bcc $\gamma(111) > \gamma(100) > \gamma(110)$ metals.

b. Critical radius is the first entity during synthesis of new materials, with judicious selection of processing parameters, as discussed in Section 1.2.6, one can control the size of it. Following nucleation, growth and Ostwald ripening results in wider particle size distribution. In the perspective of application, it is necessary to selection of shape factor (α), for example, very high α value is expected for synthesis of magnetic particles for hyperthermia treatment, whereas a relatively low α provides better packing density after consolidation of ceramic particles.

During synthesis of nanoparticles through wet chemical method, a clean glass beaker may provide high "α" value compared to old and several scratched glass beaker, as later follows relatively heterogeneous nucleation compared to former condition.

c. Consider Equations 1.23 through 1.25, and calculate all parameters.

PROBLEM 3:

a. How does the lattice strain changes with metallic and ceramic nanoparticles? Explain with a suitable example. What would be the probable method to determine the existence of such strain in nanoparticles?

b. Imagine how a Si_3N_4 cantilever beam (attached at one side to a base) is used as image processing tip, where an upward and downward deflection represent different states. The beam dimension is 10 μm long, 200 nm wide, and 100 nm thick. The modulus of elasticity of Si_3N_4 is 310 GPa and density 3440 kg/m^3. What would be the frequency of beam?

c. How do fluid dynamics depend on Reynolds numbers? Calculate and compare the settling velocity of 100 nm spherical zirconia nanoparticles in water and acetone. Given: $\rho_{ZrO2} - 5.98$ g/cc, $\rho_{acetone} = 0.784$ g/cc, $\eta_{acetone} = 0.306$ cP (1 cP = 0.1 Pa.s).

HINTS 3:
Before taking into account, please look on the topic discussed in Sections 1.2.8 through 1.2.10, and follow the steps as described in the following:

a. Please check the Section 1.2. to answer the first part of the problem. XRD is the prime technique to determine such strain, for details please consider Section 2.3.4.

b. First evaluate the spring constant with help of Equation 1.34, followed by evaluate the effective mass when beam is fixed in one side, and, finally, calculate the frequency by Equation 1.33.

c. Please check Section 1.2.10. With the help of Equation 1.37, determine the settling velocity in two fluids and compare, where, R = 50 nm, ρ_f (acetone) = 0.784 g/cc, ρ_f (water) = 1 g/cc, ρ_p (particle) = 5.98 g/cc.

PROBLEM 4:
Among several wet chemical processes, coprecipitation is an easy and economic approach and thus adopted to synthesis pure 100 g yttrium aluminum garnet (YAG) nanoparticles. In the beginning, a precipitate was obtained in presence of $Y(NO_3)_3.6H_2O$, $Al(NO_3)_3.9H_2O$, and NH_4HCO_3, followed by high temperature calcination is required to obtain pure phase with desire particle size of 40 nm. In order to achieve the pure YAG ($Y_3Al_5O_{12}$) phase instead of any other intermediate phase, judicious temperature selection is required, and it can be done by DSC-TG analysis in air at 10°C/min, and the obtained result is shown in Figure 10.1. Given, sample mass = 9.681 m, Y = 88.9, Al = 27, N = 14, O = 16, and H = 1.

Calculate, (a) amount of precursor required to make 0.5 M $Y(NO_3)_3.6H_2O$, 0.8 M $Al(NO_3)_3.9H_2O$, and 0.5 M NH_4HCO_3 solutions prior to initiate the coprecipitation synthesis, (b) the all probable phase changes and weight loss during dynamic heating, (c) crystalline

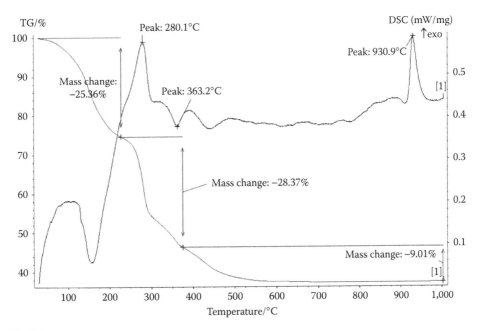

FIGURE 10.1
DSC-TG thermal analysis of precipitated particles in order to determine the crystallization temperature of YAG phase.

phase transformation temperature of YAG phase, (d) enthalpy change during crystallization of YAG, and (e) Avrami kinetics exponent and activation energy of the YAG crystallization process. Predict the morphology, and what would be the strategy to select calcination
temperature in order to obtain free flowing crystalline YAG nanoparticles without content
of any impurity phases?

HINTS 4:

Before taking into account, please look on the topic discussed in Section 2.2, and follow the
steps as described in the following:

Step 1: Write a balance equation and simply calculate the required amount of precursor from stoichiometric equation, molecular weight, targeted molarity, and weight
of the powder.

Step 2: Each endothermic and exothermic peak are related to certain phase change.
So, correlate those particular temperatures with three significant weight loss in
TG plot, verify with the chemical formula of precipitate that forms during copre-
cipitation process.

Step 3: Pick up the exothermic temperature without any weight loss to determine the
crystallization temperature, since amorphous to crystallization process does not
involve with any weight loss.

Step 4: Either equipment provide the enthalpy value, or it can be determined by
Equation 2.6. The area can be obtained from plotting the profile on graph paper or
direct through origin software.

Step 5: Consider the original raw data, plot separately within the temperature range
of 920°C–1000°C, and follow the process described in Section 2.2.4 and Equations
2.9 and 2.12. The fraction can be estimated either on graph paper or software.

PROBLEM 5:

a. Let's consider a peak $2\theta = 43.2°$ of face-centered cubic copper (Cu) is obtained using
X-ray diffraction pattern by Cu $= K\alpha$ radiation with a wavelength of $\lambda = 0.154$ nm.
Atomic radius of Cu is 1.28 Å. What would be the probable Miller indices (hkl) for this
peak?

b. Index the pattern and evaluate the lattice parameters of MgO after obtaining the
XRD pattern using Cu $= K\alpha$ radiation with a wavelength of $\lambda = 0.154$ nm. The
obtained 2θ values are 36.95, 42.91, 62.30, 74.64, 78.64, 94.06, 105.75, 109.78, 127.29, and
143.77. Assume MgO having the NaCl type structure.

c. Illustrate the method of analysis and calculate the true density of α – alumina,
t – zirconia, and m – zirconia. Given, atomic weight of Zr – 91.22, Al – 26.98, and
O – 16. α-Al_2O_3 (a $= 4.759$, c $= 12.991$), t-ZrO_2 (a $= 3.6067$, c $= 5.1758$). How does it help
to determine the relative density of synthesized particles or compacts?

d. State a simple and accurate method to quantify the phase content in multiphase
system.

e. How does crystallite size vary with particle size and grain size of materials? How
does microstrain influence the peak intensity and broadening behavior?

HINTS 5:

a. Consider Bragg's law, $\lambda = 2\,d\,Sin\theta$, so, $d = 0.209$ nm. In FCC, $a = 2r.\,(2)^{1/2}$, so, $a = 0.362$ nm. As we know, $d_{hkl} = a/(h^2 + k^2 + l^2)^{1/2}$, so, $h^2 + k^2 + l^2 = 3$. In consideration of principal diffraction planes of FCC, this $2\theta = 43.2°$ belongs to (111) of Cu.

b. Consider Table 2.1 and equation $d_{hkl} = a/(h^2 + k^2 + l^2)^{1/2}$, and follow Table 2.2 to solve it. The mean value of the lattice parameter can be evaluated after obtaining all "a" values for each plane.

c. Please refer to Section 2.3.2.

d. Please refer to Section 2.3.3.

e. Please refer to Section 2.3.4.

PROBLEM 6:

a. Top-down approach like high-energy attrition milling was employed to obtain the MgO nanoparticle from micron size particles followed by thermal treatment was carried at 800°C for 1 h to remove the lattice strain that was actually introduced during continuous impact in milling. XRD pattern using Cu = Kα radiation with a wavelength ($\lambda = 0.154$ nm) was measured for both milled and annealed samples and their data including hkl, 2θ, and full width half maxima (FWHM) are tabulated in Table 10.1. Compute the average crystallite size and lattice strain of ball milled particles.

b. After performing the TEM analysis, a representative HRTEM and SAED pattern schematic are given in Figure 2.9. Identify the hkl plane and d-spacing from TEM analysis. Estimate the probable surface area of particles from morphology. Given, $\rho_{WO_3} - 7.16$ g/cc (Section 2.5.1).

c. Wet-chemical followed by flash calcination process was adopted to synthesis of tetragonal 120 nm spherical $BaTiO_3$ for MLCC application and repeated thrice with variation of process parameters. XRD analysis confirms the presence of lattice parameter c/a ratio. Estimated particle size distribution by SMPS and BET surface area for three processes, and both Weibull statistical data and surface area are given in Table 10.2. Given, $\rho_{BaTiO_3} - 6.02$ g/cc. Justify the best process to obtain the targeted nanoparticles.

TABLE 10.1

Diffraction Data of Ball Milled and Annealed MgO

Peak No.	hkl	Ball Milled Powder		Annealed Powder	
		2θ	FWHM (degree)	2θ	FWHM (degree)
1	200	42.90	0.183	42.91	0.093
2	220	62.31	0.205	62.30	0.072
3	311	74.71	0.243	74.68	0.068
4	222	78.63	0.274	78.60	0.90
5	400	94.06	0.309	94.04	0.087

TABLE 10.2

Specific Surface Area, Three Parameter Weibull Distribution Parameters and Obtained c/a for $BaTiO_3$ Synthesis Are Given for Three Different Batches

Process	Specific Surface Area (m²/gm)	Shape Parameter (α)	Scale Parameter (β)	Location Parameter (γ)	c/a
1	6.92	3	150	5	1.007
2	7.96	4.6	124	8.6	1.009
3	12.63	6	80	2	1.000

HINTS 6:

a. Take an intense look on the obtained data, where broadening is noticed during ball milling compared to annealed powder.

Step 1: Thus, in consideration of Equation 2.22, first calculate the $B_r = (B^2_{ball\ milled} - B^2_{annealed})^{1/2}$. For an example, FWHM (degree) 0.183 for (200) ball milled powder is needed to convert in radian by the multiplication of $\pi/180$, where $\pi = 3.14$, so, result is 3.19×10^{-3} (radian).

Step 2: Calculate the corresponding $\cos\theta$ and $\sin\theta$ to evaluate the crystallite size (D) and strain (η) from Equation 2.21, $B_r\cos\theta = (\lambda/D) + 2\eta\sin\theta$.

Step 3: Plot $B_r\cos\theta$ (y-axis) and $\sin\theta$ (x-axis) and estimate the inhomogenous strain by the slope (2η) and crystallite size by the intercept (λ/D), an intercept $\sim 0.155 \times 10^{-3}$ obtained, so crystallite size becomes 100 nm. Slope provides microstrain of 2.75×10^{-3}.

b. Read Sections 2.4.1 and 2.4.2, and solve it.

c. Process 2 is best one, justify after careful reading Section 2.5.1.

PROBLEM 7:

a. Why particle size, excitation wavelength, dopant concentration, and coupling of dual semiconductor alter the photoluminescence spectral behavior?

b. Identify a characterization protocol to distinguish regular graphene from graphite, and explain such spectral behavior. Is it feasible to synthesis semiconductive graphene in order to use as transistors? (band gap enhance up to 0.5 ev of ribbon shaped graphene).

c. Suggest a method to quantify the elemental composition for nanoscale coating, describe the basic philosophy of such process with a relevant example.

d. Performance of photocatalyst depends on the band gap and effective photo degradation of pollutant, and thus explain the process to determine both band gap and unknown concentration of methylene blue after photocatalytic degradation.

e. What are the possible bond behaviors persist in alumina-zirconia composite particles? Explain the probable characteristic change of bond behavior during synthesis of such nanoparticles through sol-gel method.

HINTS 7:

Section 2.6 provides all answers.

PROBLEM 8:

a. Differentiate the Young and Wenzel concept on contact angle. How do you theoretically estimate the critical contact angle in order to differentiate the boundary between hydrophobic and hydrophilic surface? In consideration of water adhesion tension, work of spreading, hydration energy classify the surface characteristics starting from superhydrophilic to superhydrophobic. Given, water surface energy (72.8 mJ/m^2) and free energy of hydration (-113 mJ/m^2).

b. What are the critical parameters needed to encounter to achieve superhydrophobic and superhydrophilic surfaces?

c. Discuss the colloidal suspension stabilization mechanism. How do zeta potential and isoelectric point vary with particle size and pH of the solution?

d. How do surface tension and viscosity of colloidal suspension change with particle size and solid loading? Explain such phenomena with an example. How do the settling velocity and viscosity vary with the packing fraction of nanoparticles?

HINTS 8:
Answer from chapter 3.

PROBLEM 9:

a. How do you evaluate the band gap position of ceramic semiconductor for effective use as photocatalyst explain with a suitable example? What is happening during photo induced catalytic process, and how can their efficiency alter when changing the band gap and size of the particles?

b. Suggest some strategies to reduce the recombination of electron-hole pairs and photocatalytic reaction rate. Why is an optimum size of photocatalyst a smart choice for the photoexcited degradation?

c. Differentiate the photo electro chemical (PEC) cell based on the free energy change during conversion of optical energy to either electrical energy or chemical energy. How does output of PEC cell depends on band gap of semiconductors, effect of temperature, effect of light intensity, particle morphology and porosity, physical appearance, and coating thickness of electrodes?

d. Let's consider the fabrication of dye-sensitized solar cell (DSSC) by 15 nm TiO$_2$ nanoparticles and estimate their lab scale performance at $P_{in} = 83$ W/m^2, $\lambda = 520$ nm, 1/10th of sunlight illumination, to understand the efficiency of the module (refer to Section 7.2). In order to develop the electrode, titania was coated on 85% transparent and conductive ITO, effective surface area of photoanode was 0.5 cm^2. Different characteristic output was estimated and plotted in Figure 7.2d, calculate the (i) fill factor, (ii) power conversion efficiency (PCE), and (iii) incident photon to current conversion efficiency. What would be the expected incident photon to current conversion efficiency in direct sunlight during summer afternoon?

HINTS 9:
Part a, b, and c are in Chapter 4. Solve part d with help of Sections 4.3.5 and 7.2.5.

PROBLEM 10:

a. Segregate the ferrimagnetic and ferrielectric from entire family of magnetic and dielectric ceramics.

b. Differentiate the soft, hard ferrites and superparamagnetic, and compare their magnetic properties. Draw and explain the P-E hysteresis loop for single crystal, poly crystal, and thin coating of piezoelectric ceramics, and compare their strain-electric field characteristics.

c. What is the basic philosophy needed to encounter to develop a successful exchange-spring magnet? Is monosize (high α, shape factor value) beneficial to attain high SLP?

d. How do specific loss power, magnetic saturation, dielectric constant, dielectric loss, and piezoelectric constant vary with particle or grain size? Is always smallest nanostructured required to achieve highest value of such properties? If not, why?

e. Take a look on the relation within piezoelectric constants in Table 5.2 and correlate themselves judiciously.

HINTS 10:
Consider Chapter 5 and solve all questions.

PROBLEM 11:

a. How pore content, pore morphology, pore location, and grain size influence the elastic modulus of ceramics? Can we measure the stiffness, (F/δ), of ceramics?

b. Draw a competitive stress-strain plot for compressive and tensile strength of same ceramics, and explain such behavior. Why 4-point flexural strength data are considered for design of ceramics? How grain size and porosity influence the strength of ceramics?

c. A large ceramic ($30 \times 10 \times 3$ cm) slab was made of 5 wt% zirconia – 95 wt% alumina (ZTA) composite followed by 10 number of bar specimens were machined from different positions of the same slab to predict the design stress of the bar for load bearing application. The obtained 4-point flexural strength data are found to be 405, 310, 300, 365, 425, 345, 410, 345, 340, and 385 MPa. Calculate the average stress, "m" and "σ_0." Also estimate the design stress that would ensure a survival probability higher than 80% and 99.9% for two different applications, respectively.

d. Why SEVNB is more reliable than indented fracture toughness data? Why an optimum grain size window is required to achieve high fracture toughness of noncubic crystal system compared to cubic crystalline ceramics? Explain with proper examples.

e. Calculate the mean diagonal length 2a of the Vickers indentation of sintered Al_2O_3 measured at 0.5, 20, and 50 N. Estimate the depth of during nanoindentation under load of 100 μN. Elastic modulus – 380 GPa. Draw the probable indentation images including dimension after performing the hardness test and hardness-load characteristic behavior.

HINTS 11:

a. Please refer to Section 6.2.

b. Please refer to Section 6.3.

c. Read carefully Section 6.4 and solve it. Average stress = 363 MPa, Weibull modulus, m = 10.4, $\sigma_0 \sim$ 384 MPa. Use Equation 6.32, and calculate the design stress. When S = 0.8, m = 10.4, and $\sigma_0 \sim$ 384 MPa the design stress σ = 332 MPa. For S = 0.999, σ = 198 MPa. It indicates low design stress (load during application) regime has high survival probability. A close look on the average stress and σ_0 implies near to 5% difference, thus sometime an average stress is sufficient to predict the representative strength value of developed product.

d. Please refer to Section 6.5.

e. Consider Section 6.6.4, and pick up the hardness value 40 GPa at 0.5 N and 14.1 GPa at 20 N, use Equation 6.41 (hardness) and Equation 6.38 (indented toughness), calculate the 2a and 2c, and draw the probable indented image analogous to Figure 6.9. Draw a hardness-load characteristic in assumption with 20 N is the critical load to report the hardness value. Estimate the indentation depth h_c, by $H = 0.0408 \, P/(h_c)^2$ when hardness 14.1 GPa at optimum load.

PROBLEM 12:

a. Explain the working principle of DSSC, tandem cell, photovoltaically self-charging battery, solid oxide fuel cell, and nano electro mechanical system to achieve solar energy, synthesis of hydrogen gas from water, solar energy storage and electric energy, electrical energy by hydrogen gas, and electrical energy by waste mechanical energy, respectively. Use relevant schematics and equations to support all mechanisms.

b. Make a note on the synthesis protocol of TiO_2, WO_3, ZrO_2, and $BaTiO_3$ nanoparticles in order to use in highlighted renewable energy. Suggest few new materials and morphology to substitute the used materials for a definite application.

c. Why is TiO_2 most attractive ceramic semiconductor for DSSC? What is the significance of "Z-scheme" during splitting of water to hydrogen gas in presence of dual semiconductors like WO_3 and TiO_2? Why nanostructured zirconia is a lucrative solid electrolyte in solid oxide fuel cell? What are the advantages of 1D $BaTiO_3$ compared to spherical geometry in nano electro mechanical systems?

HINTS 12:
Please read the Chapter 7 thoroughly before answering it.

PROBLEM 13:

a. Explain the working principle of protective electrochromic glass to control the sunlight transmittance and solar heat, photoassisted degradation and decolorizing of industrial effluent, piezoelectrochemical assisted pollutant removal, nanofiltration of high salinity waste water, and self-cleaning coating using WO_3, WO_3-ZnO, $BaTiO_3$, ZrO_2, and TiO_2, respectively. Use relevant schematics and equations to support all mechanisms.

b. What is the basic philosophy to optimize and control the fibrous and cuboid morphology of WO_3 particles that can be used for electrochromic and dye degradation, respectively? Why is individual WO_3 or ZnO not enough good for the dye degradation, rather mixing of an optimum content say 10 wt% facilitate high rate of degradation?

c. Suggest a synthesis protocol of zirconia nanofilter to purify salinity water. How TiO_2 can be used for superhydrophobic coating on glass?

HINTS 13:
Please read the chapter 8 thoroughly before answering it.

PROBLEM 14:

a. Why nanostructured zirconia is considered as a potential dental ceramic? Describe zirconia nanoparticle synthesis and their blank fabrication protocol in the perspective of development of artificial teeth. How is biocompetitve hydroxyapatite particle gradually deposited on sintered zirconia surface in presence of simulated body fluid solution? Describe a method to characterize both the mechanical response of blank and formation of hydroxyapatite particle.

b. Why ZTA is a better choice over single matrix alumina in order to fabricate hip prosthesis? Brief the fabrication protocol and characterization techniques to establish a successful femoral head made of ZTA.

c. What are the advantages of nanoscale hydroxyapatite based scaffold for tissue engineering as well as drug delivery media? Describe a fabrication protocol of gelatin-HA porous scaffold, and how to determine the drug adsorption and release quantity by this scaffold. How does compressive strength of such scaffold vary with respect to liquid nitrogen treatment time, and why?

d. Suggest a material that enables to participate in cancer treatment through chemo and photodynamic therapy, and state their advantages over conventional chemotherapy only. Describe a method to make such nanoparticles to load the cancerous drug effectively and state the working principle of such combined therapy.

e. What is core-shell nanoparticle and their importance in hyperthermia treatment? Describe a method to make such nanoparticles and explain working principle to cure tumor. What extent this hypothesis is beneficial compared to only drug assisted treatment? Explain with experimental data.

HINTS 14:
Please read Chapter 9 thoroughly before answering it.

PROBLEM 15:

a. Make a note on the possible use of nanoscale materials in traditional ceramics like refractories, whiteware, cement, abrasives, and glass. What are the probable risk factors being necessary to encounter during nanoresearch? State their probable remedies.

b. Write a short note on the future research scope in renewable energy, environmental science, and medical diagnosis by nanostructured ceramics.

c. Calculate the hydrogen and oxygen consume rate when a solid oxide fuel cell is operated at 350°C. The mol fraction of reactant and products are estimated $H_2 = 0.85$, $O_2 = 0.25$, $H_2O = 0.15$. The total cell area is 2.0 cm², and the electrolyte thickness is 200 μm. If the cell is operated at 1.0 volts, then determine the rates at which hydrogen and oxygen are consumed. Also compare it when electrolyte thickness is 50 nm. Given, operating voltage = 1.0 volt, $\Delta G°_{rxn} = -237$ kJ/mol of H_2, $\sigma = 0.05$ cm⁻¹ohm⁻¹, F = 96485 C.mol⁻¹.

Consider the equations $I_{cell} = (E_{ner} - E_{cell})/R_{int}$, consumption rate of hydrogen R_{H2} (mol/s) = $I_{cell}/(2F)$, and consumption rate of hydrogen R_{O2} (mol/s) = $I_{cell}/(4F)$.

HINTS 15:

a. Refer Section 1.3.1 for the basic understanding and consult more literature to make a useful note. Please refer Section 1.5.

b. The probable research scope has been discussed at the end of each case study in Chapters 7 through 9. Please go through and make proper note on it. In order to strengthen the data, please take help from other literature as cited in reference list.

c. First calculate the $E_o = \Delta G^o_{rxn}/2F$ (please see Equation 7.20 and their illustration),

$$E_o = 1.228 \text{ V. So, } E_{ner} = 1.228 + \frac{8.314 \times 623}{2 \times 96485} \ln\left[\frac{0.85 \times (0.25)^{1/2}}{0.15}\right] = 1.248 \text{ V internal}$$

resistance of the cell, $R_{int} = \delta/(\sigma \, A_{cell}) = 0.02/(0.05 \times 2.0) = 0.2 \, \Omega$, and $I_{cell} = (1.248-1.0)/0.2 = 1.24$ A. Calculate the consumption of hydrogen and oxygen gas, and compare the values for thin electrolytes as well. This information will provide the amount of gas required to achieve the particular output and manipulate the electrolyte thickness.

Index

Note: Page numbers in italic and bold refer to figures and tables respectively.

Milton Keynes UK
Ingram Content Group UK Ltd.
UKHW050452071024
449327UK00015B/347

9 780367 570941